Effective Dynamics of Stochastic Partial Differential Equations

Effective Dynamics of Stochastic Partial Differential Equations

Jinqiao Duan

Illinois Institute of Technology
Chicago, USA

Wei Wang

Nanjing University
Nanjing, China

AMSTERDAM • BOSTON • HEIDELBERG • LONDON • NEW YORK • OXFORD
PARIS • SAN DIEGO • SAN FRANCISCO • SINGAPORE • SYDNEY • TOKYO

ELSEVIER

Elsevier
32 Jamestown Road, London NW1 7BY
225 Wyman Street, Waltham, MA 02451, USA

First edition 2014

British Library Cataloguing-in-Publication Data
A catalogue record for this book is available from the British Library

Library of Congress Cataloging-in-Publication Data
A catalog record for this book is available from the Library of Congress

ISBN: 978-0-12-800882-9

For information on all Elsevier publications
visit our website at store.elsevier.com

This book has been manufactured using Print On Demand technology. Each copy is produced to order and is limited to black ink. The online version of this book will show color figures where appropriate

Working together
to grow libraries in
developing countries

www.elsevier.com • www.bookaid.org

Dedication

To my wife, Yan Xiong, and my children, Victor and Jessica
—J. Duan
To my father, Yuliang Wang, and my mother, Lanxiu Liu
—W. Wang

Contents

Preface

Background

Mathematical models for spatial-temporal physical, chemical, and biological systems under random influences are often in the form of *stochastic partial differential equations* (SPDEs). Stochastic partial differential equations contain randomness such as fluctuating forces, uncertain parameters, random sources, and random boundary conditions. The importance of incorporating stochastic effects in the modeling of complex systems has been recognized. For example, there has been increasing interest in mathematical modeling of complex phenomena in the climate system, biophysics, condensed matter physics, materials sciences, information systems, mechanical and electrical engineering, and finance via SPDEs. The inclusion of stochastic effects in mathematical models has led to interesting new mathematical problems at the interface of dynamical systems, partial differential equations, and probability theory. Problems arising in the context of stochastic dynamical modeling have inspired challenging research topics about the interactions among uncertainty, nonlinearity, and multiple scales. They also motivate efficient numerical methods for simulating random phenomena.

Deterministic partial differential equations originated 200 years ago as mathematical models for various phenomena in engineering and science. Now stochastic partial differential equations have started to appear more frequently to describe complex phenomena under uncertainty. Systematic research on stochastic partial differential equations started in earnest in the 1990s, resulting in several books about well-posedness, stability and deviation, and invariant measure and ergodicity, including books by Rozovskii (1990), Da Prato and Zabczyk (1992, 1996), Prevot and Rockner (2007), and Chow (2007).

Topics and Motivation

However, complex systems not only are subject to uncertainty, but they also very often operate on multiple temporal or spatial scales. In this book, we focus on stochastic partial differential equations with slow and fast time scales or large and small spatial scales. We develop basic techniques, such as averaging, slow manifolds, and homogenization, to extract effective dynamics from these stochastic partial differential equations.

The motivation for extracting effective dynamics is twofold. On one hand, effective dynamics is often just what we desire. For example, the air temperature is a macroscopic consequence of the motion of a large number of air molecules. In order to decide what to wear in the morning, we do not need to know the velocity of these molecules, only their effective or collective effect, i.e., the temperature measured by a thermometer. On the other hand, multiscale dynamical systems are sometimes too complicated to analyze or too expensive to simulate all involved scales. To make progress in understanding these dynamical systems, it is desirable to concentrate on macroscopic scales and examine their effective evolution.

Audience

This book is intended as a reference for applied mathematicians and scientists (graduate students and professionals) who would like to understand effective dynamical behaviors of stochastic partial differential equations with multiple scales. It may also be used as a supplement in a course on stochastic partial differential equations. Each chapter has several exercises, with hints or solutions at the end of the book. Realizing that the readers of this book may have various backgrounds, we try to maintain a balance between mathematical precision and accessibility.

Prerequisites

The prerequisites for reading this book include basic knowledge of stochastic partial differential equations, such as the contents of the first three chapters of P. L. Chow's *Stochastic Partial Differential Equations* (2007) or the first three chapters of G. Da Prato and J. Zabczyk's *Stochastic Equations in Infinite Dimensions* (1992). To help readers quickly get up to this stage, these prerequisites are also reviewed in Chapters 3 and 4 of the present book.

Acknowledgments

An earlier version of this book was circulated as lecture notes in the first author's course *Stochastic Partial Differential Equations* at Illinois Institute of Technology over the last several years. We would like to thank the graduate students in the course for their feedback. The materials in Chapters 5, 6, and 7 are partly based on our recent research.

The first author is grateful to Ludwig Arnold for his many years of guidance and encouragement in the study of stochastic dynamical systems and stochastic partial differential equations. We have benefited from many years of productive research interactions with our collaborators and friends, especially Peter Bates, Dirk Blömker, Daomin Cao, Tomás Caraballo, Pao-Liu Chow, Igor Chueshov, Franco Flandoli, Hongjun Gao, Peter Imkeller, Peter E. Kloeden, Sergey V. Lototsky, Kening Lu, Anthony J. Roberts, Michael Röckner, Boris Rozovskii, Michael Scheutzow, Björn Schmalfuß, and Jerzy Zabczyk. The second author would especially like to thank Anthony J. Roberts, who provided him the opportunity to conduct research at the University of Adelaide, Australia. We would also like to thank our colleagues, visitors, and students at Illinois Institute of Technology (Chicago, Illinois, USA), Huazhong University of Science and Technology (Wuhan, China), and Nanjing University (Nanjing, China), particularly Guanggan Chen, Hongbo Fu, Xingye Kan, Yuhong Li, Yan Lv, and Wei Wu, for their constructive comments.

Mark R. Lytell proofread this book in its entirety. Hassan Allouba, Hakima Bessaih, Igor Cialenco, Peter E. Kloeden, and Björn Schmalfuß proofread parts of the book. Their comments and suggestions have greatly improved the presentation of this book. Finally, we would like to acknowledge the National Science Foundation for its generous support of our research.

Jinqiao Duan
Chicago, Illinois, USA

Wei Wang
Nanjing, China
October 2013

1 Introduction

Examples of stochastic partial differential equations; outlines of this book

1.1 Motivation

Deterministic partial differential equations arise as mathematical models for systems in engineering and science. Bernoulli, D'Alembert, and Euler derived and solved a linear wave equation for the motion of vibrating strings in the 18th century. In the early 19th century, Fourier derived a linear heat conduction equation and solved it via a series of trigonometric functions [192, Ch. 28].

Stochastic partial differential equations (SPDEs) appeared much later. The subject has started to gain momentum since the 1970s, with early representative works such as Cabana [58], Bensoussan and Temam [33], Pardoux [248], Faris [123], Walsh [295], and Doering [99,100], among others.

Scientific and engineering systems are often subject to uncertainty or random fluctuations. Randomness may have delicate or even profound impact on the overall evolution of these systems. For example, external noise could induce phase transitions [160, Ch. 6], bifurcation [61], resonance [172, Ch. 1], or pattern formation [142, Ch. 5], [236]. The interactions between uncertainty and nonlinearity also lead to interesting dynamical systems issues. Taking stochastic effects into account is of central importance for the development of mathematical models of complex phenomena under uncertainty in engineering and science. SPDEs emerge as mathematical models for randomly influenced systems that contain randomness, such as stochastic forcing, uncertain parameters, random sources, and random boundary conditions. For general background on SPDEs, see [30,63,76,94,127,152,159,218,260,271,306]. There has been some promising new developments in understanding dynamical behaviors of SPDEs—for example, via invariant measures and ergodicity [107,117,132,153,204], amplitude equations [43], numerical analysis [174], and parameter estimation [83,163,167], among others.

In addition to uncertainty, complex systems often evolve on multiple time and/or spatial scales [116]. The corresponding SPDE models thus involve multiple scales. In this book, we focus on stochastic partial differential equations with slow and fast time scales as well as large and small spatial scales. We develop basic techniques, including averaging, slow manifold reduction, and homogenization, to extract effective dynamics as described by reduced or simplified stochastic partial differential equations.

Effective dynamics are often what we desire. Multiscale dynamical systems are often too complicated to analyze or too expensive to simulate. To make progress in

Effective Dynamics of Stochastic Partial Differential Equations. http://dx.doi.org/10.1016/B978-0-12-800882-9.00001-9

understanding these dynamical systems, it is desirable to concentrate on significant scales, i.e., the macroscopic scales, and examine the effective evolution of these scales.

1.2   Examples of Stochastic Partial Differential Equations

In this section, we present a few examples of stochastic partial differential equations (SPDEs or stochastic PDEs) arising from applications.

Example 1.1 (Heat conduction in a rod with fluctuating thermal source).   The conduction of heat in a rod, subject to a random thermal source, may be described by a stochastic heat equation [123]

$$u_t = \kappa u_{xx} + \eta(x, t), \tag{1.1}$$

where $u(x, t)$ is the temperature at position x and time t, κ is the (positive) thermal diffusivity, and $\eta(x, t)$ is a noise process.

Example 1.2 (A traffic model).   A one-dimensional traffic flow may be described by a macroscopic quantity, i.e., the density. Let $R(x, t)$ be the deviation of the density from an equilibrium state at position x and time t. Then it approximately satisfies a diffusion equation with fluctuations [308]

$$R_t = K R_{xx} - c R_x + \eta(x, t), \tag{1.2}$$

where K, c are positive constants depending on the equilibrium state, and $\eta(x, t)$ is a noise process caused by environmental fluctuations.

Example 1.3 (Concentration of particles in a fluid).   The concentration of particles in a fluid, $C(x, t)$, at position x and time t approximately satisfies a diffusion equation with fluctuations [322, Sec. 1.4]

$$C_t = D \Delta C + \eta(x, t), \tag{1.3}$$

where D is the (positive) diffusivity, Δ is the three-dimensional Laplace operator, and $\eta(x, t)$ is an environmental noise process.

Example 1.4 (Vibration of a string under random forcing).   A vibrating string being struck randomly by sand particles in a dust storm [6,58] may be modeled by a stochastic wave equation

$$u_{tt} = c^2 u_{xx} + \eta(x, t), \tag{1.4}$$

where $u(x, t)$ is the string displacement at position x and time t, the positive constant c is the propagation speed of the wave, and $\eta(x, t)$ is a noise process.

Example 1.5 (A coupled system in molecular biology).   Chiral symmetry breaking is an example of spontaneous symmetry breaking affecting the chiral symmetry in nature. For example, the nucleotide links of RNA (ribonucleic acid) and DNA (deoxyribonucleic acid) incorporate exclusively dextro-rotary (D) ribose and D-deoxyribose, whereas the enzymes involve only laevo-rotary (L) enantiomers of amino acids. Two continuous fields $a(x, t)$ and $b(x, t)$, related to the annihilation for L and D, respectively, are described by a system of coupled stochastic partial differential equations [158]

$$\partial_t a = D_1 \Delta a + k_1 a - k_2 ab - k_3 a^2 + \eta_1(x, t), \tag{1.5}$$
$$\partial_t b = D_2 \Delta b + k_1 b - k_2 ab - k_3 b^2 + \eta_2(x, t), \tag{1.6}$$

where x varies in a three-dimensional spatial domain; D_1, D_2(both positive) and k_1, k_2 are real parameters; and η_1 and η_2 are noise processes. When $D_1 \ll D_2$, this is a slow-fast system of SPDEs.

Example 1.6 (A continuum limit of dynamical evolution of a group of "particles"). SPDEs may arise as continuum limits of a system of stochastic ordinary differential equations (SODEs or SDEs) describing the motion of "particles" under certain constraints on system parameters [7,195,196,207,214].

In particular, a stochastic Fisher–Kolmogorov–Petrovsky–Piscunov equation emerges in this context [102]

$$\partial_t u = Du_{xx} + \gamma u(1 - u) + \varepsilon\sqrt{u(1 - u)}\eta(x, t), \tag{1.7}$$

where $u(x, t)$ is the population density for a certain species; D, γ, and ε are parameters; and η is a noise process.

Example 1.7 (Vibration of a string and conduction of heat under random boundary conditions). Vibration of a flexible string of length l, randomly excited by a boundary force, may be modeled as [57,223]

$$u_{tt} = c^2 u_{xx}, \quad 0 < x < l, \tag{1.8}$$
$$u(0, t) = 0, \quad u_x(l, t) = \eta(t), \tag{1.9}$$

where $u(x, t)$ is the string displacement at position x and time t, the positive constant c is the propagation speed of the wave, and $\eta(t)$ is a noise process.

Evolution of the temperature distribution in a rod of length l, with fluctuating heat source at one end and random thermal flux at the other end, may be described by the following SPDE [96]:

$$u_t = \kappa u_{xx}, \quad 0 < x < l, \tag{1.10}$$
$$u(0, t) = \eta_1(t), \quad u_x(l, t) = \eta_2(t), \tag{1.11}$$

where $u(x, t)$ is the temperature at position x and time t, κ is the (positive) thermal diffusivity, and η_1 and η_2 are noise processes.

Random boundary conditions also arise in geophysical fluid modeling [50,51,226].

In some situations, a random boundary condition may also involve the time derivative of the unknown quantity, called a *dynamical random boundary condition* [55,79,297,300]. For example, dynamic boundary conditions appear in the heat transfer model of a solid in contact with a fluid [210], in chemical reactor theory [211], and in colloid and interface chemistry [293]. Noise enters these boundary conditions as thermal agitation or molecular fluctuations on a physical boundary or on an interface.

Noise will be defined as the generalized time derivative of a Wiener process (or Brownian motion) $W(t)$ in Chapter 3.

Note that partial differential equations with random coefficients are called *random partial differential equations (or random PDEs)*. They are different from *stochastic partial differential equations*, which contain noises in terms of Brownian motions. This distinction will become clear in the next chapter. Random partial differential equations have also appeared in mathematical modeling of various phenomena; see [14,279,169,175,208,216,212,228,250].

1.3   Outlines for This Book

We now briefly overview the contents of this book. Chapters 5, 6 and 7 are partly based on our recent research.

1.3.1   Chapter 2: Deterministic Partial Differential Equations

We briefly present a few examples of deterministic PDEs arising as mathematical models for time-dependent phenomena in engineering and science, together with their solutions by Fourier series or Fourier transforms. Then we recall some equalities and inequalities useful for estimating solutions of both deterministic and stochastic partial differential equations.

1.3.2   Chapter 3: Stochastic Calculus in Hilbert Space

We first recall basic probability concepts and Brownian motion in Euclidean space $\mathbb{R}^n$ and in Hilbert space, and then we review Fréchet derivatives and Gâteaux derivatives as needed for Itô's formula. Finally, we discuss stochastic calculus in Hilbert space, including a version of Itô's formula that is useful for analyzing stochastic partial differential equations.

1.3.3   Chapter 4: Stochastic Partial Differential Equations

We review some basic facts about stochastic partial differential equations, including various solution concepts such as weak, strong, mild, and martingale solutions and sufficient conditions under which these solutions exist. Moreover, we briefly discuss infinite dimensional stochastic dynamical systems through a few examples.

1.3.4   Chapter 5: Stochastic Averaging Principles

We consider averaging principles for a system of stochastic partial differential equations with slow and fast time scales:

$$du^\epsilon = \left[\Delta u^\epsilon + f(u^\epsilon, v^\epsilon)\right]dt + \sigma_1 \, dW_1(t), \tag{1.12}$$

$$dv^\epsilon = \frac{1}{\epsilon}\left[\Delta v^\epsilon + g(u^\epsilon, v^\epsilon)\right]dt + \frac{\sigma_2}{\sqrt{\epsilon}}dW_2(t), \tag{1.13}$$

where ϵ is a small positive parameter and W_1 and W_2 are mutually independent Wiener processes defined on a probability space $(\Omega, \mathcal{F}, \mathbb{P})$. The effective dynamics for this system are shown to be described by an averaged or effective system

$$du = \big[\Delta u + \bar{f}(u)\big]dt + \sigma_1\,dW_1(t), \tag{1.14}$$

where the averaged quantify $\bar{f}(u)$ is appropriately defined. The errors for the approximation of the original multiscale SPDE system by the effective system are quantified via normal deviation principles as well as large deviation principles.

Finally, averaging principles for partial differential equations with time-dependent, time-recurrent random coefficients (e.g., periodic, quasiperiodic, or ergodic) are also discussed.

1.3.5 Chapter 6: Slow Manifold Reduction

We first present a random center manifold reduction method for a class of stochastic evolutionary equations in a Hilbert space H:

$$du(t) = [Au(t) + F(u(t))]dt + u(t) \circ dW(t), \quad u(0) = u_0 \in H. \tag{1.15}$$

Here $\circ$ indicates the Stratonovich differential. A random center manifold is constructed as the graph of a random Lipschitz mapping $\bar{h}^s : H_c \to H_s$. Here $H = H_c \oplus H_s$. Then the effective dynamics are described by a reduced system on the random center manifold

$$du_c(t) = [A_c u_c(t) + F_c(u_c(t) + \bar{h}^s(u_c(t), \theta_t\omega))]dt + u_c(t) \circ dW(t), \tag{1.16}$$

where A_c and F_c are projections of A and F to H_c, respectively.

Then we consider random slow manifold reduction for a system of SPDEs with slow and fast time scales:

$$du^\epsilon = [Au^\epsilon + f(u^\epsilon, v^\epsilon)], \quad u^\epsilon(0) = u_0 \in H_1, \tag{1.17}$$

$$dv^\epsilon = \frac{1}{\epsilon}[Bv^\epsilon + g(u^\epsilon, v^\epsilon)]dt + \frac{1}{\sqrt{\epsilon}}dW(t), \quad v^\epsilon(0) = v_0 \in H_2, \tag{1.18}$$

with a small positive parameter ϵ and a Wiener process $W(t)$. The effective dynamics for this system are captured by a reduced system on the random slow manifold

$$d\bar{u}^\epsilon(t) = [A\bar{u}^\epsilon(t) + f(\bar{u}^\epsilon(t), \bar{h}^\epsilon(\bar{u}^\epsilon(t), \theta_t\omega) + \eta^\epsilon(\theta_t\omega))]dt, \tag{1.19}$$

where $\bar{h}^\epsilon(\cdot, \omega) : H_1 \to H_2$ is a Lipschitz mapping whose graph is the random slow manifold.

1.3.6 Chapter 7: Stochastic Homogenization

In this final chapter, we consider a microscopic heterogeneous system under random influences. The randomness enters the system at the physical boundary of small-scale obstacles (heterogeneities) as well as at the interior of the physical medium. This system is modeled by a stochastic partial differential equation defined on a domain D_ϵ

perforated with small holes (obstacles or heterogeneities) of "size" ϵ, together with random dynamical boundary conditions on the boundaries of these small holes

$$du_\epsilon(x, t) = \Big[\Delta u_\epsilon(x, t) + f(x, t, u_\epsilon, \nabla u_\epsilon)\Big]dt + g_1(x, t)dW_1(x, t),$$
$$\text{in } D_\epsilon \times (0, T), \qquad\qquad\qquad\qquad\qquad\qquad (1.20)$$

$$\epsilon^2 du_\epsilon(x, t) = \Big[-\frac{\partial u_\epsilon(x, t)}{\partial v_\epsilon} - \epsilon b u_\epsilon(x, t)\Big]dt + \epsilon g_2(x, t)dW_2(x, t),$$
$$\text{on } \partial S_\epsilon \times (0, T), \qquad\qquad\qquad\qquad\qquad\qquad (1.21)$$

with a small positive parameter ϵ, constant b, nonlinearity f, and noise intensities g_1 and g_2. Moreover, $W_1(x, t)$ and $W_2(x, t)$ are mutually independent Wiener processes, and v_ϵ is the outward unit normal vector on the boundary of small holes.

We derive a homogenized, macroscopic model for this heterogeneous stochastic system

$$dU = \Big[\vartheta^{-1}\text{div}_x(\bar{A}\nabla_x U) - b\lambda U + \vartheta\bar{f}\Big]dt$$
$$+ \vartheta g_1 dW_1(t) + \lambda g_2 dW_2(t), \qquad\qquad\qquad\qquad (1.22)$$

where ϑ and λ are characterized by the microscopic heterogeneities. Moreover, $\bar{A}$ and $\bar{f}$ are appropriately homogenized linear and nonlinear operators, respectively. This homogenized or effective model is a new stochastic partial differential equation defined on a unified domain D without small holes and with the usual boundary conditions only.

2 Deterministic Partial Differential Equations

Examples of partial differential equations; Fourier methods and basic analytic tools for partial differential equations

In this chapter, we first briefly present a few examples of deterministic partial differential equations (PDEs) arising as mathematical models for time-dependent phenomena in engineering and science, together with their solutions by Fourier series or Fourier transforms. Then we recall some equalities that are useful for estimating solutions of both deterministic and stochastic partial differential equations.

For elementary topics on solution methods for linear partial differential equations, see [147,239,258]. More advanced topics, such as well-posedness and solution estimates, for deterministic partial differential equations may be found in popular textbooks such as [121,176,231,264].

The basic setup and well-posedness for stochastic PDEs are discussed in Chapter 4.

2.1 Fourier Series in Hilbert Space

We recall some information about Fourier series in Hilbert space, which is related to Hilbert–Schmidt theory.

A vector space has two operations, addition and scalar multiplication, which have the usual properties we are familiar with in Euclidean space $\mathbb{R}^n$. A Hilbert space H is a vector space with a scalar product $\langle \cdot, \cdot \rangle$, with the *usual* properties we are familiar with in $\mathbb{R}^n$; see [198, p. 128] or [313, p. 40] for details. In fact, $\mathbb{R}^n$ is a vector space and also a Hilbert space.

A separable Hilbert space H has a countable orthonormal basis $\{e_n\}_{n=1}^{\infty}$, $\langle e_m, e_n \rangle = \delta_{mn}$, where δ_{mn} is the Kronecker delta function (i.e., it takes value 1 when $m = n$, and 0 otherwise). Moreover, for any $h \in H$, we have Fourier series expansion

$$h = \sum_{n=1}^{\infty} \langle h, e_n \rangle e_n. \tag{2.1}$$

In the context of solving PDEs, we choose to work in a Hilbert space with a countable orthonormal basis. Such a Hilbert space is a separable Hilbert space. This is naturally possible with the help of the Hilbert–Schmidt theorem [316, p. 232].

Effective Dynamics of Stochastic Partial Differential Equations. http://dx.doi.org/10.1016/B978-0-12-800882-9.00002-0

The Hilbert–Schmidt theorem [316, p. 232] says that a linear compact symmetric operator A on a separable Hilbert space H has a set of eigenvectors that form a complete orthonormal basis for H. Furthermore, all the eigenvalues of A are real, each nonzero eigenvalue has finite multiplicity, and two eigenvectors that correspond to different eigenvalues are orthogonal.

This theorem applies to a strong self-adjoint elliptic differential operator B,

$$Bu = \sum_{0 \leq |\alpha|, |\beta| \leq m} (-1)^{|\alpha|} D^\alpha (a_{\alpha\beta}(x) D^\beta u), \quad x \in D \subset \mathbb{R}^n,$$

where the domain of definition of B is an appropriate dense subspace of $H = L^2(D)$, depending on the boundary condition specified for u.

When B is invertible, let $A = B^{-1}$. If B is not invertible, set $A = (B + aI)^{-1}$ for some a such that $(B+aI)^{-1}$ exists. This may be necessary in order for the operator to be invertible, i.e., no zero eigenvalue, such as in the case of the Laplace operator with zero Neumann boundary conditions. Note that A is a linear symmetric compact operator in a Hilbert space, e.g., $H = L^2(D)$, the space of square-integrable functions on D.

By the Hilbert–Schmidt theorem, eigenvectors (also called *eigenfunctions* or *eigenmodes* in this context) of A form an orthonormal basis for $H = L^2(D)$. Note that A and B share the same set of eigenfunctions. So, we can claim that the strong self-adjoint elliptic operator B's eigenfunctions form an orthonormal basis for $H = L^2(D)$.

In the case of one spatial variable, the elliptic differential operator is the so-called Sturm–Liouville operator [316, p. 245],

$$Bu = -(pu')' + qu, \quad x \in (0, l),$$

where $p(x)$, $p'(x)$ and $q(x)$ are continuous on $(0, l)$. This operator arises in solving linear (deterministic) partial differential equations by the method of separating variables. Due to the Hilbert–Schmidt theorem, eigenfunctions of the Sturm–Liouville operator form an orthonormal basis for $H = L^2(0, l)$.

2.2 Solving Linear Partial Differential Equations

We now consider a few linear partial differential equations and their solutions.

Example 2.1 (Wave equation).   Consider a vibrating string of length l. The evolution of its displacement $u(x, t)$, at position x and time t, is modeled by the following wave equation:

$$u_{tt} = c^2 u_{xx}, \quad 0 < x < l, \tag{2.2}$$

$$u(0, t) = u(l, t) = 0, \tag{2.3}$$

$$u(x, 0) = f(x), \quad u_t(x, 0) = g(x), \tag{2.4}$$

where c is a positive constant (wave speed), and f, g are given initial data. By separating variables, $u = X(x)T(t)$, we arrive at an eigenvalue problem for the Laplacian ∂_{xx}

with zero Dirichlet boundary conditions on $(0, l)$, namely, $X''(x) = \lambda X, X(0) = X(l) = 0$. The eigenfunctions (which need to be nonzero by definition) are $\sin \frac{n\pi x}{l}$ and the corresponding eigenvalues are $\lambda_n = -(\frac{n\pi}{l})^2$ for $n = 1, 2, \ldots$ In fact, the set of normalized eigenfunctions (i.e., making each of them have norm 1),

$$e_n(x) = \sqrt{2/l} \sin \frac{n\pi x}{l}, \quad n = 1, 2, \ldots,$$

forms an orthonormal basis for the Hilbert space $H = L^2(0, l)$ of square-integrable functions, with the usual scalar product $\langle u, v \rangle = \int_0^l u(x)v(x)dx$. We construct the solution u by the Fourier expansion or eigenfunction expansion

$$u = \sum_{n=1}^{\infty} u_n(t)e_n(x). \tag{2.5}$$

Inserting this expansion into the PDE $u_{tt} = c^2 u_{xx}$, we obtain an infinite system of ordinary differential equations (ODEs)

$$\ddot{u}_n(t) = c^2 \lambda_n u_n, \quad n = 1, 2, \ldots \tag{2.6}$$

For each n, this second-order ordinary differential equation has general solution

$$u_n(t) = A_n \cos \frac{cn\pi}{l} t + B_n \sin \frac{cn\pi}{l} t, \tag{2.7}$$

where the constants A_n and B_n are determined by the initial conditions, to be

$$A_n = \langle f, e_n \rangle, \quad B_n = \frac{l}{cn\pi} \langle g, e_n \rangle.$$

Thus, the final solution is

$$u(x, t) = \sum_{n=1}^{\infty} \left[A_n \cos \frac{cn\pi}{l} t + B_n \sin \frac{cn\pi}{l} t \right] e_n(x). \tag{2.8}$$

For Neumann boundary conditions, $u_x(0, t) = u_x(l, t) = 0$, the corresponding eigenvalue problem for the Laplacian ∂_{xx} is $X''(x) = \lambda X, X'(0) = X'(l) = 0$. The eigenfunctions are $\cos \frac{n\pi x}{l}$ and the eigenvalues are $-(\frac{n\pi}{l})^2$, for $n = 0, 1, 2, \ldots$ The set of normalized eigenfunctions,

$$e_n(x) = \sqrt{2/l} \cos \frac{n\pi x}{l}, \quad n = 0, 1, 2, \ldots,$$

forms an orthonormal basis for the Hilbert space $H = L^2(0, l)$ of square-integrable functions.

For mixed boundary conditions, $u(0, t) = u_x(l, t) = 0$, the corresponding eigenvalue problem for the Laplacian ∂_{xx} is $X''(x) = \lambda X, X(0) = X'(l) = 0$. The eigenfunctions are $\sin \frac{(n+12)\pi x}{l}$ and the eigenvalues are $-[\frac{(n+12)\pi}{l}]^2$ for $n = 0, 1, 2, \ldots$ Again, the set of normalized eigenfunctions,

$$e_n(x) = \sqrt{2/l} \sin \frac{(n + \frac{1}{2})\pi x}{l}, \quad n = 0, 1, 2, \ldots,$$

forms an orthonormal basis for the Hilbert space $H = L^2(0, l)$.

Example 2.2 (Heat equation).   We now consider the heat equation for the temperature $u(x, t)$ of a rod of length l at position x and time t:

$$u_t = \nu u_{xx}, \ 0 < x < l, \tag{2.9}$$

$$u(0, t) = u(l, t) = 0, \tag{2.10}$$

$$u(x, 0) = f(x), \tag{2.11}$$

where ν is the thermal diffusivity and f is the initial temperature. By separating variables, $u = X(x)T(t)$, we arrive at an eigenvalue problem for the Laplacian ∂_{xx} with zero Dirichlet boundary conditions on $(0, l)$. Namely, $X''(x) = \lambda X$, $X(0) = X(l) = 0$. The eigenfunctions (which need to be nonzero) are $\sin \frac{n\pi x}{l}$ and the corresponding eigenvalues are $\lambda_n = -(\frac{n\pi}{l})^2$ for $n = 1, 2, \ldots$ The set of normalized eigenfunctions,

$$e_n(x) = \sqrt{2/l} \sin \frac{n\pi x}{l}, \quad n = 1, 2, \ldots,$$

forms an orthonormal basis for the Hilbert space $H = L^2(0, l)$. We construct the solution u by the Fourier expansion (or eigenfunction expansion)

$$u = \sum_{n=1}^{\infty} u_n(t)e_n(x). \tag{2.12}$$

Inserting this expansion into the above PDE (2.9), we get

$$\sum \dot{u}_n(t)e_n(x) = \nu \sum \lambda_n u_n(t)e_n. \tag{2.13}$$

This leads to the following system of ODEs:

$$\dot{u}_n(t) = \nu \lambda_n u_n(t), \quad n = 1, 2, 3, \ldots, \tag{2.14}$$

whose solutions are, for each n,

$$u_n(t) = u_n(0) \exp\left(\nu \lambda_n t\right), \tag{2.15}$$

where $u_n(0) = \langle f, e_n \rangle$ for $n = 1, 2, 3, \ldots$ Therefore, the final solution is

$$u(x, t) = \sum_{n=1}^{\infty} \langle f, e_n \rangle \exp\left(\nu \lambda_n t\right)e_n(x). \tag{2.16}$$

Introduce a semigroup of linear operators, $S(t) : H \to H$, by

$$S(t)h \triangleq \sum_{n=1}^{\infty} \langle h, e_n \rangle \exp\left(\nu \lambda_n t\right)e_n(x), h \in H, \tag{2.17}$$

for every $t \geq 0$. Then $S(0) = \mathrm{Id}_H$ (identity mapping in H) and $S(t+s) = S(t)S(s)$, for $t, s \geq 0$. Thus the above solution can be written as $u(x, t) = S(t)f(x)$; see [253, Ch. 7] for more details.

Example 2.3 (Heat equation on the real line).   We consider the heat equation on the real line

$$u_t = \nu u_{xx}, \; x \in \mathbb{R}^1, \tag{2.18}$$
$$u(x,0) = f(x), \tag{2.19}$$

where ν is the thermal diffusivity and f is the initial temperature. Taking the Fourier transform of u with respect to x,

$$U(k,t) \triangleq \frac{1}{\sqrt{2\pi}} \int_{-\infty}^{\infty} e^{ikx} u(x,t) dx, \tag{2.20}$$

we obtain an initial value problem for an ordinary differential equation

$$\frac{d}{dt} U = -\nu k^2 U, \tag{2.21}$$
$$U(k,0) = F(k), \tag{2.22}$$

where $F(k) = \frac{1}{\sqrt{2\pi}} \int_{-\infty}^{\infty} e^{ikx} f(x) dx$ is the Fourier transform for the initial temperature f. The solution to (2.21) and (2.22) is $U(k,t) = F(k)e^{-\nu k^2 t}$. By the inverse Fourier transform,

$$u(x,t) = \frac{1}{\sqrt{2\pi}} \int_{-\infty}^{\infty} F(k)e^{-\nu k^2 t} e^{-ikx} dk,$$

and, finally [239, Ch. 12],

$$u(x,t) = \frac{1}{2\sqrt{\pi \nu t}} \int_{-\infty}^{\infty} f(\xi) e^{-\frac{(x-\xi)^2}{4\nu t}} d\xi. \tag{2.23}$$

Usually, $G(x,t) \triangleq \frac{1}{2\sqrt{\pi \nu t}} e^{-\frac{x^2}{4\nu t}}$ is called the *heat kernel* or *Gaussian kernel*.

2.3   Integral Equalities

In this and the next two sections we recall some equalities and inequalities useful for estimating solutions of SPDEs as well as PDEs.

Let us review some integral identities. For more details, see [1, Sec. 7.3], [17, Ch. 7] or [121, Appendix C].

Green's theorem in $\mathbb{R}^2$: Normal form

$$\oint_C \mathbf{v} \cdot \mathbf{n} \, dc = \iint_D \nabla \cdot \mathbf{v} \, dA,$$

where $\mathbf{v}$ is a continuously differentiable vector field, C is a piecewise smooth closed curve that encloses a bounded region D in $\mathbb{R}^2$, and $\mathbf{n}$ is the unit outward normal vector to C. The curve C is positively oriented (i.e., if you walk along C in the positive orientation, the region D is to your left).

Green's theorem in $\mathbb{R}^2$: Tangential form

$$\oint_C \mathbf{v} \cdot \mathbf{T} dc = \oint_C M dx + N dy = \iint_D \left(\frac{\partial N}{\partial x} - \frac{\partial M}{\partial y} \right) dx\, dy,$$

where $\mathbf{v} = M(x, y)dx + N(x, y)dy$ is a continuously differentiable vector field, C is a piecewise smooth closed curve that encloses a bounded region D in $\mathbb{R}^2$, and $\mathbf{T}$ is the unit tangential vector to C. The curve C is positively oriented (i.e., if you walk along C in the positive orientation, the region D is to your left).

Divergence theorem in $\mathbb{R}^k$

$$\int_S \mathbf{v} \cdot \mathbf{n}\, ds = \int_D \nabla \cdot \mathbf{v}\, dA,$$

where $\mathbf{v}$ is a continuously differentiable vector field, S is a closed surface that encloses a bounded region D in $\mathbb{R}^k$, and $\mathbf{n}$ is the outward unit normal vector to S. In particular, taking $\mathbf{v}$ as a vector function with one component equal to a scalar function u and the rest of components zero, we have

$$\int_D u_{x_i} dx = \int_{\partial D} u n_i\, ds, \quad i = 1, \ldots, k,$$

where $\mathbf{n} = (n_1, \cdots, n_k)$ is the outward unit normal vector to the boundary ∂D of the domain D in $\mathbb{R}^k$.

Applying this theorem to uv and uv_{x_i}, we get the following two integration by parts formulas.

Integration by parts formula in $\mathbb{R}^k$

$$\int_D u_{x_i} v\, dx = - \int_D u v_{x_i}\, dx + \int_S u v n_i\, ds, \quad i = 1, \ldots, k,$$

and

$$\int_D u \Delta v\, dx = - \int_D \nabla u \cdot \nabla v\, dx + \int_{\partial D} u \frac{\partial v}{\partial \mathbf{n}} ds,$$

where $\mathbf{n} = (n_1, \ldots, n_k)$ is the outward unit normal vector to the boundary ∂D of the domain D in $\mathbb{R}^k$.

Stokes's theorem in $\mathbb{R}^3$

$$\oint_C \mathbf{v} \cdot d\mathbf{r} = \iint_S (\nabla \times \mathbf{v}) \cdot \mathbf{n}\, d\sigma,$$

where $\mathbf{v}$ is a continuously differentiable vector field and C is the boundary of an oriented surface S in $\mathbb{R}^3$. The direction of C is taken as counterclockwise with respect to the surface S's unit normal vector $\mathbf{n}$. Stokes's theorem relates the surface integral of the curl of a vector field over a surface to the line integral of the vector field over its boundary.

Green's identities in $\mathbb{R}^k$

$$\int_D \Delta u \, dx = \int_{\partial D} \frac{\partial u}{\partial \mathbf{n}} \, ds,$$

$$\int_D \nabla u \cdot \nabla v \, dx = -\int_D u \Delta v \, dx + \int_{\partial D} u \frac{\partial v}{\partial \mathbf{n}} \, ds,$$

and

$$\int_D (u\Delta v - v\Delta u) \, dx = \int_{\partial D} \left(u \frac{\partial v}{\partial \mathbf{n}} - v \frac{\partial u}{\partial \mathbf{n}} \right) ds,$$

where $\mathbf{n}$ is the unit outward normal vector to the boundary ∂D of the domain D in $\mathbb{R}^k$.

2.4 Differential and Integral Inequalities

The Gronwall inequality [88, p. 37], [146, p. 169] is often used to estimate solutions for differential or integral equations.

Gronwall inequality: Differential form

Assume that $y(t), g(t)$ and $h(t)$ are integrable real functions, and furthermore, $y(t) \geq 0$. If $\frac{dy}{dt} \leq g(t)y + h(t)$ for $t \geq t_0$, then

$$y(t) \leq y(t_0)e^{\int_{t_0}^{t} g(\tau) d\tau} + \int_{t_0}^{t} h(s) \left[e^{\int_s^t g(\tau) d\tau} \right] ds, \quad t \geq t_0.$$

If, in addition, g, h are constants and $t_0 = 0$, then

$$y(t) \leq y(0)e^{gt} - \frac{h}{g}(1 - e^{gt}), \quad t \geq 0.$$

Note that if constant $g < 0$, then $\lim_{t \to \infty} y(t) = -\frac{h}{g}$.

Gronwall inequality: Integral form

Assume that $v(t), k(t), c(t)$ are nonnegative integrable real functions, and $c(t)$ is additionally differentiable. If $v(t) \leq c(t) + \int_0^t u(s)v(s)ds$ for $t \geq t_0$, then

$$v(t) \leq v(t_0)e^{\int_{t_0}^{t} u(\tau) d\tau} + \int_{t_0}^{t} c'(s) \left[e^{\int_s^t u(\tau) d\tau} \right] ds, \quad t \geq t_0.$$

If, in addition, k, c are positive constants and $t_0 = 0$, then

$$v(t) \leq ce^{kt}, \quad t \geq 0.$$

2.5 Sobolev Inequalities

We now review some inequalities for weakly differentiable functions [264, Ch. 7], [287, Ch. II] or [144, Ch. 7]. Let D be a domain in $\mathbb{R}^n$. The Lebesgue spaces $L^p = L^p(D), p \geq 1$, are the spaces of measurable functions that are pth order Lebesgue integrable on a domain D in $\mathbb{R}^n$. The norm for f in L^p is defined by

$$\|f\|_{L^p} \triangleq \left(\int_D |f(x)|^p dx \right)^{\frac{1}{p}}.$$

In particular, $L^2(D)$ is a Hilbert space with the following scalar product $\langle \cdot, \cdot \rangle$ and norm $\| \cdot \|$:

$$\langle f, g \rangle := \int_D f \bar{g} dx, \quad \|f\| := \sqrt{\langle f, f \rangle} = \sqrt{\int_D |f(x)|^2 dx},$$

for f, g in $L^2(D)$. Now we introduce some common Sobolev spaces. For $k = 1, 2, \ldots$, define

$$H^k(D) := \{ f : \partial^\alpha f \in L^2(D), |\alpha| \le k \}.$$

Here $\alpha = (\alpha_1, \ldots, \alpha_n)$ with α_i's being nonnegative integers, $|\alpha| = \alpha_1 + \cdots + \alpha_n$, and $\partial^\alpha \triangleq \partial_{x_1}^{\alpha_1} \cdots \partial_{x_n}^{\alpha_n}$. Each of these is a Hilbert space with scalar product

$$\langle u, v \rangle_k = \int_D \sum_{|\alpha| \le k} \partial^\alpha u \partial^\alpha \bar{v} dx$$

and the norm

$$\|u\|_k = \sqrt{\langle u, u \rangle_k} = \sqrt{\int_D \sum_{|\alpha| \le k} |\partial^\alpha u|^2 dx}.$$

For $k = 1, 2, \ldots$ and $p \ge 1$, we further define another class of Sobolev spaces,

$$W^{k,p}(D) = \{ u : D^\alpha u \in L^p(D), |\alpha| \le k \},$$

with norm

$$\|u\|_{k,p} = \left(\sum_{|\alpha| \le k} \|\partial^\alpha u\|_{L^p}^p \right)^{\frac{1}{p}}.$$

Recall that $C_c^\infty(D)$ is the space of infinitely differentiable functions with compact support in the domain D. Then $H_0^k(D)$ is the closure of $C_c^\infty(D)$ in Hilbert space $H^k(D)$ (under the norm $\| \cdot \|_k$). It is a Hilbert space contained in $H^k(D)$. Similarly, $W_0^{k,p}(D)$ is the closure of $C_c^\infty(D)$ in Banach space $W^{k,p}(D)$ (under the norm $\| \cdot \|_{k,p}$). It is a Banach space contained in $W^{k,p}(D)$.

Standard abbreviations $L^2 = L^2(D)$, $H_0^k = H_0^k(D)$, $k = 1, 2, \ldots$ are used for these common Sobolev spaces.

Let us list several useful inequalities.

Cauchy–Schwarz inequality

For $f, g \in L^2(D)$,

$$\left| \int_D f(x)g(x)dx \right| \leq \sqrt{\int_D |f(x)|^2 dx} \sqrt{\int_D |g(x)|^2 dx}.$$

Hölder inequality

For $f \in L^p(D)$ and $g \in L^q(D)$ with $\frac{1}{p} + \frac{1}{q} = 1$, $p > 1$, and $q > 1$,

$$\left| \int_D f(x)g(x)dx \right| \leq \left(\int_D |f(x)|^p dx \right)^{\frac{1}{p}} \left(\int_D |g(x)|^q dx \right)^{\frac{1}{q}}.$$

Minkowski inequality

For $f, g \in L^p(D)$,

$$\left(\int_D |f(x) \pm g(x)|^p dx \right)^{\frac{1}{p}} \leq \left(\int_D |f(x)|^p dx \right)^{\frac{1}{p}} + \left(\int_D |g(x)|^p dx \right)^{\frac{1}{p}}.$$

Poincaré inequality

For $g \in H_0^1(D)$,

$$\|g\|^2 = \int_D |g(x)|^2 \, dx \leq \left(\frac{|D|}{\omega_n} \right)^{\frac{1}{n}} \int_D |\nabla g|^2 \, dx,$$

where $|D|$ is the Lebesgue measure of the domain D, and ω_n is the volume of the unit ball in $\mathbb{R}^n$ in terms of the Gamma function Γ:

$$\omega_n = \frac{\pi^{n2}}{\Gamma(\frac{n}{2} + 1)}.$$

It is clear that $\omega_1 = 2$, $\omega_2 = \pi$ and $\omega_3 = \frac{4}{3}\pi$.

Similarly, for $u \in W_0^{1,p}(D)$, $1 \leq p < \infty$, and $D \subset \mathbb{R}^n$ a bounded domain,

$$\|u\|_p \leq \left(\frac{|D|}{\omega_n} \right)^{\frac{1}{n}} \|\nabla u\|_p.$$

Let $u \in W^{1,p}(D)$, $1 \leq p < \infty$, and $D \subset \mathbb{R}^n$ a bounded convex domain. Take $S \subset D$, be any measurable subset, and define the spatial average of u over S by $u_S = \frac{1}{|S|} \int_S u \, dx$ (with $|S|$ being the Lebesgue measure of S). Then

$$\|u - u_S\|_p \leq \left(\frac{\omega_n}{|D|} \right)^{1 - \frac{1}{n}} d^n \|\nabla u\|_p,$$

where d is the diameter of D.

Agmon inequality

Let $D \subset \mathbb{R}^n$ be an open domain with piecewise smooth boundary. There exists a positive constant C, depending only on domain D, such that

$$\|u\|_{L^\infty(D)} \leq C\|u\|_{H^{\frac{n-1}{2}}(D)}^{\frac{1}{2}} \|u\|_{H^{\frac{n+1}{2}}(D)}^{\frac{1}{2}}, \quad \text{for } n \text{ odd},$$

$$\|u\|_{L^\infty(D)} \leq C\|u\|_{H^{\frac{n-2}{2}}(D)}^{\frac{1}{2}} \|u\|_{H^{\frac{n+2}{2}}(D)}^{\frac{1}{2}}, \quad \text{for } n \text{ even}.$$

In particular, for $n = 1$ and $u \in H^1(0, l)$,

$$\|u\|_{L^\infty(0,l)} \leq C\|u\|_{L^2(0,l)}^{\frac{1}{2}} \|u\|_{H^1(0,l)}^{\frac{1}{2}}.$$

Moreover, for $n = 1$ and $u \in H_0^1(0, l)$,

$$\|u\|_{L^\infty(0,l)} \leq C\|u\|_{L^2(0,l)}^{\frac{1}{2}} \|u_x\|_{L^2(0,l)}^{\frac{1}{2}}.$$

2.6   Some Nonlinear Partial Differential Equations

In this final section, we consider a class of nonlinear partial differential equations and present some results on the existence and uniqueness of their solutions. The basic idea is also useful to prove the existence and uniqueness of solutions for SPDEs, as in Chapter 4.

2.6.1   *A Class of Parabolic PDEs*

Let D be a bounded domain in $\mathbb{R}^n$ with piecewise smooth boundary ∂D. Consider the following nonlinear heat equation

$$u_t = \Delta u + f(u), \tag{2.24}$$

$$u(0) = u_0, \tag{2.25}$$

$$u|_{\partial D} = 0, \tag{2.26}$$

where Δ is the Laplace operator, f is a nonlinear term, and the initial datum $u_0 \in L^2(D)$. We recall the following well-posedness result.

Theorem 2.4.   *Assume that the nonlinearity $f : L^2(D) \to L^2(D)$ is Lipschitz, that is, there exists a positive constant L_f such that*

$$\|f(u_1) - f(u_2)\| \leq L_f \|u_1 - u_2\|, \quad \text{for all} \quad u_1, u_2 \in L^2(D).$$

Then for every $T > 0$ and every $u_0 \in L^2(D)$, there exists a unique solution $u \in C(0, T; L^2(D)) \cap L^2(0, T; H_0^1(D))$ for the equations (2.24)–(2.26).

The proof of the above result can be found in many textbooks [231,253,264]. We present an outline of the proof [253, Ch. 7].

2.6.1.1　Outline of the Proof of Theorem 2.4

For a given $T > 0$, denote by $\mathcal{X}_T$ the space $C(0, T; L^2(D)) \cap L^2(0, T; H_0^1(D))$ and by $\| \cdot \|_{\mathcal{X}_T}$ the usual norm on $\mathcal{X}_T$. Let $A = \Delta$ with zero Dirichlet boundary condition on D and denote by $S(t)$, $t \geq 0$, the C_0-semigroup generated by A [253, Ch. 7]. Then u in $\mathcal{X}_T$ is a solution of equations (2.24)–(2.26) if and only if u solves the following integral equation,

$$u(t) = S(t)u_0 + \int_0^t S(t - s) f(u(s))ds, \tag{2.27}$$

for $0 \leq t \leq T$. This is the generalization of classical variation of constant formula in Banach space [253, Ch. 6].

We first show the uniqueness of the solution. If u and v are two solutions to equation (2.24)–(2.26), then

$$\|u(t) - v(t)\| \leq \int_0^t \|S(t-s)[f(u(s)) - f(v(s))\|ds \leq L_f \int_0^t \|u(s) - v(s)\|ds.$$

By the Gronwall inequality in the integral form (see Section 2.4) with $c(t) = 0$, we have

$$u(t) - v(t) = 0, \quad \text{for all } 0 \leq t \leq T,$$

which yields the uniqueness of solution. Next, we show the existence of solution in space $\mathcal{X}_T$. For a given $u_0 \subset L^2(D)$, define a nonlinear mapping

$$\mathcal{T} : \mathcal{X}_T \to \mathcal{X}_T$$

by

$$(\mathcal{T}u)(t) = S(t)u_0 + \int_0^t S(t - s) f(u(s))ds, \quad 0 \leq t \leq T.$$

Then, by the Lipschitz property of f and the boundedness of semigroup $S(t)$ [253, Ch. 6], we have

$$\|\mathcal{T}u - \mathcal{T}v\|_{\mathcal{X}_T} \leq M L_f T \|u - v\|_{\mathcal{X}_T},$$

where M is the bound for the norm of $S(t)$ on $[0, T]$. Then, with an argument via induction on n, we have

$$\|\mathcal{T}^n u - \mathcal{T}^n v\|_{\mathcal{X}_T} \leq \frac{(M L_f T)^n}{n!} \|u - v\|_{\mathcal{X}_T}.$$

For n sufficiently large, $(M L_f T)^n < n!$. Then, by the Banach contraction mapping theorem, $\mathcal{T}$ has a unique fixed point in $C(0, T; L^2(D)) \cap L^2(0, T; H_0^1(D))$, which is the unique solution to the nonlinear heat equations (2.24)–(2.26).

In fact, a similar result holds when the nonlinearity f is only locally Lipachitz [253, Ch. 7].

2.6.2   A Class of Hyperbolic PDEs

We now consider the following nonlinear wave equation on a bounded domain D in $\mathbb{R}^n$ with piecewise smooth boundary ∂D,

$$u_{tt} + \alpha u_t = \Delta u + f(u), \tag{2.28}$$

$$u(0) = u_0 \in H_0^1(D), \quad u_t(0) = u_1 \in L^2(D), \tag{2.29}$$

$$u|_{\partial D} = 0, \tag{2.30}$$

where Δ is the Laplace operator, f is a nonlinear term, α is a positive constant, and u_0, u_1 are initial data. The term αu_t has a damping effect. We recall the following well-posedness result.

Theorem 2.5.   *Assume that the nonlinearity $f : L^2(D) \to L^2(D)$ is Lipschitz, that is, there exists a positive constant L_f such that*

$$\|f(u_1) - f(u_2)\| \le L_f \|u_1 - u_2\|, \quad \text{for all} \quad u_1, u_2 \in L^2(D).$$

Then, for every $T > 0$ and every $(u_0, u_1) \in H_0^1(D) \times L^2(D)$, there exists a unique solution $(u, u_t) \in C(0, T; H_0^1(D) \times L^2(D))$ for equations (2.28)–(2.30).

We also give an outline of the proof of this theorem [64, Ch. 4] and [231, Ch. 6].

2.6.2.1   Outline of the Proof of Theorem 2.5

Let $A = \Delta$, with zero Dirichlet boundary condition on D. Define a linear operator $\mathcal{A}$ by

$$\mathcal{A}(u, v) = (v, Au - \alpha v)$$

for $(u, v) \in \mathcal{D}(\mathcal{A}) \subset H_0^1(D) \times L^2(D)$. Then $\mathcal{A}$ is skew-adjoint and generates a C_0-semigroup $\mathcal{S}(t), t \ge 0$, as in [64]. We also define a nonlinear mapping

$$F(u, v) = (0, f(u)), \quad (u, v) \in H_0^1(D) \times L^2(D).$$

Then F is Lipschitz from $H_0^1(D) \times L^2(D)$ to itself, with the same Lipschitz constant as f. Denoting by $U = (u, u_t)$, we rewrite equations (2.28)–(2.30) in the following form:

$$U_t = \mathcal{A}U + F(U), \quad U(0) = (u_0, u_1) \in H_0^1(D) \times L^2(D). \tag{2.31}$$

For every given $T > 0$, $U(t) = (u(t), v(t)) \in C(0, T; H_0^1(D) \times L^2(D))$ is a solution of equations (2.28)–(2.30) if and only if

$$U(t) = \mathcal{S}(t)U(0) + \int_0^t \mathcal{S}(t - s)F(U(s))ds.$$

A similar discussion as in the proof of Theorem 2.4, via Banach contraction mapping theorem, yields the result.

Remark 2.6.   The damping term in equation (2.28)–(2.30) ensures the system is dissipative [287, e.g.], that is, solutions to equations (2.28)–(2.30) are uniformly bounded on $[0, \infty)$.

For the case of local Lipschitz nonlinearity, the well-posedness is more complicated. In this case, we usually get a unique solution that exists on a short time interval (local solution) and then certain a prior estimates on the solution ensure the existence and uniqueness of the solution for all time (global solution). For more details, see, e.g., [64,231,253,264,287].

2.7 Problems

2.1. Wave equation I
Solve the following initial-boundary value problem for the wave equation

$$u_{tt} = u_{xx}, \ 0 < x < 1,$$
$$u(x, 0) = x(1 - x),$$
$$u_t(x, 0) = 0,$$
$$u(0, t) = u(1, t) = 0.$$

2.2. Wave equation II
Solve the following initial-boundary value problem for the wave equation

$$u_{tt} = c^2 u_{xx}, \ 0 < x < l,$$
$$u(0, t) = u_x(l, t) = 0,$$
$$u(x, 0) = f(x), \quad u_t(x, 0) = g(x).$$

How about if the boundary conditions are $u_x(0, t) = u(l, t) = 0$?

2.3. Heat equation I
Solve the following initial-boundary value problem for the heat equation

$$u_t = 4u_{xx}, \ 0 < x < 1, \ t > 0,$$
$$u(x, 0) = x^2(1 - x),$$
$$u(0, t) = u(1, t) = 0.$$

2.4. Heat equation II
Solve the following initial-boundary value problem for the heat equation

$$u_t = 4u_{xx}, \ 0 < x < 1, \ t > 0,$$
$$u(x, 0) = x^2(1 - x),$$
$$u_x(0, t) = u_x(1, t) = 0.$$

How about if the boundary conditions are $u(0, t) = u_x(1, t) = 0$?

2.5. Wave equation on the real line
Consider the wave equation on the real line

$$u_{tt} = c^2 u_{xx}, \ x \in \mathbb{R}^1,$$
$$u(x, 0) = f(x),$$
$$u_t(x, 0) = g(x).$$

Find the solution by Fourier transform. The solution is the well-known d'Alembert's formula.

3 Stochastic Calculus in Hilbert Space

Basic probability concepts; Hilbert space; Brownian motion or Wiener process; stochastic calculus; Itô's formula

After recalling basic probability concepts, including Brownian motion and white noise, in Euclidean space $\mathbb{R}^n$, we review Fréchet derivatives and Gâteaux derivatives, which are needed for Itô's formula. Then we define random variables, especially Gaussian random variables, in Hilbert space. Finally, we discuss stochastic calculus in Hilbert space, including a version of Itô's formula, useful for analyzing stochastic partial differential equations.

3.1 Brownian Motion and White Noise in Euclidean Space

Recall that the Euclidean space $\mathbb{R}^n$ is equipped with the usual scalar product $\langle x, y \rangle = \sum_{j=1}^{n} x_j y_j$, which is also denoted by $x \cdot y$. This scalar product induces the Euclidean norm (or length) $\|x\|_{\mathbb{R}^n} \triangleq \sqrt{\sum_{j=1}^{n} x_j^2} = \sqrt{\langle x, x \rangle}$ and the Euclidean distance $d_{\mathbb{R}^n}(x, y) \triangleq \sqrt{\sum_{j=1}^{n} (x_j - y_j)^2} = \sqrt{\langle x - y, x - y \rangle}$. For convenience, we often denote this norm and this distance as $\| \cdot \|$ and $d(\cdot, \cdot)$, respectively.

With this distance, we can define the open ball, centered at x_0 with radius $r > 0$, as $B_r(x_0) \triangleq \{x \in \mathbb{R}^n : \|x - x_0\| < r\}$. The Borel σ-field of $\mathbb{R}^n$, i.e., $\mathcal{B}(\mathbb{R}^n)$, is generated via unions, intersections, and complements of all open balls in $\mathbb{R}^n$. Every element in $\mathcal{B}(\mathbb{R}^n)$ is called a *Borel measurable set* (or *Borel set*) in $\mathbb{R}^n$. A function $f : \mathbb{R}^n \to \mathbb{R}^1$ is called *Borel measurable* (or measurable) if for every Borel set A in $\mathbb{R}^1$, the preimage $f^{-1}(A) \triangleq \{x \in \mathbb{R}^n : f(x) \in A\}$ is a Borel set in $\mathbb{R}^n$.

A probability space $(\Omega, \mathcal{F}, \mathbb{P})$ consists of three ingredients: sample space Ω, σ-field $\mathcal{F}$ composed of certain subsets of Ω (also called *events*), and probability $\mathbb{P}$ (also called *probability measure*). For a collection $\mathcal{F}$ of subsets of Ω to be a σ-field, it must satisfy (i) the empty set $\varphi \in \mathcal{F}$; (ii) if $A \in \mathcal{F}$, then its complement $A^c \in \mathcal{F}$; and (iii) if $A_1, A_2, \ldots \in \mathcal{F}$, then $\bigcup_{i=1}^{\infty} A_i \in \mathcal{F}$. For example, $\mathcal{B}(\mathbb{R}^n)$, introduced above, is a σ-field of $\mathbb{R}^n$ (when we take $\mathbb{R}^n$ as a sample space).

When tossing a cubic die, the sample space is $\Omega = \{1, 2, 3, 4, 5, 6\}$. It is the set of all possible outcomes (i.e., different numbers on six faces). In a deterministic experiment, we can think of tossing a ball with the corresponding sample space $\Omega = \{1\}$, because

Effective Dynamics of Stochastic Partial Differential Equations. http://dx.doi.org/10.1016/B978-0-12-800882-9.00003-2

we could write only one number on a ball without causing confusion. Since there is only one outcome or one sample, we do not need to indicate it in a deterministic variable.

A property holds almost surely (a.s.) in the sample space Ω if the event that it holds is a sure event. In other words, the property holds on a subset $\tilde{\Omega} \subset \Omega$ of full probability measure (i.e., $\mathbb{P}(\tilde{\Omega}) = 1$).

In this book, we do not distinguish $\mathbb{P}(A)$ and $\mathbb{P}\{A\}$ for an event A.

A probability space is said to be complete if its σ-field contains all subsets of every zero-probability event [24, p. 17]. A probability space may be extended to a complete probability space.

A random variable X with values in Euclidean space $\mathbb{R}^n$ is a measurable mapping $X : \Omega \to \mathbb{R}^n$, namely, for every Borel set A (such as an open ball) in $\mathbb{R}^n$, the preimage $X^{-1}(A)$ is an event in $\mathcal{F}$.

The mean or mathematical expectation of a $\mathbb{R}^n$-valued random variable X is a vector in $\mathbb{R}^n$ defined as

$$\mathbb{E}(X) \triangleq \int_\Omega X(\omega)d\mathbb{P}(\omega),$$

whenever the integral, which is defined component by component, in the right-hand side exists. The variance of X is an $n \times n$ matrix defined as

$$\text{Var}(X) \triangleq \mathbb{E}[(X - \mathbb{E}(X))(X - \mathbb{E}(X))^T],$$

where ^{T} denotes the matrix transpose. The covariance of $\mathbb{R}^n$-valued random variables X and Y is an $n \times n$ matrix defined as

$$\text{Cov}(X, Y) = \mathbb{E}[(X - \mathbb{E}(X))(Y - \mathbb{E}(Y))^T].$$

We denote $\text{Cov}(X, X) = \text{Cov}(X)$. In fact, $\text{Cov}(X) = \text{Var}(X)$.

A random variable X in Euclidean space $\mathbb{R}^n$ induces a probability measure, $\mathbb{P}_X$, in $\mathbb{R}^n$ as

$$\mathbb{P}_X\{A\} \triangleq \mathbb{P}\{X^{-1}(A)\}, \quad A \in \mathcal{B}(\mathbb{R}^n). \tag{3.1}$$

The probability measure $\mathbb{P}_X$ is also called the law of X and is sometimes denoted as $\mathcal{L}(X)$. Thus $(\mathbb{R}^n, \mathcal{B}(\mathbb{R}^n), \mathbb{P}_X)$ is a probability space.

In fact, the probability measure $\mathbb{P}_X$ is a generalization of a more classical concept: probability distribution function. Recall that the distribution function $F_X(x)$ of X is defined as

$$F_X(x) = \mathbb{P}\{\omega \in \Omega : X(\omega) \leq x\} = \mathbb{P}\{X^{-1}(-\infty, x]\}. \tag{3.2}$$

If there exists a function $f : \mathbb{R}^n \to \mathbb{R}^1$ such that

$$F_X(x) = \int_{-\infty}^x f(\xi)d\xi, \tag{3.3}$$

then f is called the probability density function of the random variable X. The probability distribution measure $\mathbb{P}_X$ (or law $\mathcal{L}(X)$) of X is then

$$\mathbb{P}_X\{A\} = \int_A f(x)dx, \quad A \in \mathcal{B}(\mathbb{R}^n). \tag{3.4}$$

A scalar Gaussian (or normal) random variable $X : \Omega \to \mathbb{R}^1$, with $\mathbb{E}(X) = \mu$ and $\text{Var}(X) = \sigma^2$ (μ real and σ positive) is denoted as $X \sim \mathcal{N}(\mu, \sigma^2)$. That is, X has the probability density function

$$f(x) = \frac{1}{\sqrt{2\pi}\,\sigma} \exp \frac{-(x - \mu)^2}{2\sigma^2}. \tag{3.5}$$

Additionally, it can be shown that the odd central moments $\mathbb{E}(X - \mu)^{2k+1} = 0$ for $k = 0, 1, 2, \ldots$, and the even central moments $\mathbb{E}(X - \mu)^{2k} = 1 \cdot 3 \cdot 5 \cdots (2k - 1)\sigma^{2k}$ for $k = 1, 2, \ldots$ This implies that all moments of a Gaussian random variable can be expressed in terms of the first two moments, i.e., in terms of its mean and variance.

A random variable taking values in $\mathbb{R}^n$

$$X : \Omega \to \mathbb{R}^n$$

is called Gaussian if, for every $a = (a_1, \ldots, a_n) \in \mathbb{R}^n$, $X \cdot a = \langle X, a \rangle = a_1 X_1 + \cdots + a_n X_n$ is a scalar Gaussian random variable. A Gaussian random variable in $\mathbb{R}^n$ is denoted as $X \sim \mathcal{N}(m, Q)$, with mean vector $\mathbb{E}(X) = m$ and covariance matrix $\text{Cov}(X) = Q$. The covariance matrix Q is symmetric and nonnegative definite, i.e., eigenvalue $\lambda_j \geq 0$, $j = 1, \ldots, n$. The trace of Q is $\text{Tr}(Q) = \lambda_1 + \cdots + \lambda_n$. The covariance matrix is actually

$$Q = (Q_{ij}) = (\mathbb{E}[(X_i - m_i)(X_j - m_j)]).$$

The probability density function for this Gaussian random variable X in $\mathbb{R}^n$ is

$$f(x) = f(x_1, \ldots, x_n) = \frac{\sqrt{\det(A)}}{(2\pi)^{n/2}} e^{-\frac{1}{2}\sum_{j,k=1}^{n}(x_j - m_j)a_{jk}(x_k - m_k)}, \tag{3.6}$$

where $A = Q^{-1} = (a_{jk})$. In matrix form, this becomes

$$f(x) = \frac{\sqrt{\det(A)}}{(2\pi)^{n/2}} e^{-\frac{1}{2}(X-m)^T A(X-m)}. \tag{3.7}$$

For a two-dimensional Gaussian random variable $(X, Y)^T$, its components X, Y are independent if and only if they are uncorrelated (i.e., $\text{Cov}(X, Y) = \mathbb{E}[(X - \mathbb{E}X)$ $(Y - \mathbb{E}Y)] = \mathbb{E}(XY) - \mathbb{E}X\mathbb{E}Y = 0$). This property is, of course, not generally true for non-Gaussian random variables. Recall that $\mathbb{R}^n$-valued random variables $X_1, X_2, \ldots, X_n$ are said to be *independent* if

$$\mathbb{P}\{X_1 \in B_1, X_2 \in B_2 \ldots, X_n \in B_n\} = \mathbb{P}\{X_1 \in B_1\}\mathbb{P}\{X_2 \in B_2\} \cdots \mathbb{P}\{X_n \in B_n\}$$

for all $B_i \in \mathcal{B}(\mathbb{R}^n)$.

For $a, b \in \mathbb{R}^n$,

$$\mathbb{E}\langle X, a \rangle = \mathbb{E}\sum_{i=1}^{n} a_i X_i = \sum_{i=1}^{n} a_i \mathbb{E}(X_i) = \sum_{i=1}^{n} a_i m_i = \langle m, a \rangle, \tag{3.8}$$

and

$$
\mathbb{E}(\langle X - m, a \rangle \langle X - m, b \rangle) = \mathbb{E} \left(\sum_i a_i (X_i - m_i) \sum_j b_j (X_j - m_j) \right)
$$

$$
= \sum_{i,j} a_i b_j \mathbb{E}[(X_i - m_i)(X_j - m_j)]
$$

$$
= \sum_{i,j} a_i b_j Q_{ij} = \langle Qa, b \rangle. \tag{3.9}
$$

In particular, $\langle Qa, a \rangle = \mathbb{E}\langle X - m, a \rangle^2 \geq 0$, which confirms that Q is non-negative definite. Also, $\langle Qa, b \rangle = \langle a, Qb \rangle$, which implies that Q is symmetric.

Now we recall some basic properties of Brownian motion [18, Ch. 4], [24, Ch. 9.2], [115, Ch. 1], or [180, Ch. 2.5] in Euclidean space $\mathbb{R}^n$.

The Brownian motion or Wiener process $W(t)$ (also denoted as $W(t, \omega)$ or W_t) in $\mathbb{R}^n$ is a Gaussian stochastic process on a probability space $(\Omega, \mathcal{F}, \mathbb{P})$. Being a Gaussian process, $W(t)$ is characterized by its mean vector (taken to be the zero vector) and its covariance operator, which is an $n \times n$ symmetric nonnegative definite matrix (i.e., t times the identity matrix I). More specifically, $W(t)$ satisfies the following conditions [244, p. 11]:

(a) $W(0) = 0$, almost surely (a.s.)
(b) W has continuous sample paths or trajectories, a.s.
(c) W has independent increments, and
(d) $W(t) - W(s) \sim \mathcal{N}(0, (t - s)I), t \geq s \geq 0$.

Remark 3.1.

(i) The covariance matrix (or covariance operator) for $W(t)$ is $Q = t I$ (a diagonal matrix with each diagonal element being t) and $\text{Tr}(Q) = t n$. Because $\text{Cov}(W_i(t), W_j(t)) = 0$ for $i \neq j$, the components of $W(t)$ are pair-wise uncorrelated and thus are pair-wise independent.

(ii) $W(t) \sim N(0, t I)$, i.e., $W(t)$ has probability density function

$$
p_t(x) = \frac{1}{(2\pi t)^{\frac{n}{2}}} e^{-\frac{x_1^2 + \cdots + x_n^2}{2t}} = \frac{1}{(2\pi t)^{\frac{n}{2}}} e^{-\frac{1}{2t} \|x\|^2}.
$$

(iii) The paths of Brownian motion are Hölder continuous with exponent less than one half. Additionally,

$$
\lim_{t \to \infty} \frac{1}{t} \|W(t, \omega)\| = 0, \ a.s.
$$

(iv) For a scalar Brownian motion $W(t)$, the covariance $\mathbb{E}(W(t)W(s)) = t \wedge s \triangleq \min\{t, s\}$.

(v) Sample paths of Brownian motion $W(t)$ are not differentiable anywhere in the ordinary sense [24, p. 408], but its generalized time derivative exists, in the sense we usually use for generalized functions or weak solutions of partial differential equations [231, Ch. 2], [264, Ch. 5]. In fact, $\frac{d}{dt}W(t)$ is a mathematical model for white noise [18, p. 50].

(vi) From now on, we consider a two-sided Brownian motion $W(t), t \in \mathbb{R}$, defined by means of two independent usual Brownian motions $W^1(t)$ and $W^2(t)$:

$$W(t) \triangleq \begin{cases} W^1(t), & t \geq 0, \\ W^2(-t), & t < 0. \end{cases} \tag{3.10}$$

3.1.1 White Noise in Euclidean Space

A (Gaussian) noise is a special stationary stochastic process $\eta_t(\omega)$, with mean $\mathbb{E}\eta_t = 0$ and covariance $\mathbb{E}(\eta_t \eta_s) = Kc(t - s)$ for all t and s, for constant $K > 0$ and a function $c(\cdot)$. When $c(t - s)$ is the Dirac delta function $\delta(t - s)$, the noise η_t is called *white noise*; otherwise it is called *colored noise*.

Gaussian white noise may be modeled in terms of the "time derivative" of Brownian motion (i.e., Wiener process) $W(t)$. Let us only discuss this formally [199, Ch. 2.5], [247, Ch. 11.1] and refer to [18, Ch. 3] for a rigorous treatment.

Recall that a scalar Brownian motion $W(t)$ is a Gaussian process with stationary and independent increments, together with mean $\mathbb{E}W(t) = 0$ and covariance $\mathbb{E}(W(t)W(s)) = t \wedge s = \min\{t, s\}$.

The increment $W(t + \Delta t) - W(t) \approx \Delta t \dot{W}(t)$ is stationary, and formally, $\mathbb{E}\dot{W}(t) \approx \mathbb{E}\frac{W(t+\Delta t)-W(t)}{\Delta t} = \frac{0}{\Delta t} = 0$. Moreover, by the formal formula $\mathbb{E}(\dot{X}_t \dot{X}_s) = \partial^2 \mathbb{E}(X_t X_s)/\partial t \partial s$, we conclude that

$$\begin{aligned}
\mathbb{E}(\dot{W}(t)\dot{W}(s)) &= \frac{\partial^2}{\partial t \partial s}\mathbb{E}(W(t)W(s)) \\
&= \frac{\partial^2}{\partial t \partial s}(t \wedge s) \\
&= \frac{\partial}{\partial t}\frac{\partial}{\partial s}\begin{cases} t, & t - s < 0 \\ s, & t - s \geq 0 \end{cases} \\
&= \frac{\partial}{\partial t}\begin{cases} 0, & t - s < 0 \\ 1, & t - s \geq 0 \end{cases} \\
&= \frac{\partial}{\partial t}H(t - s) \\
&= \delta(t - s),
\end{aligned}$$

where $H(\xi)$ is the Heaviside function

$$H(\xi) = \begin{cases} 1, & \xi \geq 0, \\ 0, & \xi < 0. \end{cases} \tag{3.11}$$

Note that $\frac{d}{d\xi}H(\xi) = \delta(\xi)$. This, formally, says that $\dot{W}(t)$ is uncorrelated at different time instants.

Additionally, the spectral density function for $\eta_t \triangleq \dot{W}(t)$, i.e., the Fourier transform $\mathfrak{F}$ for its covariance function $\mathbb{E}(\eta_t \eta_s)$ is (resembling: "white light"),

$$\mathfrak{F}(\mathbb{E}(\eta_t \eta_s)) = \mathfrak{F}(\mathbb{E}(\dot{W}(t)\dot{W}(s))) = \mathfrak{F}(\delta(t - s)) = \frac{1}{\sqrt{2\pi}} e^{-isk}.$$

So the power spectrum is constant: $|\mathfrak{F}(\mathbb{E}(\eta_t \eta_s))|^2 = \frac{1}{2\pi}$. This resembles the white light. Thus, $\eta_t = \dot{W}(t)$ is taken as a mathematical model for white noise.

3.2   Deterministic Calculus in Hilbert Space

To develop tools to handle stochastic calculus in Hilbert space, we first recall some concepts of deterministic calculus in Hilbert space.

We work in a separable Hilbert space., i.e., a Hilbert space with a countable basis $\{e_n\}_{n=1}^{\infty}$. In fact, by Gram–Schmidt orthogonalization, we can take this basis to be orthonormal, which means that $\langle e_i, e_j \rangle = \delta_{ij}$ for all i and j.

The scalar product in Hilbert space H induces a norm $\|u\|_H = \|u\| \triangleq \sqrt{\langle u, u \rangle}$. With this norm, we can define a metric or distance $d(u, v) = \|u - v\|$. With this distance, an open ball centered at h of radius r is defined as the set $\{u \in H : d(u, h) < r\}$. The Borel σ-field of H, i.e., $\mathcal{B}(H)$, is generated via unions, intersections, and complements of all open balls in H. Every element in $\mathcal{B}(H)$ is called a Borel set in H. A mapping $f : H \to \mathbb{R}^n$ is called Borel measurable if, for every Borel set A in $\mathbb{R}^n$, the preimage $f^{-1}(A)$ is a Borel set in H.

A Banach space is a vector space with a norm, under which all Cauchy sequences are convergent [313, p. 52]. Euclidean space $\mathbb{R}^n$ and Hilbert spaces are also Banach spaces. In Euclidean, Hilbert, and Banach spaces, a distance, or metric, is induced by the norm $\| \cdot \|$ as $d(x, y) = \|x - y\|$. Thus we have concepts such as convergence, continuity, and differentiability (see §3.2).

An even more general space is called a *metric space*, which is a set with a metric (or distance). Euclidean, Hilbert, and Banach spaces are examples of metric spaces where distance (or metric) is induced by the norm, as above.

For calculus in Euclidean space $\mathbb{R}^n$, we have concepts of *derivative* and *directional derivative*. In Hilbert space, we have the corresponding *Fréchet derivative* and *Gâteaux derivative* [34, Ch. 2], [317, Ch. 4].

Let H and $\hat{H}$ be two Hilbert spaces, and let $F : U \subset H \to \hat{H}$ be a mapping whose domain of definition U is an open subset of H. Let $\mathcal{L}(H, \hat{H})$ be the set of all bounded (i.e., continuous) linear operators $A : H \to \hat{H}$. In particular, we denote $\mathcal{L}(H) \triangleq \mathcal{L}(H, H)$. We can also introduce a multilinear operator $A_1 : H \times H \to \hat{H}$. The space of all these multilinear operators is denoted as $\mathcal{L}(H \times H, \hat{H})$. A linear operator $A \in \mathcal{L}(H)$ is of trace class if there exist sequences $\{a_k\}$ and $\{b_k\}$ in H such that

$$Ax = \sum_{k=1}^{\infty} \langle x, a_k \rangle b_k, \quad x \in H$$

and

$$\sum_{k=1}^{\infty} \|a_k\| \|b_k\| < \infty.$$

It is known that every trace class operator A is compact. In the following we denote by $\mathcal{L}^1(H) \triangleq \mathcal{L}^1(H, H)$ the set of all trace class operators in $\mathcal{L}(H)$, which is a Banach space endowed with the norm

$$\|A\|_{\mathcal{L}^1} = \inf \left\{ \sum_{k=1}^{\infty} \|a_k\| \|b_k\| : Ax = \sum_{k=1}^{\infty} \langle x, a_k \rangle b_k, \ x \in H, \{a_k\}, \{b_k\} \subset H \right\}.$$

For a linear operator $A \in \mathcal{L}^1(H)$, its trace $\mathrm{Tr}(A)$ is defined as

$$\mathrm{Tr}(A) \triangleq \sum_{k=1}^{\infty} \langle Ae_k, e_k \rangle,$$

where $\{e_k\}$ is a complete orthonormal basis on H. The definition is independent of the choice of a complete orthonormal basis on H.

Definition 3.2. The mapping F is Fréchet differentiable at $u_0 \in U$ if there is a linear bounded operator $A : H \to \hat{H}$ such that

$$\lim_{h \to 0} \frac{\|F(u_0 + h) - F(u_0) - Ah\|_{\hat{H}}}{\|h\|_H} = 0,$$

or

$$\|F(u_0 + h) - F(u_0) - Ah\|_{\hat{H}} = o(\|h\|_H).$$

When there is no confusion, we use $\| \cdot \|$ to denote both norms in H and $\hat{H}$. The linear bounded operator A is called the Fréchet derivative of F at u_0 and is denoted as $F_u(u_0)$. Sometimes it is also denoted as $F'(u_0)$ or $dF(u_0)$.

In other words, if F is Fréchet differentiable at u_0, then

$$F(u_0 + h) - F(u_0) = F_u(u_0)h + R(u_0, h),$$

where the remainder $R(u_0, h)$ satisfies the condition $\|R(u_0, h)\| = o(\|h\|)$, i.e., $\lim_{h \to 0} \frac{\|R(u_0, h)\|}{\|h\|} = 0$.

For a nonlinear mapping $F : U \subset H \to \hat{H}$, its Fréchet derivative $F'(u_0)$ is a linear operator, i.e., $F'(u_0) \in \mathcal{L}(H, \hat{H})$. If F is linear, its Fréchet derivative is the mapping F itself.

Definition 3.3. The directional derivative of F at $u_0 \in U$ in the direction $h \in H$ is defined by the limit

$$dF(u_0, h) \triangleq \lim_{\varepsilon \to 0} \frac{F(u_0 + \varepsilon h) - F(u_0)}{\varepsilon} = \frac{d}{d\varepsilon} F(u_0 + \varepsilon h)|_{\varepsilon=0}.$$

If this limit exists for every $h \in H$, and $dF(u_0, h)$ is a linear mapping in h, then we say that F is Gâteaux differentiable at u_0 and this linear mapping $dF(u_0, h)$ is called the Gâteaux derivative of F at u_0.

If F is Gâteaux differentiable at $u_0 + \varepsilon h$ for $0 \leq \varepsilon \leq 1$, then [34, p. 68],

$$F(u_0 + \varepsilon h) - F(u_0) = \int_0^1 dF(u_0 + \varepsilon h, h) d\varepsilon.$$

The Fréchet and Gâteaux derivatives are unique. Furthermore, if F is Fréchet differentiable at u_0, then it is also Gâteaux differentiable at u_0. Conversely, if the Gâteaux derivative $dF(u_0, h)$ is a continuous mapping in u_0, then F is also Fréchet differentiable. In either case, we have the formula [34, Theorem 2.1.13], [317, Problem 4.9]

$$F_u(u_0)h = dF(u_0, h).$$

Similarly, we can define higher-order Fréchet derivatives. Each of these derivatives is a multilinear operator. For example,

$$F_{uu}(u_0) : H \times H \to \hat{H}, \tag{3.12}$$
$$(h_1, h_2) \longmapsto F_{uu}(u_0)(h_1, h_2). \tag{3.13}$$

The second-order Gâteaux derivative is defined as

$$d^2 F(u_0, h_1, h_2) = d(dF(u_0, h_1), h_2) = \frac{d}{d\varepsilon} dF(u_0 + \varepsilon h_1)|_{\varepsilon=0}$$

$$= \frac{\partial^2}{\partial \varepsilon_1 \partial \varepsilon_2} F(u_0 + \varepsilon_1 h_1 + \varepsilon_2 h_2)|_{\varepsilon_1 = \varepsilon_2 = 0}. \tag{3.14}$$

If F is second-order Fréchet differentiable at u_0, then it is also second-order Gâteaux differentiable at u_0. Conversely, if the second-order Gâteaux derivative $d^2 F(u_0, h_1, h_2)$ is a continuous mapping in u_0, then F is also second-order Fréchet differentiable. In either case, we have the formula [34, Theorem 2.1.27]

$$F_{uu}(u_0)(h_1, h_2) = d^2 F(u_0, h_1, h_2).$$

We denote

$$F''(u_0)h^2 \triangleq F_{uu}(u_0)(h, h),$$
$$F'''(u_0)h^3 \triangleq F_{uuu}(u_0)(h, h, h),$$

and similarly for higher-order Fréchet derivatives.

If F is nth order Fréchet differentiable in a neighborhood of u_0, then

$$F(u_0 + h) = F(u_0) + \frac{1}{1!} F_u(u_0)h + \frac{1}{2!} F_{uu}(u_0)h^2$$

$$+ \cdots + \frac{1}{n!} F_{u...u}(u_0)h^n + R(u_0, h),$$

or

$$F(u_0 + h) = F(u_0) + \frac{1}{1!}F'(u_0)h + \frac{1}{2!}F''(u_0)h^2$$

$$+ \cdots + \frac{1}{n!}F^{(n)}(u_0)h^n + R(u_0, h),$$

where the remainder $R(u_0, h)$ satisfies the condition $\|R(u_0, h)\| = o(\|h\|^n)$, i.e., $\lim_{h \to 0} \frac{\|R(u_0,h)\|}{\|h\|^n} = 0$.

More specifically, we have the Taylor expansion in Hilbert space

$$F(u + h) = F(u) + F'(u)h + \frac{1}{2!}F''(u)h^2$$

$$+ \cdots + \frac{1}{m!}F^{(m)}(u)h^m + R_{m+1}(u, h),$$

where the remainder

$$R_{m+1}(u, h) = \frac{1}{(m+1)!} \int_0^1 (1 - s)^m F^{(m+1)}(u + sh)h^{m+1} ds.$$

Let us consider several examples.

Example 3.4. Let $F : \mathbb{R}^n \to \mathbb{R}^1$ be a real-valued function of variable $\mathbf{x} = (x_1, \ldots, x_n)$. The first-order Fréchet derivative at $\mathbf{x}_0$ is

$$F_{\mathbf{x}}(\mathbf{x}_0) : \mathbb{R}^n \to \mathbb{R}^1,$$

$$F_{\mathbf{x}}(\mathbf{x}_0)h = (DF(\mathbf{x}_0))^T h, \quad h \in \mathbb{R}^n,$$

where $DF = (\partial F / \partial x_1, \ldots, \partial F / \partial x_n)^T$ is the Jacobian vector or gradient of F. In calculus, we usually use D instead of d to denote first-order derivative (or gradient). The second-order Fréchet derivative at $\mathbf{x}_0$ is

$$F_{\mathbf{xx}}(\mathbf{x}_0) : \mathbb{R}^n \times \mathbb{R}^n \to \mathbb{R}^1,$$

$$F_{\mathbf{xx}}(\mathbf{x}_0)(h, k) = h^T H(F(\mathbf{x}_0))k, \quad h, k \in \mathbb{R}^n,$$

where the matrix $H(F) = \left(\frac{\partial^2 F}{\partial x_i \partial x_j} \right)_{n \times n}$ is also called the *Hessian matrix* of F. Moreover, the Gâteaux derivative at $\mathbf{x}_0$ is

$$dF(\mathbf{x}_0, h) = (DF(\mathbf{x}_0))^T h, \quad h \in \mathbb{R}^n.$$

Example 3.5. Let H be a Hilbert space with scalar product $\langle \cdot, \cdot \rangle$ and norm $\| \cdot \|^2 = \langle \cdot, \cdot \rangle$, as in [76, p. 34].

Consider a functional $F(u) = \|u\|^{2p}$ for $p \in [1, \infty)$. In this case, $F_u(u_0)h = 2p\|u_0\|^{2p-2}\langle u_0, h \rangle$ and

$$F_{uu}(u_0)(h, k) = 2p\|u_0\|^{2p-2}\langle h, k \rangle + 4p(p-1)\|u_0\|^{2p-4}\langle u_0, h \rangle \langle u_0, k \rangle$$

$$= 2p\|u_0\|^{2p-2}\langle h, k \rangle + 4p(p-1)\|u_0\|^{2p-4}\langle (u_0 \otimes u_0)h, k \rangle,$$

where $(a \otimes b)h := a\langle b, h \rangle$ as introduced in §3.3.

We list two particular cases.

(i) $p = \frac{1}{2}$:

$$F_u(u_0)h = \frac{1}{\|u_0\|}\langle u_0, h\rangle,$$

and

$$F_{uu}(u_0)(h, k) = \frac{1}{\|u_0\|}\langle h, k\rangle - \frac{1}{\|u_0\|^3}\langle u_0, h\rangle\langle u_0, k\rangle.$$

(ii) $p = 1$:

$$F_u(u_0)h = 2\langle u_0, h\rangle,$$

and

$$F_{uu}(u_0)(h, k) = 2\langle h, k\rangle.$$

Example 3.6.   It is interesting to see the relation between the Fréchet derivative and the variational derivative in the context of calculus of variations. The variational derivative is usually considered for functionals defined as spatial integrals, such as a Lagrange functional in mechanics. For example,

$$F(u) = \int_0^l G(u(x), u_x(x))dx,$$

where u is defined for $x \in [0, l]$ and satisfies zero Dirichlet boundary conditions at $x = 0, l$. Then it is known [165, p. 415] that

$$F_u(u)h = \int_0^l \frac{\delta F}{\delta u}h(x)dx \tag{3.15}$$

for h in the Hilbert space $H_0^1(0, l)$. The quantity $\frac{\delta F}{\delta u}$ is the classical variational derivative of F. The equation (3.15) above gives the relation between Fréchet derivative and variational derivative.

One difficulty for problems in infinite dimensional space is that a bounded and closed set may not be compact, contrary to the case in finite dimensional Euclidean space. We need the following result on compactness.

For a Banach space $\mathcal{S}$, let $\mathcal{S}'$ be its dual space. We can also define the double dual space $\mathcal{S}''$. If S coincides with its double dual space S'', then it is called a *reflexive Banach space*.

Let $\mathcal{X} \subset \mathcal{Y} \subset \mathcal{Z}$ be three reflexive Banach spaces, and let $\mathcal{X} \subset \mathcal{Y}$ with compact and dense embedding. Define a new Banach space

$$G = \left\{v : v \in L^2(0, T; \mathcal{X}), \frac{dv}{dt} \in L^2(0, T; \mathcal{Z})\right\}$$

with norm

$$\|v\|_G^2 = \int_0^T |v(s)|_{\mathcal{X}}^2 \, ds + \int_0^T \left|\frac{dv}{ds}(s)\right|_{\mathcal{Z}}^2 ds, \quad v \in G.$$

We recall the following result [215, Theorem 5.1].

Lemma 3.7. *If B is bounded in G, then it is precompact in $L^2(0, T; \mathcal{Y})$.*

We will be concerned with the limit of a family of functions u^ϵ, which are parameterized by a small parameter ϵ, as $\epsilon \to 0$. The compactness of u^ϵ refers to the compactness of the set $\{u^\epsilon : \epsilon > 0\}$. Compactness of u^ϵ in some space is needed because passing the limit $\epsilon \to 0$ is usually a difficult issue. But a clever and easy approach is to consider compactness in a weak sense [313, Ch. V]. Here we recall convergence in some weak senses that are needed in later chapters.

We recall the definitions and some properties of weak convergence and weak* convergence [313, Ch. V.1].

Definition 3.8. A sequence $\{s_n\}$ in Banach space S is said to converge weakly to $s \in S$ if, for every $s' \in S'$,

$$\lim_{n \to \infty} (s', s_n)_{S', S} = (s', s)_{S', S}.$$

This is written as $s_n \rightharpoonup s$ weakly in S. Note that (s', s) denotes the value of the continuous linear functional s' at the point s.

By the Riesz representation theorem [313, Ch. III. 6], every continuous linear functional in Hilbert space H is represented in scalar product $\langle \cdot, \cdot \rangle$. Therefore, a sequence $\{s_n\}$ in Hilbert space H converges weakly to s in H, if

$$\langle s_n, h \rangle \to \langle s, h \rangle, \quad \text{for every } h \in H.$$

The Riesz representation theorem also implies that a Hilbert space is a reflexive Banach space.

Lemma 3.9 (Eberlein–Shmulyan). *Assume that Banach space S is reflexive, and let $\{s_n\}$ be a bounded sequence in S. Then there exists a subsequence $\{s_{n_k}\}$ and $s \in S$ such that $s_{n_k} \rightharpoonup s$ weakly in S as $k \to \infty$. Moreover, if all weakly convergent subsequences of $\{s_n\}$ have the same limit s in S, then the sequence $\{s_n\}$ itself weakly converges to s.*

Since a Hilbert space is a reflexive Banach space, a bounded sequence in a Hilbert space has a weakly convergent subsequence.

Definition 3.10. A sequence $\{s_n'\}$ in S' is said to converge *weakly** to $s' \in S'$ if, for every $s \in S$,

$$\lim_{n \to \infty} (s_n', s)_{S', S} = (s', s)_{S', S}.$$

This is written as $s_n' \rightharpoonup s'$ *weakly** in S'.

By the previous lemma, we have the following result.

Lemma 3.11.   *Assume that the dual space S' is reflexive, and let $\{s'_n\}$ be a bounded sequence in S'. Then there exists a subsequence $\{s'_{n_k}\}$ and $s' \in S'$ such that $s'_{n_k} \rightharpoonup s'$ weakly* in S' as $k \to \infty$. Moreover, if all weakly* convergent subsequences of $\{s'_n\}$ have the same limit s' in S', the sequence $\{s'_n\}$ itself weakly* converges to s'.*

3.3   Random Variables in Hilbert Space

Let $(\Omega, \mathcal{F}, \mathbb{P})$ be a probability space, with sample space Ω, σ-field $\mathcal{F}$, and probability measure $\mathbb{P}$. The Borel σ-field, $\mathcal{B}(H)$, in Hilbert space H, the smallest σ-field containing all open balls. Every element of $\mathcal{B}(H)$ is called a Borel set in H. A random variable in Hilbert space H (i.e., taking values in H) is a Borel measurable mapping

$$X : \Omega \to H.$$

That is, for every Borel set A in H, the preimage $X^{-1}(A) \triangleq \{\omega \in \Omega : X(\omega) \in A\}$ is an event in $\mathcal{F}$.

Similar to real-valued or vector-valued random variables, mathematical expectation of X is defined in terms of the integral with respect to the probability measure $\mathbb{P}$

$$\mathbb{E}(X) = \int_{\Omega} X(\omega)d\mathbb{P}(\omega).$$

The variance of X is

$$\mathrm{Var}(X) = \mathbb{E}\langle X - \mathbb{E}(X), X - \mathbb{E}(X)\rangle = \mathbb{E}\|X - \mathbb{E}(X)\|^2 = \mathbb{E}\|X\|^2 - \|\mathbb{E}(X)\|^2.$$

In the special case when $\mathbb{E}(X) = 0$, the variance and the second moment are the same, i.e., $\mathrm{Var}(X) = \mathbb{E}\|X\|^2$.

The covariance operator of X is defined as

$$\mathrm{Cov}(X) = \mathbb{E}[(X - \mathbb{E}(X)) \otimes (X - \mathbb{E}(X))], \tag{3.16}$$

where for every pair $a, b \in H$, $a \otimes b$ is a bilinear operator (called the *tensor product*), defined by

$$a \otimes b : H \to H, \tag{3.17}$$

$$(a \otimes b)h = a\langle b, h\rangle, \quad h \in H. \tag{3.18}$$

Let X and Y be two random variables taking values in Hilbert space H. The correlation operator of X and Y is defined by

$$\mathrm{Corr}(X, Y) = \mathbb{E}[(X - \mathbb{E}(X)) \otimes (Y - \mathbb{E}(Y))]. \tag{3.19}$$

Remark 3.12.   The covariance operator $\mathrm{Cov}(X)$ is a symmetric, positive-definite and trace-class linear operator with trace

$$\mathrm{Tr}(\mathrm{Cov}(X)) = \mathbb{E}\langle X - \mathbb{E}(X), X - \mathbb{E}(X)\rangle = \mathbb{E}\|X - \mathbb{E}(X)\|^2. \tag{3.20}$$

Also,

$$\mathrm{Tr}(\mathrm{Corr}(X, Y)) = \mathbb{E}\langle X - \mathbb{E}(X), Y - \mathbb{E}(Y)\rangle. \tag{3.21}$$

To study a sequence of random variables $\{X_n\}$ in an infinite dimensional space $\mathcal{M}$, we need the following definitions. We also often denote $\{X_n\}$ simply by X_n. So we will not distinguish these two notations.

A random variable X in Hilbert space H induces a probability measure, $\mathbb{P}_X$, in H as follows:

$$\mathbb{P}_X\{A\} \triangleq \mathbb{P}\{X^{-1}(A)\}, \quad A \in \mathcal{B}(H). \tag{3.22}$$

The probability measure $\mathbb{P}_X$ is also called the law of X and is sometimes denoted as $\mathcal{L}(X)$. Thus $(H, \mathcal{B}(H), \mathbb{P}_X)$ is a probability space.

A sequence of probability measures μ_n on $(H, \mathcal{B}(H))$ converges weakly to a probability measure μ if

$$\lim_{n \to \infty} \int_H \varphi(x)\mu_n(dx) = \int_H \varphi(x)\mu(dx) \tag{3.23}$$

for every bounded continuous function φ on H. A sequence of random variables X_n, taking values in H, converges weakly to a random variable X if the sequence of laws $\mathcal{L}(X_n)$ converges weakly to the law $\mathcal{L}(X)$.

Definition 3.13 (Tightness). A sequence of random variables $\{X_n\}$ with values in metric space $\mathcal{M}$ is called *tight* if, for every $\delta > 0$, there exists a compact set $K_\delta \subset \mathcal{M}$ such that

$$\mathbb{P}\{X_n \in K_\delta\} \geq 1 - \delta, \quad \text{for all } n.$$

Here the metric space $\mathcal{M}$ may be a Hilbert or Banach space. An example is $\mathcal{M} = C(0, T; H)$, with $T > 0$ and H being a Hilbert space. Moreover, tightness yields a compactness of the random variable in the sense of distribution. Let $\mathcal{L}(X_n)$ denote the distribution of X_n in space $\mathcal{M}$. We have the following compactness result [94, Theorem 2.3].

Theorem 3.14 (Prohorov theorem). *Assume that $\mathcal{M}$ is a separable Banach space. The set of probability measures $\{\mathcal{L}(X_n)\}$ on $(\mathcal{M}, \mathcal{B}(\mathcal{M}))$ is relatively compact if and only if $\{X_n\}$ is tight.*

By the Prohorov theorem, if the set of random variables $\{X_n\}$ is tight, then there is a subsequence $\{X_{n_k}\}$ and some probability measure μ such that

$$\mathcal{L}(X_{n_k}) \to \mu, \quad \text{weakly as } k \to \infty.$$

The Prohorov theorem has: "counterparts" in other parts of mathematics. The Heine-Borel theorem [17, p. 59] says that every bounded sequence in Euclidean space has a convergent subsequence. The Eberlein–Shmulyan lemma above (Lemma 3.9) implies that every bounded sequence in Hilbert space has a weakly convergent subsequence. The well-known Ascoli-Arzela theorem [313, p. 85] for $C(S)$, the space of continuous functions on a compact metric space S, states that every equi-bounded and equi-continuous sequence has a convergent subsequence.

Tightness usually does not imply the almost sure convergence of the sequence of random variables. However, the following theorem links weak convergence of probability measures and almost sure convergence of random variables [94, Theorem 2.4].

Theorem 3.15 (Skorohod theorem). *If a sequence of probability measures $\{\mu_n\}$ on $(\mathcal{M}, \mathcal{B}(\mathcal{M}))$ weakly converges to a probability measure μ, then there exists a probability space $(\Omega, \mathcal{F}, \mathbb{P})$ and random variables, $X_1, X_2, \ldots$ and X such that the law $\mathcal{L}(X_n) = \mu_n$ and the law $\mathcal{L}(X) = \mu$ and*

$$\lim_{n \to \infty} X_n = X, \quad \mathbb{P}\text{-}a{:}s{:}$$

3.4 Gaussian Random Variables in Hilbert Space

Inspired by the definition of Gaussian random variables in Euclidean space (§3.1), we now introduce Gaussian random variables taking values in Hilbert space H with scalar product $\langle \cdot, \cdot \rangle$. See [94, Ch. 1] for more information.

Definition 3.16. A random variable $X : \Omega \to H$ in Hilbert space H is called a Gaussian random variable if, for every a in H, the scalar random variable $\langle X, a \rangle$ is a scalar Gaussian random variable.

Remark 3.17. If X is a Gaussian random variable taking values in Hilbert space H, then there exist a vector m in H and a symmetric, positive-definite operator $Q : H \to H$ such that, for all $a, b \in H$,

 (i) $\mathbb{E}\langle X, a \rangle = \langle m, a \rangle$, and
 (ii) $\mathbb{E}(\langle X - m, a \rangle \langle X - m, b \rangle) = \langle Qa, b \rangle$.

We call m the mean vector and Q the covariance operator for X. We denote these as $\mathbb{E}(X) \triangleq m$ and $\text{Cov}(X) \triangleq Q$, respectively. A Gaussian random variable X, with mean vector m and covariance operator Q, is also symbolically denoted by $X \sim \mathcal{N}(m, Q)$.

Remark 3.18. The Borel probability measure μ on $(H, \mathcal{B}(H))$, induced by a Gaussian random variable X taking values in Hilbert space H, is called a Gaussian measure in H. For this Gaussian measure, there exist an element $m \in H$ and a positive-definite symmetric continuous linear operator $Q : H \to H$ such that for all $h, h_1, h_2 \in H$,

 (i) Mean vector $m : \int_H \langle x, h \rangle d\mu(x) = \langle m, h \rangle$, and
 (ii) Covariance operator $Q : \int_H \langle x, h_1 \rangle \langle x, h_2 \rangle d\mu(x) - \langle m, h_1 \rangle \langle m, h_2 \rangle = \langle Qh_1, h_2 \rangle$.

Since the covariance operator Q is positive-definite and symmetric, the eigenvalues of Q are positive and the eigenvectors, $e_n, n = 1, 2, \ldots$, form an orthonormal basis (after an appropriate normalization) for Hilbert space H. Let q_n be the eigenvalue corresponding to the eigenvector e_n, i.e.,

$$Qe_n = q_n e_n, \quad n = 1, 2, \ldots$$

Then the trace $\text{Tr}(Q) = \sum_{n=1}^{\infty} q_n$.
 Note that

$$X - m = \sum_{n=1}^{\infty} X_n e_n, \tag{3.24}$$

with coefficients $X_n = \langle X - m, e_n \rangle$ and

$$
\begin{aligned}
\mathbb{E}X_n^2 &= \mathbb{E}(\langle X - m, e_n \rangle \langle X - m, e_n \rangle) \\
&= \langle Qe_n, e_n \rangle = \langle q_n e_n, e_n \rangle = q_n.
\end{aligned}
\tag{3.25}
$$

Therefore,

$$
\| X - m \|^2 = \sum_{n=1}^{\infty} X_n^2,
\tag{3.26}
$$

and

$$
\mathbb{E}\| X - m \|^2 = \sum_{n=1}^{\infty} \mathbb{E}X_n^2 = \sum_{n=1}^{\infty} q_n = \mathrm{Tr}(Q).
\tag{3.27}
$$

We use $L^2(\Omega, H)$, or just $L^2(\Omega)$, to denote the Hilbert space of square-integrable random variables $X : \Omega \to H$. In this Hilbert space, the scalar product is

$$
\langle X, Y \rangle = \mathbb{E}\langle X(\omega), Y(\omega) \rangle,
$$

where $\mathbb{E}$ denotes the expectation with respect to probability $\mathbb{P}$. This scalar product induces the usual root mean-square norm

$$
\| X \|_{L^2(\Omega)} \triangleq \sqrt{\mathbb{E}\| X(\omega) \|_H^2}.
$$

When there is no confusion, we also denote this norm as $\| \cdot \|$. This norm provides an appropriate convergence concept: convergence in mean square or convergence in $L^2(\Omega)$. This limit is usually denoted by lim in m.s.

3.5 Brownian Motion and White Noise in Hilbert Space

Let $(\Omega, \mathcal{F}, \mathbb{P})$ be a probability space. We define Brownian motion or Wiener process [e.g., 94, Ch. 4.1], in Hilbert space H. We consider a symmetric positive linear operator Q in H. If the trace $\mathrm{Tr}(Q) < +\infty$, we say Q is a trace class (or nuclear) operator. Then there exists an orthonormal basis (formed by eigenfunctions of Q) $\{e_k\}$ for H and a (bounded) sequence of nonnegative real numbers (eigenvalues of Q) q_k such that

$$
Qe_k = q_k e_k, \quad k = 1, 2, \ldots
$$

We can "imagine" the covariance operator Q as an $\infty \times \infty$ diagonal matrix with diagonal elements $q_1, q_2, \ldots, q_n, \ldots$

A stochastic process $W(t)$, or W_t, taking values in H, for $t \geq 0$, is called a Wiener process with covariance operator Q if

(a) $W(0) = 0$ a.s.,

(b) W has continuous sample paths a.s.,

(c) W has independent increments, and

(d) $W(t) - W(s) \sim \mathcal{N}(0, (t-s)Q), t \geq s.$

Hence, $W(t) \sim \mathcal{N}(0, tQ)$, i.e., $\mathbb{E}W(t) = 0$ and $\mathrm{Cov}(W(t)) = tQ$. Sometimes, $W(t)$ is also called a Q-Wiener process.

The two-sided Wiener process, $W(t)$ for $t \in \mathbb{R}$, is defined as in §3.2.

For every $a \in H$, we have

$$a = \sum_n \langle a, e_n \rangle e_n$$

and

$$Qa = \sum_n \langle a, e_n \rangle Q e_n = \sum_n q_n \langle a, e_n \rangle e_n.$$

We define the fractional power of the operator Q as follows: For $\gamma \in (0, 1)$,

$$Q^\gamma a \triangleq \sum_n q_n^\gamma \langle a, e_n \rangle e_n \tag{3.28}$$

whenever the right-hand side is well defined. Furthermore, for symmetric and positive-definite Q, we can also define Q^γ as in (3.28) for all $\gamma \in \mathbb{R}$. In fact, for a given function $h : \mathbb{R}^1 \to \mathbb{R}^1$, we define the operator $h(Q)$ through the following natural formula [316, p. 293],

$$h(Q)u = \sum_n h(q_n) \langle u, e_n \rangle e_n,$$

as long as the right-hand side is well defined.

Representations of Brownian motion in Hilbert space

An H-valued Brownian motion $W(t)$ has an infinite series representation [94, Ch. 4.1]

$$W(t) = \sum_{n=1}^{\infty} \sqrt{q_n} W_n(t) e_n, \tag{3.29}$$

where

$$W_n(t) := \begin{cases} \frac{\langle W(t), e_n \rangle}{\sqrt{q_n}}, & q_n > 0, \\ 0, & q_n = 0, \end{cases} \tag{3.30}$$

are standard scalar independent Brownian motions, that is, $W_n(t) \sim \mathcal{N}(0, t)$, $\mathbb{E}W_n(t) = 0$, $\mathbb{E}W_n(t)^2 = t$ and $\mathbb{E}(W_n(t)W_n(s)) = \min\{t, s\}$. The infinite series (3.29) converges in $L^2(\Omega)$ as long as $\mathrm{Tr}(Q) = \sum q_n < \infty$. From this representation, we conclude

$$\lim_{t \to \infty} \frac{1}{t} \|W(t)\| = 0, \quad a.s.$$

Note that we may imagine that $W(t)$ to be a Brownian motion with countably many components, $\langle W(t), e_n \rangle, n = 1, 2, \ldots$, with the covariance being an infinity by infinity diagonal "matrix" $t Q$ of diagonal elements $tq_1, tq_2, \ldots$ Thus the components $\langle W, e_i \rangle$ and $\langle W, e_j \rangle$ are uncorrelated and hence independent for $i \neq j$. Therefore, $W_n(t)$'s are pair-wise independent standard scalar Brownian motions.

For $a, b \in H$, we have the following identities:

$$\mathbb{E}\langle W(t), W(t) \rangle = \mathbb{E}\|W(t)\|^2 = \mathbb{E}\left\langle \sum_{n=1}^{\infty} \sqrt{q_n} W_n(t) e_n, \sum_{n=1}^{\infty} \sqrt{q_n} W_n(t) e_n \right\rangle$$

$$= \sum_{n=1}^{\infty} q_n \mathbb{E}[W_n(t) W_n(t)]$$

$$= t \sum_{n=1}^{\infty} q_n = t \mathrm{Tr}(Q), \tag{3.31}$$

$$\mathbb{E}\langle W(t), a \rangle = \langle 0, a \rangle = 0, \tag{3.32}$$

and

$$\mathbb{E}(\langle W(t), a \rangle \langle W(t), b \rangle) = \mathbb{E}\left[\left\langle \sum_{n=1}^{\infty} \sqrt{q_n} W_n(t) e_n, a \right\rangle \left\langle \sum_{n=1}^{\infty} \sqrt{q_n} W_n(t) e_n, b \right\rangle \right]$$

$$= \mathbb{E} \sum_{m,n} \sqrt{q_m q_n} W_m(t) W_n(t) \langle e_m, a \rangle \langle e_n, b \rangle$$

$$= t \sum_{n} \langle e_n, a \rangle \langle q_n e_n, b \rangle = t \sum_{n} \langle Q \langle e_n, a \rangle e_n, b \rangle$$

$$= t \left\langle Q \sum_{n} \langle e_n, a \rangle e_n, b \right\rangle = t \langle Qa, b \rangle,$$

where we have used the fact that $a = \sum_n \langle e_n, a \rangle e_n$ in the final step.

In particular, taking $a = b$, we obtain

$$\mathbb{E}\langle W(t), a \rangle^2 = t \langle Qa, a \rangle \tag{3.33}$$

and

$$\mathrm{Var}(\langle W(t), a \rangle) = t \langle Qa, a \rangle. \tag{3.34}$$

More generally,

$$\mathbb{E}(\langle W(t), a \rangle \langle W(s), b \rangle) = \min\{t, s\} \langle Qa, b \rangle. \tag{3.35}$$

In the context of SPDEs, the basis $\{e_n\}$ of Hilbert space H often depends on a spatial variable, say x. Thus, we often write $W(t)$ as $W(x, t)$, which gives

$$\mathbb{E}[W(x, t)W(y, s)] = \mathbb{E}\left\{\sum_{n=1}^{\infty} \sqrt{q_n}W_n(t)e_n(x) \sum_{m=1}^{\infty} \sqrt{q_m}W_m(s)e_m(y)\right\}$$

$$= \sum_{n,m=1}^{\infty} \sqrt{q_n q_m}\mathbb{E}[W_n(t)W_m(s)]e_n(x)e_m(y)$$

$$= \min\{t, s\}\sum_{n=1}^{\infty} q_n e_n(x)e_n(y)$$

$$= \min\{t, s\}q(x, y), \tag{3.36}$$

where the kernel function, or spatial correlation, is

$$q(x, y) = \sum_{n=1}^{\infty} q_n e_n(x)e_n(y). \tag{3.37}$$

The smoothness of $q(x, y)$ depends on the decaying property of q_n's.

The following example shows that the covariance operator Q can be represented in terms of the spatial correlation $q(x, y)$. In fact, it is an integral operator with kernel function $q(x, y)$.

Example 3.19. Consider $H = L^2(0, 1)$ with orthonormal basis $e_n = \sqrt{2}\sin(n\pi x)$. In the above infinite series representation (3.37), taking the derivative with respect to x, we get

$$\partial_x W(t) = \sum_{n=1}^{\infty} \sqrt{2}\,n\pi\,\sqrt{q_n}\,W_n(t)\cos(n\pi x). \tag{3.38}$$

In order for this series to converge, we need $n\pi\sqrt{q_n}$ to converge to zero sufficiently fast as $n \to \infty$. So q_n being small helps. In this sense, the trace $\text{Tr}(Q) = \sum q_n$ may be seen as a measurement (or quantification) for spatial regularity of white noise $\dot{W}(t)$. This is one reason for introducing the concept of a trace class operator $Q : \text{Tr}(Q) < \infty$.

In this case, the covariance operator Q can be calculated explicitly as follows:

$$Qa = Q\sum_n \langle e_n, a\rangle e_n = \sum_n \langle e_n, a\rangle Qe_n$$

$$= \sum_n \langle e_n, a\rangle q_n e_n$$

$$= \sum_n \int_0^1 a(y)e_n(y)dy\, q_n\, e_n(x)$$

$$= \int_0^1 q(x, y)a(y)dy, \tag{3.39}$$

where the kernel $q(x, y) = \sum_{n=1}^{\infty} q_n e_n(x)e_n(y)$.

3.5.1 White Noise in Hilbert Space

Motivated by §3.1, the Gaussian white noise in Hilbert space H is modeled as

$$\dot{W}(t) = \sum_{n=1}^{\infty} \sqrt{q_n}\, \dot{W}_n(t)e_n, \tag{3.40}$$

where $\dot{W}_n(t), n = 1, 2, \ldots$ are standard scalar independent Brownian motions.

3.6 Stochastic Calculus in Hilbert Space

We first recall the definition of Itô integral in Euclidean space $\mathbb{R}^1$

$$\int_0^T f(t, \omega) dB_t(\omega), \tag{3.41}$$

where $f(t, \omega)$ is called the integrand, and scalar Brownian motion B_t is called the integrator. Here we distinguish B_t, a Brownian motion in Euclidean space $\mathbb{R}^1$, from W_t, a Brownian motion in Hilbert space. Assume that the integrand f is measurable in (t, ω), i.e., it is measurable with respect to the σ-field $\mathcal{B}([0, T]) \otimes \mathcal{F}$. Also assume that f is adapted to the filtration $\mathcal{F}_t^B \triangleq \sigma(B_s : s \leq t)$ generated by Brownian motion B_t. Namely, $f(t, \cdot)$ is measurable with respect to $\mathcal{F}_t^B$ for each t.

We partition the time interval $[0, T]$ into small subintervals with the maximal length $\delta : 0 = t_0 < t_1 < \cdots < t_i < t_{i+1} < \cdots < t_n = T$ and then consider the sum

$$\sum_{i=0}^{n-1} f(\tau_i, \omega)(B_{t_{i+1}}(\omega) - B_{t_i}(\omega)), \tag{3.42}$$

where $\tau_i \in [t_i, t_{i+1}]$, and its convergence in an appropriate sense. However, the limit of these sums as $\delta \to 0$, even when it exists, depends on the choice of τ_i. Thus τ_i's can not be chosen arbitrarily, unlike the case of Riemann-Stieltjes integrals. In fact, it turns out that τ_i's have to be chosen in a fixed fashion in every subinterval $[t_i, t_{i+1}]$. The integral value depends on this specific fashion in the choices of τ_i. For example, for $\tau_i = t_i$ we obtain the Itô integral, whereas for $\tau = \frac{1}{2}(t_i + t_{i+1})$, we have the Stratonovich integral.

When the integrand is mean-square-Riemann integrable (with respect to time), $\mathbb{E}\int_0^T f^2(t, \omega)dt < \infty$, the Itô integral is defined as

$$\int_0^T f(t, \omega) dB_t(\omega) \triangleq \lim_{\delta \to 0} \text{ in m.s. } \sum_{i=0}^{n-1} f(t_i, \omega)(B_{t_{i+1}}(\omega) - B_{t_i}(\omega)), \tag{3.43}$$

where the limit is taken in mean square (m.s.).

The Stratonovich integral of f with respect to B_t, $\int_0^T f(t, \omega) \circ dB_t(\omega)$, is defined similarly except taking $\tau_i = \frac{1}{2}(t_i + t_{i+1})$. When the integrand is mean-square-Riemann integrable (with respect to time), $\mathbb{E} \int_0^T f^2(t, \omega) dt < \infty$, the Stratonovich integral is defined as

$$\int_0^T f(t, \omega) \circ dB_t(\omega) \triangleq \lim_{\delta \to 0} \text{in m.s.} \sum_{i=0}^{n-1} f\left(\tfrac{1}{2}(t_i + t_{i+1}), \omega\right) (B_{t_{i+1}}(\omega) - B_{t_i}(\omega)).$$

(3.44)

We now introduce stochastic integrals in Hilbert space. Given a probability space $(\Omega, \mathcal{F}, \mathbb{P})$, let $W(t)$ be a Brownian motion (or Wiener process) taking values in a Hilbert space U. We now define the Itô stochastic integral

$$\int_0^T \Phi(t, \omega) dW(t),$$

where the integrand $\Phi(t, \omega)$ is usually a linear operator

$$\Phi : U \to H.$$

The integrand is assumed to be jointly measurable in (t, ω). It is also assumed to be adapted to the filtration $\mathcal{F}_t^W \triangleq \sigma(W_s : s \leq t)$, generated by Brownian motion W_t, namely, $\Phi(t, \cdot)$ is measurable with respect to $\mathcal{F}_t^W$, for each t.

Assume that the covariance operator $Q : U \to U$ for $W(t)$ is symmetric and nonnegative definite. We further assume that $\text{Tr}(Q) < \infty$, i.e., Q is of trace class. From the previous section, §3.5, there exists an orthonormal basis $\{e_n\}$ (formed by eigenvectors of Q) for U, and a sequence of nonnegative real numbers q_n (eigenvalues of Q) such that

$$Qe_n = q_n e_n, \quad n = 1, 2, \ldots$$

For simplicity, let us assume that $q_n > 0$ for all n. Moreover, $W(t)$ has the following expansion:

$$W(t) = \sum_{n=1}^{\infty} \sqrt{q_n} \, W_n(t) e_n,$$

(3.45)

where W_n's are standard scalar independent Brownian motions.

With the expansion for the Brownian motion W in (3.45), we define the Itô integral $\int_0^T \Phi(t, \omega) dW(t)$ as in [314] by

$$\int_0^T \Phi(t, \omega) dW(t) = \sum_{n=1}^{\infty} \sqrt{q_n} \int_0^T \Phi(t, \omega) e_n dW_n(t).$$

(3.46)

Note that each term, $\int_0^T \Phi(t, \omega) e_n dW_n(t)$, is an Itô integral in Euclidean space $\mathbb{R}^1$, as defined above or in [244, Ch. 3.1]. This definition is sufficient for our purpose.

The Stratonovich integral in Hilbert space will be defined in §4.5. The conversion between the Stratonovich integral and Itô integral will be discussed there too.

Here we list a few properties of the Itô integral [94, Ch. 4.2] and [314]. In the following, the integrands Φ, F, and G satisfy the measurability and adaptedness conditions specified above.

Zero-mean property

$$\mathbb{E} \int_0^T \Phi(r, \omega) dW(r) = 0. \tag{3.47}$$

Itô isometry

$$\mathbb{E} \left\| \int_0^T \Phi(r, \omega) dW(r) \right\|^2 = \mathbb{E} \int_0^T \mathrm{Tr} \left[\left(\Phi(r) Q^{\frac{1}{2}} \right)^* \left(\Phi(r) Q^{\frac{1}{2}} \right) \right] dr. \tag{3.48}$$

Generalized Itô isometry

$$\mathbb{E} \left\langle \int_0^a F(r, \omega) dW(r), \int_0^b G(r, \omega) dW(r) \right\rangle$$
$$= \mathbb{E} \int_0^{a \wedge b} \mathrm{Tr} \left[\left(G(r, \omega) Q^{\frac{1}{2}} \right)^* \left(F(r, \omega) Q^{\frac{1}{2}} \right) \right] dr, \tag{3.49}$$

where $a \wedge b \triangleq \min\{a, b\}$.

By the properties of stochastic integral, it is natural to introduce the space $U^0 = Q^{1/2}U$ for a symmetric, positive-definite operator Q on Hilbert space U. Here $Q^{1/2}$ is defined in (3.28). For simplicity we assume Q is positive-definite, because if not, we can just consider Q restricted on the orthogonal complement of the kernel space of Q in U. Let $Q^{-1/2}$ be the inverse of $Q^{1/2}$. Now define a scalar product

$$\langle u, v \rangle_0 = \left\langle Q^{-1/2} u, Q^{-1/2} v \right\rangle, \quad u, v \in U_0.$$

Then U^0 with this scalar product is a Hilbert space.

In the construction of the stochastic integral, an important process is the Hilbert–Schmidt operator-valued process [94, Ch. 4.2]. A linear operator Φ in $\mathcal{L}(U^0, H)$ is called Hilbert–Schmidt from U^0 to H if, for every complete orthonormal basis $\{e_k^0\}$ of U^0,

$$\sum_{k=1}^\infty \| \Phi e_k^0 \|^2 < \infty.$$

The value of the series is independent of the choice of $\{e_k^0\}$. The space of all Hilbert–Schmidt operators is denoted by $\mathcal{L}^2(U^0, H)$, which is a separable Hilbert space with scalar product

$$\langle \Phi, \Psi \rangle_{\mathcal{L}^2} = \sum_{k=1}^\infty \left\langle \Phi e_k^0, \Psi e_k^0 \right\rangle, \quad \Phi, \Psi \in \mathcal{L}^2(U^0, H).$$

Furthermore, by the definition of U^0, every complete orthonormal basis of U^0 can be represented by $\{\sqrt{q_k} e_k\}$ for some complete orthonormal basis $\{e_k\}$ of U. Then the norm on $\mathcal{L}^2(U^0, H)$ is

$$\|\Phi\|^2_{\mathcal{L}^2(U^0,H)} = \langle \Phi, \Phi \rangle_{\mathcal{L}^2} = \mathrm{Tr}\left[\left(\Phi Q^{1/2}\right)^* \left(\Phi Q^{1/2}\right)\right], \quad \Phi \in \mathcal{L}^2(U^0, H).$$

Now for an $\mathcal{L}^2(U^0, H)$-valued process $\Phi(t, \omega)$, $0 \le t \le T$, the stochastic integral $\int_0^T \Phi(t, \omega)dW(t)$ is well defined if [94, p. 94]:

$$\mathbb{E}\int_0^T \|\Phi(t)\|^2_{\mathcal{L}^2(U^0,H)}dt = \mathbb{E}\int_0^T \mathrm{Tr}\left[\left(\Phi(t)Q^{1/2}\right)^* \left(\Phi(t)Q^{1/2}\right)\right]dt < \infty.$$

3.7 Itô's Formula in Hilbert Space

We continue with the stochastic calculus in Hilbert space H and discuss a useful version of Itô's formula [76, p. 10], [94, p. 105], [260, p. 75].

Theorem 3.20 (Itô's formula). *Let u be the solution of the* SPDE:

$$du = b(u)dt + \Phi(u)dW(t), \quad u(0) = u_0, \tag{3.50}$$

where $b : H \to H$ *and* $\Phi : H \to \mathcal{L}^2(U_0, H)$ *are bounded and continuous, and* $W(t)$ *is a* U*-valued* Q*-Wiener process. Assume that* F *is a smooth, deterministic function*

$$F : [0, \infty) \times H \to \mathbb{R}^1.$$

Then:

(i) *Itô's formula: Differential form*

$$dF(t, u(t)) = F_u(t, u(t))(\Phi(u(t))dW(t))$$
$$+ \left\{ F_t(t, u(t)) + F_u(t, u(t))(b(u(t))) \right.$$
$$\left. + \frac{1}{2}\mathrm{Tr}\left[F_{uu}(t, u(t))(\Phi(u(t))Q^{\frac{1}{2}})(\Phi(u(t))Q^{\frac{1}{2}})^* \right] \right\}dt, \tag{3.51}$$

where F_u *and* F_{uu} *are Fréchet derivatives,* F_t *is the usual partial derivative with respect to time, and* $*$ *denotes adjoint operation. This formula is understood with the following symbolic operations in mind:*

$$\langle dt, dW(t) \rangle = \langle dW(t), dt \rangle = 0, \quad \langle dW(t), dW(t) \rangle = \mathrm{Tr}(Q)dt.$$

(ii) *Itô's formula: Integral form*

$$F(t, u(t)) = F(0, u(0)) + \int_0^t F_u(s, u(s))(\Phi(u(s))dW(s))$$
$$+ \int_0^t \left\{ F_t(s, u(s)) + F_u(s, u(s))(b(u(s))) \right.$$
$$\left. + \frac{1}{2}\mathrm{Tr}\left[F_{uu}(s, u(s))(\Phi(u(s))Q^{\frac{1}{2}})(\Phi(u(s))Q^{\frac{1}{2}})^* \right] \right\}ds, \tag{3.52}$$

where F_u and F_{uu} are Fréchet derivatives, and F_t is the usual partial derivative with respect to time. The stochastic integral in the right-hand side is interpreted as

$$\int_0^t F_u(s, u(s))(\Phi(u(s))dW(s)) = \int_0^t \tilde{\Phi}(u(s))dW(s),$$

where the integrand $\tilde{\Phi}(u(s))$ is defined by

$$\tilde{\Phi}(u(s))(v) \triangleq F_u(s, u(s))(\Phi(u(s))v)$$

for all $s > 0, v \in H, \omega \in \Omega$.

Remark 3.21. Note that [260, Proposition B.0.10]:

$$\text{Tr}\left[F_{uu}(s, u(s))\left(\Phi(u(s))Q^{\frac{1}{2}}\right)\left(\Phi(u(s))Q^{\frac{1}{2}}\right)^*\right]$$
$$= \text{Tr}\left[\left(\Phi(u(s))Q^{\frac{1}{2}}\right)^* F_{uu}(s, u(s))\left(\Phi(u(s))Q^{\frac{1}{2}}\right)\right].$$

Remark 3.22. The trace term $\text{Tr}[F_{uu}(t, u(t))(\Phi(u(t))Q^{\frac{1}{2}})(\Phi(u(t))Q^{\frac{1}{2}})^*]$ comes from the second-order Fréchet derivative F_{uu}. In fact, let $W(t)$ be a Brownian motion in U with covariance operator Q; for example,

$$W(t) = \sum_{n=1}^{\infty} \sqrt{q_n} W_n(t)e_n,$$
$$Qe_n = q_n e_n,$$

where $W_n(t), n = 1, 2, \ldots$, are scalar independent Brownian motions and $\{e_n\}$ is an orthonormal basis of U. Then

$$dW_n(t)dW_m(t) = \begin{cases} dt, & n = m, \\ 0, & n \neq m, \end{cases}$$

which means that for every scalar integrable random process $G(t)$

$$\int_{t_0}^t G(s)dW_n(s)dW_m(s) = \begin{cases} \int_{t_0}^t G(s)ds, & n = m, \\ 0, & n \neq m. \end{cases}$$

For brevity, we write $F(t, u(t)) = F(u(t))$. Therefore,

$$F_{uu}(u(t))(\Phi(u(t))dW(t), \Phi(u(t))dW(t))$$
$$= F_{uu}(u(t))\left(\Phi(u(t))\Sigma_{n=1}^{\infty}\sqrt{q_n}dW_n(t)e_n, \Phi(u(t))\Sigma_{m=1}^{\infty}\sqrt{q_m}dW_m(t)e_m\right)$$
$$= \Sigma_{n=1}^{\infty}\Sigma_{m=1}^{\infty}\sqrt{q_n}dW_n(t)\sqrt{q_m}dW_m(t)F_{uu}(u(t))(\Phi(u(t))e_n, \Phi(u(t))e_m)$$
$$= \left(\Sigma_{n=1}^{\infty}q_n F_{uu}(u(t))(\Phi(u(t))e_n, \Phi(u(t))e_n)\right)dt$$
$$= \left(\Sigma_{n=1}^{\infty}F_{uu}(u(t))\left(\Phi(u(t))Q^{1/2}e_n, \Phi(u(t))Q^{1/2}e_n\right)\right)dt$$
$$= \text{Tr}\left[\left(\Phi(u(s))Q^{\frac{1}{2}}\right)^* F_{uu}(s, u(s))\left(\Phi(u(s))Q^{\frac{1}{2}}\right)\right]dt$$
$$= \text{Tr}\left[F_{uu}(u(t))\left(\Phi(u(t))Q^{1/2}\right)\left(\Phi(u(t))Q^{1/2}\right)^*\right]dt.$$

If $F_{uu}(u(t))$ is a linear operator from $H \times H$ to H, that is, $F_{uu}(u(t)) \in \mathcal{L}(H \times H, H)$, then for $\Phi(u(t)) \in \mathcal{L}^2(U^0, H)$ [260, Proposition B.0.10],

$$\Phi(u(t))^* F_{uu}(u(t)) \Phi(u(t)) \in \mathcal{L}^1(U_0, U_0),$$

$$F_{uu}(u(t)) \Phi(u(t)) \Phi(u(t))^* \in \mathcal{L}^1(H, H).$$

By a similar analysis as above, we again have

$$
\begin{aligned}
F_{uu}&(u(t))(\Phi(u(t))dW(t), \Phi(u(t))dW(t)) \\
&= \left(\Phi(u(t))^* F_{uu}(u(t)) \Phi(u(t))dW(t), dW(t) \right) \\
&= \sum_{n=1}^{\infty} \left(\left(Q^{1/2} \right)^* \Phi(u(t))^* F_{uu}(u(t)) \Phi(u(t)) Q^{1/2} e_n, e_n \right) dt \\
&= \mathrm{Tr} \left[\left(Q^{1/2} \right)^* \Phi(u(t))^* F_{uu}(u(t)) \Phi(u(t)) Q^{1/2} \right] dt \\
&= \mathrm{Tr} \left[F_{uu}(u(t)) \left(\Phi(u(t)) Q^{1/2} \right) \left(\Phi(u(t)) Q^{1/2} \right)^* \right] dt.
\end{aligned}
$$

Example 3.23. Let us look at a typical application of Itô's formula for SPDEs. Consider the SPDE

$$du = b(u)dt + \Phi(u)dW(t), \quad u(0) = u_0, \tag{3.53}$$

in Hilbert space $H = L^2(D)$, $D \subset \mathbb{R}^n$, with the usual scalar product $\langle u, v \rangle = \int_D uv dx$. Define an energy functional $F(u) = \frac{1}{2} \int_D u^2 dx = \frac{1}{2} \|u\|^2$. In this case, $F_u(u)(h) = \int_D uh dx$ and $F_{uu}(u)(h, k) = \int_D h(x)k(x)dx$. By Itô's formula in differential form,

$$\frac{1}{2} d\|u\|^2 = \left\{ \langle u, b(u) \rangle + \frac{1}{2} \mathrm{Tr} \left[\left(\Phi(u) Q^{\frac{1}{2}} \right) \left(\Phi(u) Q^{\frac{1}{2}} \right)^* \right] \right\} dt + \langle u, \Phi(u)dW(t) \rangle.$$

Then the corresponding integral form becomes

$$
\begin{aligned}
\frac{1}{2} \mathbb{E} \|u(t)\|^2 = {}& \frac{1}{2} \mathbb{E} \|u(0)\|^2 + \mathbb{E} \int_0^t \langle u(r), b(u(r)) \rangle dr \\
& + \frac{1}{2} \mathbb{E} \int_0^t \mathrm{Tr} \left[\left(\Phi(u(r)) Q^{\frac{1}{2}} \right) \left(\Phi(u(r)) Q^{\frac{1}{2}} \right)^* \right] dr.
\end{aligned}
$$

Note that in this special case, F_u is a bounded operator in $L(H, \mathbb{R})$, which can be identified with H itself due to the Riesz representation theorem [313, Ch. III.6].

For a given U-valued Wiener process $W(t)$, the following Burkholder–Davis–Gundy type inequality [94, Lemma 7.2] for stochastic Itô integrals in Hilbert space is important for estimating solutions of SPDEs.

Lemma 3.24 (Burkholder–Davis–Gundy). *For* $r \geq 1$ *and* $\Phi(t) \in \mathcal{L}^2(U^0, H)$, $t \in [0, T]$, *the following estimate holds*:

$$\mathbb{E} \left[\sup_{0 \leq s \leq t} \left\| \int_0^s \Phi(\tau)dW(\tau) \right\|^{2r} \right] \leq C_r \mathbb{E} \left[\int_0^t \|\Phi(s)\|^2_{\mathcal{L}^2(U^0, H)} ds \right]^r, \quad 0 \leq t \leq T,$$

where

$$C_r = (r(2r-1))^r \left(\frac{2r}{2r-1}\right)^{2r^2}.$$

3.8 Problems

3.1. A random variable in $\mathbb{R}^n$

Let $X : \Omega \to \mathbb{R}^n$ be a random variable in $\mathbb{R}^n$, and let h be a given vector in $\mathbb{R}^n$. Define $a(\omega) \triangleq \langle X(\omega), h \rangle = h_1 X_1(\omega) + \cdots + h_n X_n(\omega)$. Is this a random variable in $\mathbb{R}^1$?

3.2. A random variable in a Hilbert space

Let $X : \Omega \to H$ be a random variable in a Hilbert space H. For every $h \in H$, is $a(\omega) := \langle X(\omega), h \rangle$ a scalar random variable in $\mathbb{R}^1$?

3.3. A simple Hilbert space

Consider the Hilbert space $l^2 = \{x = (x_1, x_2, \ldots): \sum_{n=1}^{\infty} x_n^2 < \infty\}$ with scalar product $\langle x, y \rangle = \sum_{n=1}^{\infty} x_n y_n$.

(i) Describe the Borel σ-field $\mathcal{B}(l^2)$.

(ii) Define a linear operator $Q : l^2 \to l^2$ by

$$Q e_n = \frac{1}{n^2} e_n, \quad n = 1, 2, \ldots,$$

where $e_1 = (1, 0, \ldots), e_2 = (0, 1, 0, \ldots), \ldots$ form an orthonormal basis for l^2. What is Qx for $x \in l^2$? What is $\mathrm{Tr}(Q)$? Is Q a trace class operator (i.e., $\mathrm{Tr}(Q) < \infty$)?

(iii) What is the Brownian motion $W(t)$ taking values in the Hilbert space l^2 with covariance operator Q? What is the mean and variance for $W(t) - W(s)$ for $t > s$?

(iv) Can you think of a way to visualize the sample paths of this Brownian motion $W(t)$?

3.4. Another Hilbert space

Consider the Hilbert space $L^2(0, 1) = \{u(x) : \int_0^1 |u(x)|^2 dx < \infty\}$ with scalar product $\langle u, v \rangle = \int_0^1 u(x)v(x)dx$ and orthonormal basis $e_n(x) = \sqrt{2} \sin(n\pi x)$, $n = 1, 2, \ldots$ Note that this space is different from the Hilbert space $L^2(\Omega)$ of random variables of finite variance.

(i) Describe the Borel σ-field $\mathcal{B}(L^2(0, 1))$.

(ii) Consider the linear operator $Q = (-\partial_{xx})^{-1} : L^2(0, 1) \to L^2(0, 1)$, the inverse operator of the (unbounded) operator $-\Delta = -\partial_{xx}$, with domain of definition $\mathrm{Dom}(-\Delta) = H_0^2(0, 1)$. What are eigenvalues and eigenvectors for Q? What is $\mathrm{Tr}(Q)$?

(iii) What is the Brownian motion $W(t)$ taking values in $L^2(0, 1)$ with covariance operator Q?

Note: $H_0^k(0, 1)$ is a Sobolev space of functions u, for which u and all its up to k th order derivatives are all in $L^2(0, 1)$, and "have compact support, or have zero boundary conditions" (thus simplifying the integration by parts). For the rigorous definition of Sobolev spaces, see §2.5 or [264, Ch. 7], for example.

3.5. Measurable mapping of a random variable

Let $X : \Omega \to H$ be a random variable taking values in a Hilbert space H, and $f : H \to \mathbb{R}^n$ be a Borel measurable function. Is $f(X)$ a random variable taking values in $\mathbb{R}^n$? If so, why?

3.6. Derivatives in Euclidean space

Consider a function $F : \mathbb{R}^2 \to \mathbb{R}^1$ defined by $F(\mathbf{x}) = F(x_1, x_2) = x_1 + x_2^3 + \sin(x_1)$. What is the first-order (Fréchet) derivative $F_{\mathbf{x}}(\mathbf{x}_0)$ and the second-order (Fréchet) derivative $F_{\mathbf{xx}}(\mathbf{x}_0)$ at every given $\mathbf{x}_0 \in \mathbb{R}^2$, particularly at $\mathbf{x}_0 = (1, 2)^T$?

3.7. Derivatives in Hilbert space

Consider a functional $F : H_0^1(0, 1) \to \mathbb{R}^1$ defined by $F(u) = \frac{1}{2} \int_0^1 (u^2 + u_x^2) dx$. What is the first-order Frechét derivative $F_u(u_0)$ and the second-order Frechét derivative $F_{uu}(u_0)$ at every given $u_0 \in H_0^1(0, 1)$, particularly at $u_0 = \sin^2(\pi x)$?

3.8. Mean of a stochastic integral involving scalar product

Is it true that for $f \in H$,

$$\mathbb{E} \int_0^T \langle f(t, \omega), dW(t, \omega) \rangle = 0?$$

Hint: Use the series expansion for $W(t)$ and calculate the scalar product, then integrate, and finally take the mean.

3.9. Brownian motion

For $s < t$, compute $\mathbb{E} \langle W(t) - W(s), W(s) \rangle$.

3.10. Brownian motion again

Compute $\mathbb{E} (\langle W(t), a \rangle \langle W(s), b \rangle)$.

3.11. Itô's formula in Hilbert space

Consider the SPDE on the interval $0 < x < l$,

$$du = (u_{xx} + u - u^3) dt + \epsilon dW(t), u(0, t) = u(l, t) = 0,$$

where ϵ is a real parameter and $W(t)$ is a Brownian motion. We assume that the covariance operator Q for the Brownian motion $W(t)$ has the same eigenfunctions as those of the linear operator ∂_{xx} with zero Dirichlet boundary conditions on $(0, l)$ and has the corresponding eigenvalues $q_n = \frac{1}{n^2}, n = 1, 2, \ldots$ Introduce the Hilbert space $H = L^2(0, l)$ with scalar product $\langle u, v \rangle = \int_0^l uv dx$ and norm $\|u\| = \sqrt{\langle u, u \rangle}$.

(a) Apply Itô's formula in differential form to obtain $d\|u\|^2$.
(b) Apply Itô's formula in integral form to obtain $\|u\|^2 = \|u(x, 0)\|^2 + \cdots$.

3.12. Itô's formula in Hilbert space again

Consider the SPDE

$$u_t = \nu u_{xx} + u u_x + \sin(u) + g(u)\dot{w}(t), \quad u(0, t) = u(l, t) = 0, \quad 0 < x < l,$$

where $w(t)$ is a scalar Brownian motion and $g(u)$ is a given function. Define an energy functional $F(u) = \frac{1}{2} \int_0^l u^2 dx = \frac{1}{2} \|u\|^2$ and verify that the Fréchet derivatives are $F_u(u_0)h = \int_0^l u_0 h \, dx$ and $F_{uu}(u_0)(h, k) = \int_0^l hk \, dx$.

(a) Apply Itô's formula in differential form to obtain $d\|u\|^2$.

(b) Apply Itô's formula in integral form to obtain $\|u\|^2 = \|u(x, 0)\|^2 + \cdots$.

3.13. Concentration of particles in a fluid

Consider again the SPDE in Example 1.3. The concentration of particles in a fluid, $C(x, t)$, at position x and time t, approximately satisfies a diffusion equation with fluctuations [322, Ch. 1.4]

$$C_t = \kappa\, C_{xx} + \dot{W}(x, t), \quad C(0, t) = C(l, t) = 0, \quad 0 < x < l, t > 0,$$

where κ is the diffusivity (a positive parameter) and W is a Brownian motion. We assume that the covariance operator Q for the Brownian motion $W(t)$ has the same eigenfunctions as those of the linear operator ∂_{xx} with zero Dirichlet boundary conditions on $(0, l)$ and has the corresponding eigenvalues $q_n = \frac{1}{n^2}, n = 1, 2, \ldots$ Introduce the Hilbert space $H = L^2(0, l)$ with the usual scalar product $\langle u, v \rangle = \int_0^l uv\,dx$ and norm $\|u\| = \sqrt{\langle u, u \rangle}$.

(a) Apply Itô's formula in differential form to obtain $d\|C\|^2$.

(b) Apply Itô's formula in integral form to obtain $\|C\|^2 = \|C(x, 0)\|^2 + \cdots$.

4 Stochastic Partial Differential Equations

Linear and nonlinear SPDEs; solution concepts; impact of noise; and infinite dimensional stochastic dynamical systems

In this chapter we review some basic facts about stochastic partial differential equations, includingvarious solution concepts such as weak, strong, mild, and martingale solutions, and sufficient conditions under which these solutions exist. Additionally, we briefly discuss infinite dimensional stochastic dynamical systems through a few examples.

For general background on SPDEs, such as well-posedness and basic properties of solutions, see books [76,94,257,260,271] or other references [9,10,11,12,13,63,145, 152,184,185,201,218,232,237,295].

For SPDEs, it is a common practice to denote a solution process $u(x, t)$ by u or $u(t)$. Similarly W_t, $W(t)$, or $W(x, t)$ all refer to a Wiener process in a Hilbert space. We also call a solution process an *orbit* or a *trajectory* when regarded as a function of time alone, as in dynamical systems theory.

4.1 Basic Setup

In order to consider well-posedness for nonlinear stochastic partial differential equations, we first recall various solution concepts, and then discuss well-posedness results. For further details, refer to [76,94,260]. Note that strong and weak solutions have different meanings for stochastic and deterministic partial differential equations.

Let H be a separable Hilbert space and $A : \mathcal{D}(A) \to H$ be the infinitesimal generator of a C_0-semigroup $S(t)$, $t \geq 0$, on H. Let U be another separable Hilbert space and let $W(t)$, $t \geq 0$, be a U-valued Q-Wiener process defined on a complete probability space $(\Omega, \mathcal{F}, \mathbb{P})$. Consider the following SPDE in Hilbert space H:

$$du(t) = [Au(t) + f(t, u(t))]dt + \Phi(t, u(t))dW(t), \quad u(0) = u_0 \in H, \qquad (4.1)$$

where A and Φ are linear differential operators and f is a linear or nonlinear mapping. This includes the special cases of a linear SPDE with additive noise

$$du(t) = [Au(t) + f(t)]dt + \Phi \, dW(t) \qquad (4.2)$$

Effective Dynamics of Stochastic Partial Differential Equations. http://dx.doi.org/10.1016/B978-0-12-800882-9.00004-4

and a linear SPDE with multiplicative noise

$$du(t) = [Au(t) + f(t)]dt + \Phi u(t)dW(t). \tag{4.3}$$

We impose the following assumptions on f and Φ in SPDE (4.1). For a given $T > 0$ and $U^0 \triangleq Q^{1/2}U$,

(A_1) $f : [0, T] \times H \to H$ is $(\mathcal{B}([0, T]) \times \mathcal{B}(H), \mathcal{B}(H))$ measurable.
(A_2) $\Phi : [0, T] \times H \to \mathcal{L}^2(U^0, H)$ is $(\mathcal{B}([0, T]) \times \mathcal{B}(H), \mathcal{B}(\mathcal{L}^2(U^0, H)))$ measurable.

Furthermore, we assume that there is a positive constant Lip such that for all t in $[0, T]$,

(A_3) Lipschitz condition:

$$\|f(t, u) - f(t, v)\| + \|\Phi(t, u) - \Phi(t, v)\|_{\mathcal{L}^2(U^0, H)} \le Lip\|u - v\|, \quad u, v \in H.$$

(A_4) Sublinear growth condition:

$$\|f(t, u)\|^2 + \|\Phi(t, u)\|^2_{\mathcal{L}^2(U^0, H)} \le Lip(1 + \|u\|^2), \quad u \in H.$$

Remark 4.1. For a nonautonomous SPDE, i.e., when f and/or Φ in SPDE (4.1) explicitly depend on time t, the global Lipschitz condition (A_3) does not necessarily imply the sublinear growth condition (A_4).

Now we consider the solution to (4.1) in different senses and various spaces. We introduce the following two function spaces. For $T > 0$, let $L^2(\Omega; C([0, T]; H))$ be the space of all H-valued adapted processes $u(t, \omega)$, defined on $[0, T] \times \Omega$, which are continuous in t for almost every (a.e.) fixed sample $\omega \in \Omega$ and which satisfy the following condition:

$$\|u\|_T \triangleq \left\{ \mathbb{E} \sup_{0 \le t \le T} \|u(t, \omega)\|^2 \right\}^{\frac{1}{2}} < \infty.$$

Then $L^2(\Omega; C([0, T]; H))$ is a Banach space with norm $\| \cdot \|_T$.

A predictable σ-field is the smallest σ-field on the product space $[0, T] \times \Omega$ for which all continuous $\mathcal{F}$-adapted processes are measurable. A predictable process is a process measurable with respect to the predictable σ-field [94, p. 76], [206, p. 57].

Let $L^2(\Omega; L^2([0, T]; H))$ be the space of all H-valued predictable processes $u(t, \omega)$, defined on $[0, T] \times \Omega$, having the property

$$\|u\|_{2,T} = \left\{ \mathbb{E} \int_0^T \|u(t, \omega)\|^2 dt \right\}^{\frac{1}{2}} < \infty.$$

Then $L^2(\Omega; L^2([0, T]; H))$ is also a Banach space with norm $\| \cdot \|_{2,T}$.

4.2   Strong and Weak Solutions

We are now ready to define strong and weak solutions for SPDE (4.1).

Definition 4.2 (Strong solution).   A $\mathcal{D}(A)$-valued predictable process $u(t)$ is called a strong solution of SPDE (4.1) if

$$\mathbb{P}\left\{\int_0^T \left[\|u(s)\| + \|Au(s)\| + \|f(s, u(s))\|\right]ds < \infty\right\} = 1,$$

$$\mathbb{P}\left\{\int_0^T \|\Phi(s, u(s))\|_{\mathcal{L}^2(U^0, H)}^2 ds < \infty\right\} = 1,$$

and

$$u(t) = u_0 + \int_0^t [Au(s) + f(s, u(s))]ds + \int_0^t \Phi(s, u(s))dW(s), \quad \mathbb{P}\text{-a.s.}$$

for $t \in [0, T]$.

Definition 4.3 (Weak solution).   An H-valued predictable process $u(t)$ is called a weak solution of SPDE (4.1) if

$$\mathbb{P}\left\{\int_0^T \|u(s)\| ds < \infty\right\} = 1$$

and for any $\xi \in \mathcal{D}(A^*)$, with A^* the adjoint operator of A,

$$\langle u(t), \xi \rangle = \langle u_0, \xi \rangle + \int_0^t [\langle u(s), A^*\xi \rangle + \langle f(s, u(s)), \xi \rangle]ds$$

$$+ \int_0^t \langle \Phi(s, u(s))dW(s), \xi \rangle, \quad \mathbb{P}\text{-a.s.}$$

for $t \in [0, T]$, provided that the integrals on the right-hand side are well defined.

The relation between strong and weak solutions is described in the following theorem.

Theorem 4.4 (Relation between strong and weak solutions).

(i) Every strong solution of SPDE *(4.1) is also a weak solution.*
(ii) Let $u(t)$ *be a weak solution of* SPDE *(4.1) with values in* $\mathcal{D}(A)$ *such that*

$$\mathbb{P}\left\{\int_0^T \left[\|Au(s)\| + \|f(s, u(s))\|\right]ds < \infty\right\} = 1,$$

and

$$\mathbb{P}\left\{\int_0^T \|\Phi(s, u(s))\|_{\mathcal{L}^2(U^0, H)}^2 ds < \infty\right\} = 1.$$

Then the weak solution $u(t)$ *is also a strong solution of* SPDE *(4.1).*

The following result provides a sufficient condition for the existence of a weak solution.

Theorem 4.5 (Existence and uniqueness of weak solution of nonautonomous SPDE). *Assume that u_0 in H is $\mathcal{F}_0$-measurable and conditions $(A_1) - (A_4)$ are satisfied. Then the nonautonomous SPDE (4.1) has a unique weak solution $u \in L^2(\Omega; C([0, T]; H))$.*

Remark 4.6. The conditions for the existence of a strong solution are complicated, but some special cases are considered in [94, Ch. 5.6, Ch. 6.5 and Ch. 7.4].

Remark 4.7. Here the *weak* and *strong* solutions are similar in meaning to those of deterministic PDEs. In some references e.g., [76, p. 168] *weak* and *strong* solutions may have different meanings.

4.3   Mild Solutions

We now define the following *mild solution* concept for SPDE (4.1).

Definition 4.8 (Mild solution). An H-valued predictable process $u(t)$ is called a mild solution of SPDE (4.1) if

$$\mathbb{P}\left\{ \int_0^T \|u(s)\|^2 ds < \infty \right\} = 1,$$

$$\mathbb{P}\left\{ \int_0^t \left[\|S(t - s)f(s, u(s))\| + \|S(t - s)\Phi(s, u(s))\|^2_{\mathcal{L}^2(U^0, H)} \right] ds < \infty \right\} = 1,$$

and

$$u(t) = S(t)u_0 + \int_0^t S(t - s)f(s, u(s))ds$$

$$+ \int_0^t S(t - s)\Phi(s, u(s))\, dW(s), \quad \mathbb{P}\text{-a.s.}$$

for all t in $[0, T]$.

The relation between mild and weak solutions is described in the following theorem.

Theorem 4.9 (Relation between mild and weak solutions).

  (i) Let $u(t)$ be a weak solution of SPDE (4.1) such that

$$\mathbb{P}\left\{ \int_0^T \left[\|u(s)\| + \|f(s, u(s))\| \right] ds < \infty \right\} = 1,$$

  and

$$\mathbb{P}\left\{ \int_0^T \|\Phi(s, u(s))\|^2_{\mathcal{L}^2(U^0, H)} ds < \infty \right\} = 1.$$

Then $u(t)$ is also a mild solution.

(ii) Conversely, suppose that u(t) is a mild solution of SPDE *(4.1) such that*

$$(t, \omega) \rightarrow \int_0^t S(t-s)f(s, u(s, \omega))ds$$

and

$$(t, \omega) \rightarrow \int_0^t S(t-s)\Phi(s, u(s, \omega))dW(s)$$

have predictable versions. Also suppose that

$$\mathbb{P}\left\{ \int_0^T \|f(s, u(s))\|ds < \infty \right\} = 1,$$

and

$$\int_0^T \mathbb{E}\left(\int_0^t \|\langle S(t-s)\Phi(s, u(s)), A^*\xi \rangle\|_{\mathcal{L}^2(U^0, H)}^2 ds \right) dt < \infty$$

for all $\xi \in \mathcal{D}(A^*)$. Then u(t) is also a weak solution.

4.3.1   Mild Solutions of Nonautonomous SPDES

For the existence and uniqueness of mild solutions of the non-autonomous SPDE (4.1), we state the following theorem [94, Theorem 7.4].

Theorem 4.10 (Existence and uniqueness of mild solution of nonautonomous SPDE).   *Assume that u_0 in H is $\mathcal{F}_0$-measurable and the conditions (A_1)–(A_4) are satisfied. Then the nonautonomous* SPDE *(4.1) has a unique mild solution u in $L^2(\Omega; C([0, T]; H)) \cap L^2(\Omega; L^2([0, T]; H))$.*

If A generates an analytic semigroup $S(t)$ in a Hilbert space H, functions f, Φ are locally Lipschitz in u from a space V to H (here V is continuously embedded in H), and f, Φ are locally sublinear, then SPDE (4.1) has a unique local mild solution for every initial value u_0 in V [76, p. 165]. If this SPDE is autonomous, i.e., f and Φ do not depend on time explicitly, then the local sublinearity condition is a consequence of local Lipschitz condition and thus can be dropped.

Regularity in space for solutions of SPDEs is also an important issue [e.g., 127].

4.3.2   Mild Solutions of Autonomous SPDES

To facilitate a random dynamical systems approach for (autonomous) stochastic partial differential equations arising as mathematical models for complex systems under fluctuations, global existence and uniqueness (i.e., a unique solution exists for all time $t > 0$) of mild solutions are desirable.

In the rest of this section, we present a semigroup approach for global existence and uniqueness of mild solutions for a class of autonomous stochastic partial differential equations with either local or global Lipschitz coefficients, e.g., [76, Ch. 6], [94, Ch.7].

A sufficient condition for global existence and uniqueness of mild solutions is provided and an example is presented to demonstrate the result [139]. In the case of local Lipschitz coefficients, to guarantee global well-posedness, an *a priori* estimate on the solution u under norm $\| \cdot \|_\gamma$, for some $\gamma \in (0, 1)$ (see details below) is often needed.

For deterministic partial differential equations, the semigroup approach for well-posedness is presented in, for example, [156, Ch. 3], [231, Ch. 9] and [253, Ch. 4–8].

Let $D \subset \mathbb{R}^d$ be a bounded domain with smooth boundary ∂D. We consider the following nonlinear SPDE:

$$\frac{\partial}{\partial t} u(x, t) = (\kappa \Delta - \alpha) u(x, t) + f(u(x, t)) + \sigma(u(x, t)) \frac{\partial}{\partial t} W(x, t), \qquad (4.4)$$

$$u|_{\partial D} = 0, \quad u(x, 0) = h(x), \qquad (4.5)$$

where $t > 0, x \in D, \kappa$, and α are positive constants; $\Delta = \sum_{i=1}^{d} \frac{\partial^2}{\partial x_i^2}$ is the Laplace operator, and $h(x)$ is a given function in $L^2(D)$. Define $H = L^2(D)$ with the usual scalar product $\langle u, v \rangle = \int_D uv dx$ and the induced norm $\| \cdot \|$. The coefficients f, σ are given measurable functions and $W(x, t)$ is an H-valued, Q-Wiener process to be defined below. Such a SPDE models a variety of phenomena in biology, quantum field theory, and neurophysiology [225,248,277].

We work in a complete probability space $(\Omega, \mathcal{F}, \mathbb{P})$.

Define a linear operator $A \triangleq -\kappa \Delta + \alpha$ with zero Dirichlet boundary condition on D and denote by $S(t)$ the C_0-semigroup generated by $-A$ on H e.g., [253, Theorem 2.5]. Denote by $\{e_k(x)\}_{k \geq 1}$ the complete orthonormal system of eigenfunctions in H for the linear operator A, i.e., for $k = 1, 2, \cdots$,

$$(-\kappa \Delta + \alpha) e_k = \lambda_k e_k, \quad e_k|_{\partial D} = 0, \qquad (4.6)$$

with $\alpha < \lambda_1 \leq \lambda_2 \leq \cdots \leq \lambda_k \leq \cdots$.

Let us specify the Q-Wiener process $W(x, t)$ [76, p.38]. Let Q be given by the following linear integral operator:

$$Q\varphi(x) = \int_D q(x, y) \varphi(y) dy, \quad \varphi \in H,$$

where the kernel q is positive, symmetric (i.e., $q(x, y) = q(y, x)$) and square integrable, namely, $\int_D \int_D |q(x, y)|^2 dx \, dy < \infty$. Then the eigenvalues $\{\mu_k\}_{k \geq 1}$ of Q are positive. For simplicity, we assume that the covariance operator Q commutes with operator A, and thus operator Q also has eigenfunctions $\{e_k(x)\}_{k \geq 1}$. By the formula (3.37), we calculate $\mathrm{Tr}(Q) = \sum \mu_k = \sum < \mu_k e_k, e_k > = \int_D \sum \mu_k e_k(x) e_k(x) dx = \int_D q(x, x) dx$. When $\int_D q(x, x) dx < \infty$ (i.e., Q is a trace class operator), W has a representation via an infinite series in H

$$W(x, t) = \sum_{k=1}^{\infty} \sqrt{\mu_k} \, w_k(t) e_k(x),$$

where $\{w_k(t)\}_{k\geq 1}$ is a sequence of independent scalar Wiener processes. For more information regarding Q-Wiener processes, see §3.5.

Now we seek a mild solution for SPDE (4.4), that is, an H-valued predictable process $u(t)$ such that

$$u(t) = S(t)h + \int_0^t S(t-s)f(u(s))ds + \int_0^t S(t-s)\sigma(u(s))dW(s), \quad \mathbb{P}\text{-}a.s. \quad (4.7)$$

for $t \geq 0$.

We consider mild solutions of the autonomous SPDE (4.4) under either a global or local Lipschitz condition on the coefficients f and σ. Here we present a semigroup approach, mainly following [76, Theorem 3.6.5] but with a different sufficient condition to guarantee global existence [139]; see Assumption (**B**) below.

After some preliminaries, we prove global well-posedness results under either a global or local Lipschitz condition and finally present an example.

4.3.2.1   Formulation

Recall the definition of fractional power e.g., [269, Ch. 3.10] of A, based on (4.6). For $\gamma > 0$, define A^γ by

$$A^\gamma u = \sum_{i=1}^\infty \lambda_i^\gamma \langle u, e_i \rangle e_i$$

whenever the right-hand side is well defined. For more information on fractional power operators, we refer to [156, Ch. 1.4] or [253, Ch. 2.6]. This definition is consistent with what know about fractional power of a positive-definite symmetric matrix in linear algebra. Denote by H^γ the domain of $A^{\gamma/2}$ in H, that is,

$$H^\gamma \triangleq \left\{ u \in H : \sum_{k=1}^\infty \lambda_k^{\gamma/2} \langle u, e_k \rangle e_k \quad \text{converges in} \quad H \right\}.$$

Then H^γ is a Banach space endowed with norm

$$\|u\|_\gamma \triangleq \left\{ \sum_{k=1}^\infty \lambda_k^\gamma \langle u, e_k \rangle^2 \right\}^{\frac{1}{2}}.$$

Moreover, the embedding $H^{\gamma_1} \subset H^{\gamma_2}$ is continuous for $0 < \gamma_2 < \gamma_1$.

By the definition of covariance operator Q of the Wiener process $W(t)$, for a predictable process $\Sigma(t)$ in H with $\Sigma(t) \in \mathcal{L}^2(U^0, H)$,

$$\|\Sigma(t)\|_{\mathcal{L}^2(U^0, H)} = \left(\int_D q(x, x) \Sigma^2(t)(x) dx \right)^{\frac{1}{2}},$$

where $U^0 = Q^{1/2} H$ and $q(x, y)$ is the kernel of the covariance operator Q. Moreover, assume that $q(x, y) \leq r_0$ for some positive number r_0.

Before starting to prove our main theorem, we need some fundamental inequalities that are formulated in the following lemmas. Similar inequalities are also used for the existence of mild solution of stochastic parabolic equations by Chow [76, Lemma 3.5.1].

First, because $S(t)$, $t \geq 0$, is a contraction semigroup in H, we have the following estimate.

Lemma 4.11. *Assume that $h \in H$ and $F(t)$ is a predictable process in H such that* $\mathbb{E} \int_0^T \|F(t)\|^2 dt < \infty$. *Then*

$$\sup_{0 \leq t \leq T} \|S(t)h\| \leq \|h\|, \text{ and} \tag{4.8}$$

$$\mathbb{E} \sup_{0 \leq t \leq T} \left\| \int_0^t S(t-s)F(s)ds \right\|^2 \leq T\mathbb{E} \int_0^T \|F(t)\|^2 dt. \tag{4.9}$$

By Lemma 3.24 with $r = 1$, we have the following result.

Lemma 4.12. *Assume that $\Sigma(t)$ is a predictable process in H such that* $\mathbb{E} \int_0^T \int_D q(x,x)\Sigma^2(t)(x)dxdt = \mathbb{E} \int_0^T \|\Sigma(t)\|^2_{\mathcal{L}^2(U^0,H)} dt < \infty$. *Then*

$$\mathbb{E} \sup_{0 \leq t \leq T} \left\| \int_0^t S(t-s)\Sigma(s)dW(s) \right\|^2 \leq 4\mathbb{E} \int_0^T \|\Sigma(t)\|^2_{\mathcal{L}^2(U^0,H)} dt. \tag{4.10}$$

We also need the following two Lemmas for $\gamma \in (0, 1]$.

Lemma 4.13. *Assume that $h \in H$ and $F(t)$ is a predictable process in H such that* $\mathbb{E} \int_0^T \|F(t)\|^2 dt < \infty$. *Then, for $t > 0$, $S(t)h \in H^1$ and $\int_0^t S(t-s)F(s)ds$ is an H^1-valued process. Furthermore,*

$$\int_0^T \|S(t)h\|^2_\gamma dt \leq \frac{1}{2\alpha^{1-\gamma}} \|h\|^2, \text{ and} \tag{4.11}$$

$$\mathbb{E} \int_0^T \left\| \int_0^t S(t-s)F(s)ds \right\|^2_\gamma dt \leq \frac{T}{2\alpha^{1-\gamma}} \mathbb{E} \int_0^T \|F(t)\|^2 dt. \tag{4.12}$$

Proof. Expand $h \in H$ in terms of a complete orthonormal basis $\{e_k\}_{k=1}^\infty$ as

$$h = \sum_{k=1}^\infty h_k e_k,$$

where $h_k = \langle h, e_k \rangle$. Then, for $t > 0$,

$$S(t)h = \sum_{k=1}^\infty h_k e^{-\lambda_k t} e_k$$

and direct calculation yields that $S(t)h \in H^1$ and for $0 < \gamma \leq 1$,

$$\int_0^T \|S(t)h\|_\gamma^2 \, dt = \sum_{k=1}^\infty \lambda_k^\gamma h_k^2 \int_0^T e^{-2\lambda_k t} \, dt \leq \frac{1}{2\alpha^{1-\gamma}} \|h\|^2,$$

where we have used the fact that $\lambda_k > \alpha$. This gives (4.11). The other estimate, (4.12), can be obtained similarly. $\qquad\square$

Lemma 4.14. *Assume that $\Sigma(t)$ is a predictable process in H such that $\mathbb{E} \int_0^T \int_D q(x, x) \Sigma^2(t)(x) dx dt = \mathbb{E} \int_0^T \|\Sigma(t)\|_{\mathcal{L}^2(U^0, H)}^2 dt < \infty$. Then $\int_0^t S(t - s) \Sigma(s) dW(s)$ is an H^1-valued process and*

$$\mathbb{E} \int_0^T \left\| \int_0^t S(t - s) \Sigma(s) dW(s) \right\|_\gamma^2 dt$$

$$\leq \frac{1}{2\alpha^{1-\gamma}} \mathbb{E} \int_0^T \|\Sigma(t)\|_{\mathcal{L}^2(U^0, H)}^2 dt. \tag{4.13}$$

Proof. Expanding $\Sigma(t)$ in terms of a complete orthonormal basis $\{e_k\}_{k=1}^\infty$ as follows

$$\Sigma(t) = \sum_{k=1}^\infty \Sigma_k(t) e_k(x).$$

Then

$$\mathbb{E} \left\| \int_0^t S(t - s) \Sigma(s) dW(s) \right\|_\gamma^2 = \mathbb{E} \sum_{k=1}^\infty \lambda_k^\gamma \int_0^t e^{-2\lambda_k(t-s)} \Sigma_k^2(s) \mu_k \, ds.$$

Noticing that

$$\|\Sigma(s)\|_{\mathcal{L}^2(U^0, H)}^2 = \sum_{k=1}^\infty \mu_k \Sigma_k^2(s),$$

using Lemma 3.24 and $\lambda_k > \alpha$, we have (4.13). $\qquad\square$

4.3.2.2   Well-Posedness Under Global Lipschitz Condition

We consider the autonomous SPDE (4.4) or its equivalent mild form (4.7) in two separate cases: *global Lipschitz* coefficients and *local Lipschitz* coefficients.

First we consider the global Lipschitz case with the following assumption on coefficients f and σ in equation (4.7):

(H) Global Lipschitz condition: The nonlinearity f and noise intensity σ are measurable functions, and there exist positive constants β, r_1, and r_2 such that

$$\|f(u) - f(v)\|^2 \leq \beta \|u - v\|^2 + r_1 \|u - v\|_\gamma^2$$

and

$$\|\sigma(u) - \sigma(v)\|_{\mathcal{L}^2(U^0, H)}^2 \leq \beta \|u - v\|^2 + r_2 \|u - v\|_\gamma^2$$

for all $u, v \in H^\gamma$.

Remark 4.15.   In fact, in this autonomous case (i.e., f and σ do not depend on time t explicitly), the *global Lipschitz condition* implies the *sublinear growth condition*: there exists a constant $C > 0$ such that

$$\|f(u)\|^2 + \|\sigma(u)\|^2_{\mathcal{L}^2(U^0, H)} \leq C(1 + \|u\|^2 + \|u\|^2_\gamma) \tag{4.14}$$

for every $u \in H^\gamma$. The situation is different for nonautonomous SPDEs; see Remark 4.1.

Remark 4.16.   When $\gamma = 1$, it is proved in [76, Ch. 3.3.7] that the assumption **(H)** on f and σ ensure the global existence and uniqueness of the solution of Equation (4.7), provided that $r_1 + r_2 < 1$.

From now on, we shall restrict ourselves to the case $0 < \gamma < 1$. We present the following theorem for SPDE (4.4) on a finite time interval $[0, T]$ for any fixed $T > 0$. The following result is essentially in [76, Ch. 3] and [94, Ch. 7], but since we consider more regular mild solutions, the proof is thus modified.

Theorem 4.17 (Well-posedness under global Lipschitz condition).   *Assume that the global Lipschitz condition* **(H)** *holds and consider the autonomous* SPDE *(4.4) with initial data* $h \in H$. *Then there exists a unique solution u as an adapted, continuous process in H. Moreover, for every $T > 0$, u belongs to $L^2(\Omega; C([0, T]; H)) \bigcap L^2(\Omega; L^2([0, T]; H^\gamma))$ and the following property holds:*

$$\mathbb{E}\left\{ \sup_{0 \leq t \leq T} \|u(t)\|^2 + \int_0^T \|u(t)\|^2_\gamma \, dt \right\} < \infty.$$

Proof.   We choose a sufficiently small $T_0 < T$ and denote by Y_{T_0} the set of predictable random processes $\{u(t)\}_{0 \leq t \leq T}$ in space

$$L^2(\Omega; C([0, T_0]; H)) \cap L^2(\Omega; L^2([0, T_0]; H^\gamma))$$

such that

$$\|u\|_{T_0} = \left\{ \mathbb{E}\left(\sup_{0 \leq t \leq T_0} \|u(t)\|^2 + \int_0^{T_0} \|u(t)\|^2_\gamma \, dt \right) \right\}^{\frac{1}{2}} < \infty.$$

Then Y_{T_0} is a Banach space with the norm $\| \cdot \|_{T_0}$.

Let Γ be a nonlinear mapping on Y_{T_0} defined by

$$\Gamma(u)(t) \triangleq S(t)h + \int_0^t S(t - s)f(u(s))ds$$

$$+ \int_0^t S(t - s)\sigma(u(s))dW(s), \quad t \in [0, T_0].$$

We first verify that $\Gamma : Y_{T_0} \to Y_{T_0}$ is well defined and bounded. In the following, C' denotes a positive constant whose value may change from line to line.

It follows from (4.8) and (4.11) that

$$\mathbb{E} \sup_{0 \le t \le T_0} \|S(t)h\|^2 + \mathbb{E} \int_0^{T_0} \|S(t)h\|_\gamma^2 dt \le C' \|h\|^2. \tag{4.15}$$

Let $\{u(s)\}_{s \in [0, T_0]}$ be in Y_{T_0}. Then $F(t) = f(u(t))$ and $\Sigma(t) = \sigma(u(t))$ are predictable random processes in H. Hence, by (4.9), (4.12), and (4.14), we obtain

$$\mathbb{E} \sup_{0 \le t \le T_0} \left\| \int_0^t S(t - s) f(u(s)) ds \right\|^2 + \mathbb{E} \int_0^{T_0} \left\| \int_0^t S(t - s) f(u(s)) ds \right\|_\gamma^2 dt$$

$$\le C' \mathbb{E} \int_0^{T_0} \|f(u(s))\|^2 ds$$

$$\le C' \mathbb{E} \int_0^{T_0} (1 + \|u(s)\|^2 + \|u(s)\|_\gamma^2) ds$$

$$\le C'(1 + \|u(s)\|_{T_0}^2). \tag{4.16}$$

Similarly, by making use of (4.10), (4.13), and (4.14), we have

$$\mathbb{E} \sup_{0 \le t \le T_0} \left\| \int_0^t S(t - s) \sigma(u(s)) dW(s) \right\|^2$$

$$+ \mathbb{E} \int_0^{T_0} \left\| \int_0^t S(t - s) \sigma(u(s)) dW(s) \right\|_\gamma^2 dt \le C'(1 + \|u(s)\|_{T_0}^2). \tag{4.17}$$

From (4.15), (4.16), and (4.17), it follows that $\Gamma : Y_{T_0} \to Y_{T_0}$ is well defined and bounded. To show that Γ is a contraction mapping in Y_{T_0}, we introduce an equivalent norm $\| \cdot \|_{\mu, T_0}$ in Y_{T_0} as

$$\|u\|_{\mu, T_0} = \left\{ \mathbb{E} \left(\sup_{0 \le t \le T_0} \|u(t)\|^2 + \mu \int_0^{T_0} \|u(t)\|_\gamma^2 dt \right) \right\}^{\frac{1}{2}},$$

where μ is a positive parameter. Note that for $u, v \in Y_{T_0}$,

$$\|\Gamma(u) - \Gamma(v)\|_{\mu, T_0}^2$$

$$= \mathbb{E} \left\{ \sup_{0 \le t \le T_0} \|\Gamma(u)(t) - \Gamma(v)(t)\|^2 + \mu \int_0^{T_0} \|\Gamma(u)(t) - \Gamma(v)(t)\|_\gamma^2 dt \right\}.$$

By making use of (4.9), (4.10) and a simple inequality $(a + b)^2 \leq C_\epsilon a^2 + (1 + \epsilon)b^2$ with $C_\epsilon = (1 + \epsilon)/\epsilon$ for any $\epsilon > 0$, we get

$$\mathbb{E} \sup_{0 \leq t \leq T_0} \|\Gamma(u)(t) - \Gamma(v)(t)\|^2$$

$$= \mathbb{E} \sup_{0 \leq t \leq T_0} \left\| \int_0^t S(t - s)(f(u(s)) - f(v(s)))ds \right.$$

$$\left. + \int_0^t S(t - s)(\sigma(u(s)) - \sigma(v(s)))dW(s) \right\|^2$$

$$\leq C_\epsilon T_0 \mathbb{E} \int_0^{T_0} \|f(u(s)) - f(v(s))\|^2 dt$$

$$+ 4(1 + \epsilon)\mathbb{E} \int_0^{T_0} \|\sigma(u(s)) - \sigma(v(s))\|_{\mathcal{L}^2(U^0, H)}^2 \, dt.$$

Similarly, by making use of (4.12) and (4.13), we obtain

$$\mathbb{E} \int_0^{T_0} \|\Gamma(u)(t) - \Gamma(v)(t)\|_\gamma^2 dt \leq \frac{T_0 C_\epsilon}{2\alpha^{1-\gamma}} \mathbb{E} \int_0^{T_0} \|f(u(s)) - f(v(s))\|^2 \, dt$$

$$+ \frac{1 + \epsilon}{2\alpha^{1-\gamma}} \mathbb{E} \int_0^{T_0} \|\sigma(u(s)) - \sigma(v(s))\|_{\mathcal{L}^2(U^0, H)}^2 dt.$$

Hence, by assumption **(H)**, we get

$$\|\Gamma(u) - \Gamma(v)\|_{\mu, T_0}^2$$

$$\leq \left(C_\epsilon T_0 + \frac{C_\epsilon T_0 \mu}{2\alpha^{1-\gamma}} \right) \mathbb{E} \int_0^{T_0} (\beta \|u(s) - v(s)\|^2 + r_1 \|u(s) - v(s)\|_\gamma^2)ds$$

$$+ \left(4(1 + \epsilon) + \frac{(1 + \epsilon)\mu}{2\alpha^{1-\gamma}} \right) \mathbb{E} \int_0^{T_0} (\beta \|u(s) - v(s)\|^2 + r_2 \|u(s) - v(s)\|_\gamma^2)ds$$

$$\leq \rho_1 \mathbb{E} \sup_{0 \leq t \leq T_0} \|u(t) - v(t)\|^2 + \rho_2 \mu \mathbb{E} \int_0^{T_0} \|u(s) - v(s)\|_\gamma^2 \, ds,$$

where

$$\rho_1 = \beta(1 + \epsilon)T_0 \left(4 + \frac{T_0}{\epsilon} + \frac{T_0 \mu'}{2\epsilon} + \frac{\mu'}{2} \right),$$

$$\rho_2 = (1 + \epsilon) \left(\frac{1}{2} + \frac{T_0}{\epsilon \mu'} + \frac{T_0}{2\epsilon} + \frac{4}{\mu'} \right) \frac{r_1 + r_2}{\alpha^{1-\gamma}},$$

with $\mu' = \frac{\mu}{\alpha^{1-\gamma}}$.

Note that we can always assume that $\frac{r_1 + r_2}{\alpha^{1-\gamma}} < 1$. If this is not the case, choose $M > 0$ such that $\frac{r_1 + r_2}{(\alpha + M)^{1-\gamma}} < 1$, and rewrite Equation (4.4) as

$$\frac{\partial}{\partial t}u(x, t) = [\kappa \Delta - (\alpha + M)]u(x, t) + [f(u(x, t)) + M u(x, t)]$$

$$+ \sigma(u(x, t))\frac{\partial}{\partial t}W(x, t).$$

Then **(H)** holds, with β replaced by $\beta + M^2$. So, it is possible to choose μ' sufficiently large and ϵ, T_0 sufficiently small that $\rho = \rho_1 \vee \rho_2 < 1$, which implies that Γ is a contraction mapping in Y_{T_0}. Thus, there exists a unique local solution of Equation (4.7) over $(0, T_0)$, and the solution can be extended to the whole finite interval $[0, T]$. This completes the proof. $\qquad\qquad\square$

Remark 4.18. To show the possible choice of the parameters μ', ϵ, and T_0 to make $\rho < 1$, we first take $\frac{T_0}{\epsilon} < \frac{1}{16}$ and $\mu' = 16^2$, then $\rho_2 \leq (1 + \epsilon)(\frac{1}{2} + \frac{1}{16^3} + \frac{1}{32} + \frac{4}{16^2})$, which yields $\rho_2 < \frac{3}{4}$ providing $\epsilon = \frac{1}{3}$. On the other hand, ρ_1 can be made less than $\frac{3}{4}$ by taking T_0 sufficiently small.

4.3.2.3   Well-Posedness Under Local Lipschitz Condition

In Theorem 4.17, if the global Lipschitz condition on the coefficients is relaxed to hold locally, then we only obtain a local solution that may blow up in finite time. To get a global solution, we impose the following conditions:

(Hn) Local Lipschitz condition: The nonlinearity f and noise intensity σ are measurable functions, and there exist constants $r_n > 0$ such that

$$\|f(u) - f(v)\|^2 + \|\sigma(u) - \sigma(v)\|^2 \leq r_n \|u - v\|^2,$$

for all u, $v \in H^\gamma$ with $\|u\|_\gamma + \|v\|_\gamma \leq n$, $n = 1, 2, \cdots$.

(B) A priori estimate: The solution u satisfies the *a priori* estimate

$$\mathbb{E}\|u(t)\|_\gamma^2 \leq K(t), \quad 0 \leq t < \infty,$$

where $K(t)$ is defined and finite for all $t > 0$.

Theorem 4.19 (Well-posedness under local Lipschitz condition).   *Assume that the local Lipschitz condition* **(Hn)** *and the* a priori *estimate* **(B)** *hold. Consider the autonomous* SPDE *(4.4) with initial datum $h \in H$. Then there exists a unique solution u as an adapted, continuous process in H. Moreover, for every $T > 0$, u is in $L^2(\Omega; C([0, T]; H)) \bigcap L^2(\Omega; L^2([0, T]; H^\gamma))$ and the following property holds:*

$$\mathbb{E}\left\{ \sup_{0 \leq t \leq T} \|u(t)\|^2 + \int_0^T \|u(t)\|_\gamma^2 \, dt \right\} < \infty.$$

Proof.   For any integer $n \geq 1$, let $\chi_n : [0, \infty) \to [0, 1]$ be a C^∞-function such that

$$\chi_n(r) = \begin{cases} 1, & 0 \leq r \leq n, \\ 0, & r \geq 2n. \end{cases}$$

Consider the truncated system

$$\frac{\partial}{\partial t}u(x, t) = (\kappa \Delta - \alpha)u(x, t) + f_n(u(x, t)) + \sigma_n(u(x, t))\frac{\partial}{\partial t}W(x, t), \qquad (4.18)$$

$$u|_{\partial D} = 0, \quad u(x, 0) = h(x), \qquad (4.19)$$

where $f_n(u) = \chi_n(\|u\|_\gamma) f(u)$ and $\sigma_n(u) = \chi_n(\|u\|_\gamma)\sigma(u)$. The assumption (**Hn**) implies that f_n and σ_n satisfy the global conditions (**H**). Hence, by Theorem 4.17, the system (4.18)–(4.19) has a unique solution $u^n(t)$ in $L^2(\Omega; C([0, T]; H)) \bigcap L^2(\Omega; L^2([0, T]; H^\gamma))$. Define an increasing sequence of stopping times $\{\tau_n\}_{n\geq 1}$ by

$$\tau_n \triangleq \inf \{t > 0 : \|u^n(t)\|_\gamma \geq n\}$$

when it exists, and $\tau_n = \infty$ otherwise. Let $\tau_\infty = \lim_{n\to\infty} \tau_n$, a.s., and set $u^{\tau_n}(t) = u^n(t \wedge \tau_n)$. Then $u^{\tau_n}(t)$ is a local solution of Equation (4.4) for $0 \leq t \leq \tau_n$. By assumption (**B**), for any $t \in [0, T \wedge \tau_n]$,

$$\mathbb{E}\|u^{\tau_n}(t)\|_\gamma^2 \leq K(t). \tag{4.20}$$

Also note that

$$\begin{aligned}
\mathbb{E}\|u^{\tau_n}(T)\|_\gamma^2 &= \mathbb{E}\|u^n(T \wedge \tau_n)\|_\gamma^2 \\
&\geq \mathbb{E}\left\{1_{\{\tau_n \leq T\}}\|u^n(T \wedge \tau_n)\|_\gamma^2\right\} \\
&\geq \mathbb{P}\{\tau_n \leq T\} n^2.
\end{aligned} \tag{4.21}$$

In view of (4.20) and (4.21), we get $\mathbb{P}\{\tau_n \leq T\} \leq \frac{K(T)}{n^2}$, which, by the Borel–Cantelli Lemma, implies

$$\mathbb{P}\{\tau_\infty > T\} = 1.$$

Hence, $u(t) := \lim_{n\to\infty} u^n(t)$ is a global solution as claimed. This completes the proof. $\quad\square$

4.3.2.4   An Example

Let us look at an example.

Example 4.20.   Let $D \subset \mathbb{R}^2$ be a bounded domain with smooth boundary ∂D and denote $H \triangleq L^2(D)$. Consider the following SPDE on D for $t > 0$:

$$\frac{\partial}{\partial t}u(x, t) = (\Delta - 1)u(x, t) + \sin u(x, t) + \cos u(x, t)\,\frac{\partial}{\partial t}W(x, t), \tag{4.22}$$

$$u|_{\partial D} = 0, \quad u(x, 0) = h(x). \tag{4.23}$$

The global Lipschitz condition (**H**) holds for this equation. Therefore, by Theorem 4.17, there is a unique mild solution $\{u(x, t)\}_{t\geq 0}$, with a given initial datum $h \in H$.

4.4   Martingale Solutions

In this section, we consider even weaker solutions, which are called *martingale solutions* [36], [94, Ch. 8]. In some literature, these kinds of solutions are also called "weak" solutions, whereas the weak solutions of the last section are called "strong" solutions [232, Ch. V]. Some new variant of the martingale solution approach is introduced in recent research in SPDEs (e.g., [12]), which will not be discussed here.

We start with a brief introduction to martingales [180, Ch. 1.3], [270, Ch. 4 and 5]. Particularly, we consider martingales in Hilbert space e.g., [94, Ch. 3.4].

Let

$$V \subset H \subset V'$$

be a Gelfand triple of separable Hilbert spaces with continuous embeddings. Let $(\Omega, \mathcal{F}, \mathbb{P})$ be a complete probability space with a filtration $\mathcal{F}_t, t \in \mathbb{R}$, satisfying $\mathcal{F}_t \subset \mathcal{F}$ and

$$\mathcal{F}_{t_1} \subset \mathcal{F}_{t_2},$$

for $t_1 \leq t_2$.

We recall the concept of conditional expectation. Let X be an H-valued random variable and $\mathfrak{g}$ be a σ-field contained in $\mathcal{F}$. Then the conditional expectation of X given $\mathfrak{g}$, $\mathbb{E}(X|\mathfrak{g})$, is defined as an H-valued $\mathfrak{g}$-measurable random variable such that

$$\int_A X d\mathbb{P} = \int_A \mathbb{E}(X|\mathfrak{g}) d\mathbb{P}, \quad \text{for every } A \in \mathfrak{g}. \tag{4.24}$$

Consider an integrable H-valued adapted process $\{M(t) : t \in \mathbb{R}\}$, that is, $M(t)$ is $\mathcal{F}_t$-measurable with $\mathbb{E}\|M(t)\| < \infty$. The process $M(t)$ is called a *martingale* or $\mathcal{F}_t$-*martingale* if

$$\mathbb{E}\{M(t)|\mathcal{F}_s\} = M(s), \quad \text{a.s.},$$

for any $t \geq s$. Let $T > 0$ be fixed and denote by $\mathcal{M}_T^2(H)$ the space of all H-valued, continuous, square integrable martingales $\{M(t) : 0 \leq t \leq T\}$. Then the space $\mathcal{M}_T^2(H)$, equipped with norm

$$\|M\|_{\mathcal{M}_T^2(H)} = \left(\mathbb{E} \sup_{0 \leq t \leq T} \|M(t)\|^2 \right)^{1/2},$$

is a Banach space [94, Proposition 3.9].

For $M \in \mathcal{M}_T^2(H)$, there exists a unique, increasing, and nuclear operator [94, Proposition 3.12] valued adapted process $V(t)$ with $V(0) = 0$, such that the process

$$\langle M(t), a \rangle \langle M(t), b \rangle - \langle V(t)a, b \rangle, \quad a, b \in H, \quad 0 \leq t \leq T,$$

is an $\mathcal{F}_t$-martingale. Equivalently, the process

$$M(t) \otimes M(t) - V(t), \quad 0 \leq t \leq T$$

is an $\mathcal{F}_t$-martingale. The process $V(t)$ is also denoted by $\ll M(\cdot) \gg$, which is called the quadratic variation process of M.

For $M \in \mathcal{M}_T^2(H)$ the following representation theorem is important in the construction of martingale solutions [94, Theorem 8.2].

Theorem 4.21 (Representation theorem).   *Let $M \in \mathcal{M}_T^2(H)$ with*

$$\ll M(t) \gg = \int_0^t (\Phi(s)Q^{1/2})^*(\Phi(s)Q^{1/2})ds, \quad 0 \leq t \leq T,$$

where Φ is a predictable $\mathcal{L}^2(U^0, H)$-valued process [94, p.76], and Q is a given bounded nonnegative symmetric operator on Hilbert space U. Then there exists a probability space $(\widetilde{\Omega}, \widetilde{\mathcal{F}}, \widetilde{\mathcal{F}}_t, \widetilde{\mathbb{P}})$ and an H-valued Q-Wiener process, $\widetilde{W}$ defined on $(\Omega \times \widetilde{\Omega}, \mathcal{F} \times \widetilde{\mathcal{F}}, \mathbb{P} \times \widetilde{\mathbb{P}})$ such that

$$M(t, \omega, \tilde{\omega}) = \int_0^t \Phi(s, \omega, \tilde{\omega})d\widetilde{W}(s, \omega, \tilde{\omega}), \quad 0 \leq t \leq T$$

where $M(t, \omega, \tilde{\omega}) = M(t, \omega)$ and $\Phi(t, \omega, \tilde{\omega}) = \Phi(t, \omega)$ for all $(\omega, \tilde{\omega}) \in \Omega \times \widetilde{\Omega}$.

Now let $A : \mathcal{D}(A) \to H$ be the infinitesimal generator of a C_0-semigroup $S(t)$, $t \geq 0$ on H. Assume that $F(\cdot) : H \to H$ is measurable, Q is a symmetric, positive-definite operator on a Hilbert space U and $\Phi(\cdot) : H \to \mathcal{L}^2(U^0, H)$. If there exist a probability space $(\Omega, \mathcal{F}, \mathbb{P})$ with a filtration $\mathcal{F}_t$, a Q-Wiener process W, and a random process $u(t)$ such that $u(t)$ is a weak solution to the SPDE,

$$du(t) = [Au(t) + F(u(t))]dt + \Phi(u(t))dW(t), \quad u(0) = u_0, \tag{4.25}$$

then we say that the SPDE (4.25) has a *martingale solution*. The sequence $(\Omega, \mathcal{F}, \mathbb{P}, \{\mathcal{F}_t\}, W^Q, u)$ is called a martingale solution to (4.25). We impose the following assumptions:

(A_1') Semigroup $S(t), t > 0$, is compact.

(A_2') $F : H \to H$ and $\Phi : H \to \mathcal{L}^2(U^0, H)$ are globally Lipschitz continuous in u.

Then we have the following existence result [94, Theorem 8.1].

Theorem 4.22 (Existence of martingale solutions).   *Assume that (A_1') and (A_2') hold. Then SPDE (4.25) has a martingale solution.*

The martingale solution is related to the following martingale problem. For any given $T > 0$, let $\Omega_c = C(0, T; V')$, $\mathcal{C}$ be the Borel σ-field, and $Pr(\Omega_c)$ be the set of all probability measures on $(\Omega_c, \mathcal{C})$. Define the canonical process $\xi : \Omega_c \to V'$ by

$$\xi_t(\omega) = \omega(t), \quad \omega \in \Omega_c.$$

Additionally, define the σ-field $\mathcal{C}_t \triangleq \sigma\{\xi_s : 0 \leq s \leq t\}$ for $0 \leq t \leq T$.

Definition 4.23 (Martingale problem).   Given $H, V, V', U, A, F, \Phi, Q$ as above and $u_0 \in H$. The Martingale problem $\mathcal{M}^c(H, V, V', U, A, F, \Phi, Q, u_0)$ is to find a probability measure P on Ω_c such that

(M_1) $P\{\xi_0 = u_0\} = 1$,

(M_2) $P\{C(0, T; H) \cap L^2(0, T; V)\} = 1$, and

(M_3) For any $\varphi \in C^2(H)$,

$$M_t^\varphi \triangleq \varphi(\xi_t) - \varphi(\xi_0) - \int_0^t \mathcal{L}\varphi(\xi(s))ds$$

is a square integrable, continuous P-martingale. Here the second-order differential operator $\mathcal{L}$ is defined by

$$\mathcal{L}\varphi(u) \triangleq \tfrac{1}{2}\text{Tr}\left[(\Phi(u)Q^{1/2})^* D^2\varphi(u)(\Phi(u)Q^{1/2})\right] + \langle Au + F(u), D\varphi(u)\rangle,$$

where $D\varphi$ and $D^2\varphi$ are the first-order and second-order Fréchet derivatives, respectively.

The differential operator $\mathcal{L}$ is called the *Kolmogorov operator* [265].

We call the probability measure P on Ω_c satisfying (M_1)–(M_3) a solution of the martingale problem $\mathcal{M}^c(H, V, V', U, A, F, \Phi, Q, u_0)$.

Remark 4.24. The martingale problem $\mathcal{M}^c(H, V, V', U, A, F, \Phi, Q, u_0)$ is also called the martingale problem related to $\mathcal{L}$.

Now we recall the following result for the martingale problem $\mathcal{M}^c(H, V, V', U, A, F, \Phi, Q, u_0)$; see [315].

Theorem 4.25. *Assume that* (A'_1) *and* (A'_2) *hold. Then the martingale problem* $\mathcal{M}^c(H, V, V', U, A, F, \Phi, Q, u_0)$ *has a unique solution.*

For the relation of the martingale solution and the solution to the martingale problem, we state the following theorem [232, Theorem V.1]:

Theorem 4.26. *The probability measure P is a solution to the martingale problem* $\mathcal{M}^c(H, V, V', U, A, F, \Phi, Q, u_0)$ *if and only if there exists a martingale solution* $(\Omega, \mathcal{F}, \mathbb{P}, W^Q, u)$ *to (4.25) with initial value $u_0 \in H$ such that P is the image measure* $\mathbb{P}$ *under the mapping $\omega \to u(\cdot, \omega)$.*

4.5 Conversion Between Itô and Stratonovich SPDEs

In this section we convert Stratonovich SPDEs to Itô SPDEs, and vice versa. We only do this formally. Rigorous verifications can be worked out in the setup in [94, Ch. 4.2] for stochastic integrals, but following a similar discussion for stochastic ordinary differential equations [193, Ch. 6.1].

The Stratonovich integral and Itô integral (introduced in §3.6) are both defined as a limit in mean square (m.s.) of a sum, but they differ only in the selection of time instants used in subintervals $[t_j, t_{j+1}]$ of a partition of the time interval $[0, t]$. In the Itô integral, the left end point t_j is always used, whereas in the Stratonovich integral, the middle point $\frac{t_j + t_{j+1}}{2}$ is taken.

4.5.1 Case of Scalar Multiplicative Noise

We start our investigation on the following Stratonovich SPDE in Hilbert space H, with a scalar multiplicative noise

$$du = Au\, dt + f(t, u)dt + \Phi(t, u) \circ dW(t), \tag{4.26}$$

where A is a linear partial differential operator, $\Phi(t,)$: $H \to H$ is a Fréchet differentiable (nonlinear) mapping, $W(t)$ is a standard *scalar* Wiener process, and the nonlinear mapping f is locally Lipschitz continuous in u. Here $\circ dW$ is interpreted as a Stratonovich differential.

We now verify that the equivalent Itô formulation for SPDE (4.26) is

$$du = Au\, dt + f(t, u)dt + \frac{1}{2}\Phi(t, u)\Phi_u(t, u)dt + \Phi(t, u)dW(t), \qquad (4.27)$$

where Φ_u is the Fréchet derivative of Φ with respect to u. Thus a Stratonovich SPDE can be converted to an Itô SPDE with an extra correction term. Similarly, we can convert an Itô SPDE to a Stratonovich SPDE by subtracting this correction term $\frac{1}{2}\Phi(t, u)\Phi_u(t, u)$.

Remark 4.27. If the noise intensity Φ does not depend on u (additive noise), then the Fréchet derivative Φ_u is zero, and thus the correction term is absent. This implies that Stratonovich SPDEs and Itô SPDEs are identical in the case of additive noise.

Let $0 = t_0 < t_1 < \cdots < t_n = t$ be a partition of $[0, t]$. Denote $\Delta W(t_j) \triangleq W(t_{j+1}) - W(t_j)$ and $\Delta t_j \triangleq t_{j+1} - t_j$, for $j = 0, \cdots, n - 1$. In fact, we formally calculate, by the definition of Stratonovich integral,

$$\int_0^t \Phi(s, u(s)) \circ dW(s)$$

$$= \lim_{\max \Delta t_j \to 0} \text{ in m.s. } \sum_{j=0}^{n-1} \Phi\left(\frac{t_j + t_{j+1}}{2}, u\left(\frac{t_j + t_{j+1}}{2}\right)\right) \Delta W(t_j)$$

$$= \lim_{\max \Delta t_j \to 0} \text{ in m.s. } \sum_{j=0}^{n-1} \Phi\left(t_j, \frac{1}{2}[u(t_j) + u(t_{j+1})]\right) \Delta W(t_j), \qquad (4.28)$$

where lim in m.s. is the limit in the sense of mean square. The final step above is derived as follows.

In fact, note that

$$\Phi\left(\frac{t_j + t_{j+1}}{2}, u\left(\frac{t_j + t_{j+1}}{2}\right)\right) \Delta W(t_j) = \Phi\left(t_j + \frac{\Delta t_j}{2}, u\left(t_j + \frac{\Delta t_j}{2}\right)\right) \Delta W(t_j),$$

$$u\left(\frac{t_j + t_{j+1}}{2}\right) = u\left(t_j + \frac{\Delta t_j}{2}\right) = u(t_j) + \frac{du}{dt} \cdot \frac{\Delta t_j}{2} + O((\Delta t_j)^2),$$

and

$$u\left(\frac{t_j + t_{j+1}}{2}\right) = u\left(t_{j+1} - \frac{\Delta t_j}{2}\right) = u(t_{j+1}) - \frac{du}{dt} \cdot \frac{\Delta t_j}{2} + O((\Delta t_j)^2).$$

Thus,

$$u\left(\frac{t_j + t_{j+1}}{2}\right) = \frac{1}{2} \cdot \left(u\left(t_j + \frac{\Delta t_j}{2}\right) + u\left(t_{j+1} - \frac{\Delta t_j}{2}\right)\right)$$

$$= \frac{1}{2} \cdot (u(t_{j+1}) + u(t_j)) + O((\Delta t_j)^2).$$

Hence, we have

$$\Phi\left(\tfrac{t_j+t_{j+1}}{2}, u\left(\tfrac{t_j+t_{j+1}}{2}\right)\right)\Delta W(t_j)$$

$$= \left[\Phi\left(t_j, u\left(t_j + \tfrac{\Delta t_j}{2}\right)\right) + \Phi_t\left(t_j, u\left(t_j + \tfrac{\Delta t_j}{2}\right)\right)\tfrac{\Delta t_j}{2} + O((\Delta t_j)^2)\right]\Delta W(t_j)$$

$$= \Phi\left(t_j, \tfrac{1}{2}\left[u\left(t_j + \tfrac{\Delta t_j}{2}\right) + u\left(t_{j+1} - \tfrac{\Delta t_j}{2}\right)\right]\right)\Delta W(t_j) + O((\Delta t_j)^{3/2})$$

$$= \Phi\left(t_j, \tfrac{1}{2}[u(t_j) + u(t_{j+1})]\right)\Delta W(t_j) + O((\Delta t_j)^{3/2}).$$

This verifies (4.28).

Now let us move on to further examine the right-hand side of (4.28). By the Taylor expansion and the mean value theorem, we have

$$\sum_{j=0}^{n-1}\Phi\left(t_j, \tfrac{1}{2}[u(t_j) + u(t_{j+1})]\right)\Delta W(t_j)$$

$$= \sum_{j=0}^{n-1}\Phi\left(t_j, u(t_j) + \tfrac{1}{2}[u(t_{j+1}) - u(t_j)]\right)\Delta W(t_j)$$

$$= \sum_{j=0}^{n-1}\Phi(t_j, u(t_j))\Delta W(t_j)$$

$$+ \tfrac{1}{2}\sum_{j=0}^{n-1}\Phi_u(t_j, u(t_j))\Delta u(t_j)\Delta W(t_j) + O((\Delta t_j)^2), \tag{4.29}$$

where, by the definition of a Stratonovich integral,

$$\Delta u(t_j) \triangleq u(t_{j+1}) - u(t_j) = [Au(t_j) + f(t_j, u(t_j))]\Delta t_j$$

$$+ \Phi\left(t_j, u\left(\tfrac{t_j+t_{j+1}}{2}\right)\right)\Delta W(t_j).$$

Using (4.29) and the following two facts:

$$\lim_{\max \Delta t_j \to 0} \text{ in m.s. } \sum_{j=0}^{n-1}\Delta t_j \Delta W(t_j) = 0,$$

$$\lim_{\max \Delta t_j \to 0} \text{ in m.s. } \sum_{j=0}^{n-1}\Delta W(t_j)\Delta W(t_j) = t,$$

while passing the limit $\max \Delta t_j \to 0$ in (4.28), we get

$$\int_0^t \Phi(s, u(s)) \circ dW(s) = \int_0^t \Phi(s, u(s)) dW(s) + \tfrac{1}{2}\int_0^t \Phi(s, u(s))\Phi_u(s, u(s)) ds.$$

This yields (4.27).

4.5.2   Case of General Multiplicative Noise

Consider a Stratonovich SPDE in Hilbert space H

$$du = Au\,dt + F(u)dt + \Phi(u) \circ dW(t), \tag{4.30}$$

where A is a second-order linear partial differential operator, F is a locally Lipschitz continuous mapping, $\Phi(\,\cdot\,)$ is a Fréchet differentiable mapping defined from Hilbert space H to $\mathcal{L}^2(U^0, H)$, and $W(t)$ is a Brownian motion (or Wiener process) taking values in a Hilbert space U with covariance operator Q. Recall that $U^0 = Q^{1/2}U$. Note that $\Phi(u) \circ dW(t)$ is interpreted as a Stratonovich differential. We want to convert this Stratonovich SPDE to an Itô SPDE, and vice versa.

The Brownian motion $W(t)$ taking values in Hilbert space U can be represented as

$$W(t) = \sum_{n=1}^{\infty} \sqrt{q_n} W_n(t) e_n, \tag{4.31}$$

where $W_n(t)$'s are independent standard scalar Brownian motions, $\{e_n\}$ is a complete orthonormal basis of U, and $\{q_n\}$ is a sequence of nonnegative real numbers such that

$$Qe_n = q_n e_n, \quad n = 1, 2, \cdots.$$

Substituting (4.31) into (4.30), we get

$$
\begin{aligned}
du &= Au\,dt + F(u)dt + \Phi(u) \circ dW(t) \\
&= Au\,dt + F(u)dt + \Phi(u) \sum_{n=1}^{\infty} \sqrt{q_n} e_n \circ dW_n(t) \\
&= Au\,dt + F(u)dt + \sum_{n=1}^{\infty} \sqrt{q_n} \Phi(u)(e_n) \circ dW_n(t) \\
&= Au\,dt + F(u)dt + \sum_{n=1}^{\infty} \sqrt{q_n}\, \tilde{\Phi}_n(u) \circ dW_n(t),
\end{aligned}
\tag{4.32}
$$

where

$$
\begin{aligned}
\tilde{\Phi}_n(\,\cdot\,) &: H \to H \\
u &\mapsto \Phi(u)(e_n).
\end{aligned}
$$

Now we prove that the stochastic differential term in the last expression of (4.32) has the following property:

$$\tilde{\Phi}_n(u) \circ dW_n(t) = \tfrac{1}{2}\tilde{\Phi}_n'(u)(\tilde{\Phi}_n(u))dt + \tilde{\Phi}_n(u)dW_n(t), \tag{4.33}$$

where $\tilde{\Phi}_n'(u)$ denotes the Fréchet derivative of $\tilde{\Phi}_n$ with respect to u, and $\tilde{\Phi}_n(u)dW_n(t)$ is an Itô differential.

In fact, for fixed $n \in N$, let $0 < t_0^n < t_1^n < \cdots < t_m^n = t$ be a partition of $[0, t]$. Denote by $\Delta t_j^n = t_{j+1}^n - t_j^n$, $\Delta W_n(t_j^n) = W_n(t_{j+1}^n) - W_n(t_j^n)$, and $\delta_m = \max\{\Delta t_0^n, \Delta t_1^n, \cdots, \Delta t_m^n\}$.

According to the definition of Stratonovich integral, we have

$$\int_0^t \tilde{\Phi}_n(u) \circ dW_n(t)$$

$$= \lim_{\max \Delta t_j \to 0} \text{ in m.s. } \sum_{j=0}^{m-1} \tilde{\Phi}_n\left(u\left(t_j^n + \frac{\Delta t_j^n}{2}\right)\right) \Delta W_n(t_j^n)$$

$$= \lim_{\max \Delta t_j \to 0} \text{ in m.s. } \sum_{j=0}^{m-1} \tilde{\Phi}_n\left(u(t_j^n) + \int_{t_j^n}^{t_j^n + \frac{\Delta t_j^n}{2}} (Au + F(u))dt\right.$$

$$\left. + \int_{t_j^n}^{t_j^n + \frac{\Delta t_j^n}{2}} \tilde{\Phi}_n(u) \circ dW_n(t)\right) \Delta W_n(t_j^n)$$

since u satisfies Equation (4.30) :

$$= \lim_{\max \Delta t_j \to 0} \text{ in m.s. } \sum_{j=0}^{m-1} \tilde{\Phi}_n\left(u(t_j^n)\right.$$

$$+ (Au\left(t_j^n\right) + F(u\left(t_j^n\right)))\frac{\Delta t_j^n}{2} + \tilde{\Phi}_n\left(u\left(t_j^n + \frac{\Delta t_j^n}{4}\right)\right) \Delta \tilde{W}_n\left(t_j^n\right)\right) \Delta W_n\left(t_j^n\right)$$

where $\tilde{W}_n\left(t_j^n\right) = W_n\left(t_j^n + \frac{\Delta t_j^n}{2}\right) - W_n\left(t_j^n\right)$

$$= \lim_{\max \Delta t_j \to 0} \text{ in m.s. } \sum_{j=0}^{m-1}\left(\tilde{\Phi}_n\left(u\left(t_j^n\right)\right)\right.$$

$$+ \tilde{\Phi}_n'\left(\left(Au\left(t_j^n\right) + F\left(u\left(t_j^n\right)\right)\right)\frac{\Delta t_j^n}{2} + \tilde{\Phi}_n\left(u\left(t_j^n + \frac{\Delta t_j^n}{4}\right)\right) \Delta \tilde{W}_n\left(t_j^n\right)\right)$$

$$+ \text{(higher-order terms with } \Delta t_j^n)\right) \Delta W_n\left(t_j^n\right)$$

Since $\tilde{\Phi}_n(u)$ is Fréchet differentiable:

$$= \int_0^t \tilde{\Phi}_n(u)d_n W(t) + \lim_{\delta_m \to 0} \sum_{j=0}^{m-1} \tilde{\Phi}_n'(u(t_j^n))$$

$$\times \left[\tilde{\Phi}_n\left(u\left(t_j^n + \frac{\Delta t_j^n}{4}\right)\right) \Delta \tilde{W}_n(t_j^n)\right] \Delta W_n\left(t_j^n\right)$$

$$= \int_0^t \tilde{\Phi}_n(u)dW_n(t) + \frac{1}{2}\int_0^t \tilde{\Phi}_n'(u)(\tilde{\Phi}_n(u))dt.$$

In the final step, we have used the facts that $\lim_{\delta_m \to 0} \sum_{j=0}^{m-1} \Delta t_j^n \Delta W_n(t_j^n) = 0$ and $\lim_{\delta_m \to 0} \sum_{j=0}^{m-1} \Delta \tilde{W}_n(t_j^n) \Delta W_n(t_j^n) = \frac{1}{2} \Delta t_j^n$. This proves (4.33).

Then, inserting (4.33) into (4.32), we have

$$
\begin{aligned}
du &= Au\,dt + F(u)dt + \Phi(u) \circ dW(t) \\
&= Au\,dt + F(u)dt + \sum_{n=1}^{\infty} \sqrt{q_n} \left(\frac{1}{2} \tilde{\Phi}_n'(u)(\tilde{\Phi}_n)dt + \tilde{\Phi}_n(u)dW_n(t) \right) \\
&= Au\,dt + F(u)dt + \Phi(u)dW(t) + \frac{1}{2} \sum_{n=1}^{\infty} \sqrt{q_n} \tilde{\Phi}_n'(u)(\tilde{\Phi}_n(u))dt \\
&= Au\,dt + F(u)dt + \Phi(u)dW(t) + \frac{1}{2} \sum_{n=1}^{\infty} \sqrt{q_n} \left(\Phi'(u)(\Phi(u)e_n) \right) e_n\,dt.
\end{aligned}
$$

Note that $\Phi(u)e_n \in H$ and $\Phi'(u)(\cdot) \in \mathcal{L}^2(U^0, H)$.

Thus, the Stratonovich SPDE

$$
du = [Au + F(u)]dt + \Phi(u) \circ dW(t) \tag{4.34}
$$

converts to the following Itô SPDE:

$$
du = \left[Au + F(u) + \frac{1}{2} \sum_{n=1}^{\infty} \sqrt{q_n} \left(\Phi'(u)(\Phi(u)e_n) \right) e_n \right] dt + \Phi(u)\,dW(t), \tag{4.35}
$$

where $\Phi' = \Phi_u$ is the Fréchet derivative of $\Phi(u)$ with respect to u.

Remark 4.28. The term $\sum_{n=1}^{\infty} \frac{1}{2} \sqrt{q_n} \left(\Phi'(u)(\Phi(u)e_n) \right) e_n\,dt$ is called a correction term. If the noise is an additive noise, i.e., $\Phi(u)$ does not depend on the state variable u, then the Fréchet derivative $\Phi_u \equiv 0$, the correction term is zero, and in this case, the Stratonovich SPDE is the same as the Itô SPDE. This is also true for the case of scalar multiplicative noise discussed in the previous subsection.

An Itô SPDE can also be converted to a Stratonovich SPDE by subtracting the correction term.

4.5.3   Examples

Example 4.29 (Stratonovich SPDE to Itô SPDE).   Consider a Stratonovich SPDE

$$
u_t = u_{xx} + \sin(u) + u \circ \dot{W}(t),
$$

where $W(t)$ is a scalar Brownian motion. Then

$$
Au = \partial_{xx} u, \quad f(u) = \sin(u), \quad \Phi(t, u) = u, \quad \text{and} \quad \Phi_u = I.
$$

The correction term is

$$
\frac{1}{2} \Phi_u \Phi(u)dt = \frac{1}{2} u\,dt.
$$

Thus the corresponding Itô SPDE is

$$u_t = u_{xx} + \sin(u) + \tfrac{1}{2}u + u\dot{W}(t).$$

Example 4.30 (Itô SPDE to Stratonovich SPDE). Consider an Itô SPDE

$$u_t = u_{xx} + u - u^3 + \epsilon u \dot{W}(t),$$

where $W(t)$ is a scalar Brownian motion. Now

$$Au = \partial_{xx} u, \; f(u) = u - u^3 \quad \text{and} \quad \Phi(u) = \epsilon u.$$

The correction term is

$$\tfrac{1}{2}\Phi_u(u)\Phi(u)dt = \tfrac{1}{2}\epsilon^2 u \, dt.$$

Thus, the Stratonovich SPDE, corresponding to the above Itô SPDE, is

$$u_t = u_{xx} + u - u^3 - \tfrac{1}{2}\epsilon^2 u + \epsilon u \circ \dot{W}(t).$$

4.6 Linear Stochastic Partial Differential Equations

We now consider the stochastic wave and heat equations on bounded intervals.

4.6.1 Wave Equation with Additive Noise

We consider the wave equation for the displacement $u(x, t)$ of a vibrating string at position x and time t, subject to additive noise

$$u_{tt} = c^2 u_{xx} + \epsilon \dot{W}, \quad 0 < x < l, \tag{4.36}$$

$$u(0, t) = u(l, t) = 0, \tag{4.37}$$

$$u(x, 0) = f(x), \quad u_t(x, 0) = g(x), \tag{4.38}$$

where c is a positive constant (wave speed), ϵ is a positive real parameter modeling the noise intensity, and $W(t)$ is a Brownian motion taking values in Hilbert space $H = L^2(0, l)$. For simplicity, we assume that the initial data f and g are deterministic.

We discuss the solution and its correlation at different time instants. For simplicity, we assume the existence of a basis of common eigenfunctions for the covariance operator Q of $W(t)$ and for the Laplacian ∂_{xx} with the zero Dirichlet boundary conditions on $(0, l)$. This assumption is equivalent to Q commuting with ∂_{xx} [42].

Consider the Fourier expansion or eigenfunction expansion

$$u(x, t) = \sum_{n=1}^{\infty} u_n(t) e_n(x), \tag{4.39}$$

$$W(x, t) = \sum_{n=1}^{\infty} \sqrt{q_n} W_n(t) e_n(x), \tag{4.40}$$

where $q_n (\geq 0)$ are eigenvalues of Q, and

$$e_n(x) = \sqrt{2/l} \sin \frac{n\pi x}{l}, \quad n = 1, 2, \cdots$$

are eigenfunctions of ∂_{xx} with corresponding eigenvalues $\lambda_n = -(n\pi/l)^2$. In fact, $\{e_n\}_{n=1}^{\infty}$ forms an orthonomal basis for H.

Inserting these expansions into SPDE (4.36), we obtain

$$\ddot{u}_n(t) = c^2 \lambda_n u_n + \epsilon \sqrt{q_n} \dot{W}_n(t), \quad n = 1, 2, \cdots. \tag{4.41}$$

For each n, this second-order SDE can be solved by converting to a linear system of first order SDEs [244, Exercise 5.13], and we obtain (Problem 4.7)

$$u_n(t) = \left[A_n - \epsilon \frac{l}{cn\pi} \sqrt{q_n} \int_0^t \sin \frac{cn\pi}{l} s \, dW_n(s) \right] \cos \frac{cn\pi}{l} t$$

$$+ \left[B_n + \epsilon \frac{l}{cn\pi} \sqrt{q_n} \int_0^t \cos \frac{cn\pi}{l} s \, dW_n(s) \right] \sin \frac{cn\pi}{l} t, \tag{4.42}$$

where A_n and B_n are constants.

The final solution is

$$u(x, t) = \sum_{n=1}^{\infty} \left\{ \left[A_n - \epsilon \frac{l}{cn\pi} \sqrt{q_n} \int_0^t \sin \frac{cn\pi}{l} s \, dW_n(s) \right] \cos \frac{cn\pi}{l} t \right.$$

$$\left. + \left[B_n + \epsilon \frac{l}{cn\pi} \sqrt{q_n} \int_0^t \cos \frac{cn\pi}{l} s \, dW_n(s) \right] \sin \frac{cn\pi}{l} t \right\} e_n(x), \tag{4.43}$$

where the constants A_n and B_n are determined by the initial conditions as

$$A_n = \langle f, e_n \rangle, \quad B_n = \frac{l}{cn\pi} \langle g, e_n \rangle.$$

When the noise is at one mode, say, at the first mode $e_1(x)$ (i.e., $q_1 > 0$ but $q_n = 0$, $n = 2, 3, \ldots$), we see that the solution contains randomness only at that mode. So, for the linear stochastic diffusion system, there are no interactions between modes. In other words, if we randomly force a few high-frequency modes, then there is no impact of randomness on low-frequency modes.

The mean value for the solution is

$$\mathbb{E}u(x, t) = \sum_{n=1}^{\infty} \left[A_n \cos \frac{cn\pi t}{l} + B_n \sin \frac{cn\pi t}{l} \right] e_n(x). \tag{4.44}$$

Now we calculate the covariance of the solution u at different time instants t and s, i.e., $\mathbb{E}\langle u(x, t) - \mathbb{E}u(x, t), u(x, s) - \mathbb{E}u(x, s)\rangle$.

Using Itô isometry, we get

$$
\mathbb{E}\langle u(x, t) - \mathbb{E}u(x, t), u(x, s) - \mathbb{E}u(x, s)\rangle
$$

$$
= \sum_{n=1}^{\infty} \frac{\epsilon^2 l^2 q_n}{c^2 n^2 \pi^2} \left[\int_0^{t \wedge s} \sin^2 \frac{cn\pi r}{l} dr \cos \frac{cn\pi t}{l} \cos \frac{cn\pi s}{l} \right.
$$

$$
+ \int_0^{t \wedge s} \cos^2 \frac{cn\pi r}{l} dr \sin \frac{cn\pi t}{l} \sin \frac{cn\pi s}{l}
$$

$$
\left. - \int_0^{t \wedge s} \sin \frac{cn\pi r}{l} \cos \frac{cn\pi r}{l} dr \left(\cos \frac{cn\pi t}{l} \sin \frac{cn\pi s}{l} + \cos \frac{cn\pi s}{l} \sin \frac{cn\pi t}{l} \right) \right].
$$

After integrating, we obtain the covariance as

$$
\mathrm{Cov}(u(x, t), u(x, s)) = \mathbb{E}\langle u(x, t) - \mathbb{E}u(x, t), u(x, s) - \mathbb{E}u(x, s)\rangle
$$

$$
= \sum_{n=1}^{\infty} \frac{\epsilon^2 l^2 q_n}{2c^2 n^2 \pi^2} \left[(t \wedge s) \cos \frac{cn\pi(t - s)}{l} - \frac{l}{2cn\pi} \sin \frac{2cn\pi(t \wedge s)}{l} \cos \frac{cn\pi(t + s)}{l} \right.
$$

$$
\left. + \frac{l}{2cn\pi} \cos \frac{2cn\pi(t \wedge s)}{l} \sin \frac{cn\pi(t + s)}{l} - \frac{l}{2cn\pi} \sin \frac{cn\pi(t + s)}{l} \right]
$$

$$
= \sum_{n=1}^{\infty} \frac{\epsilon^2 l^2 q_n}{2c^2 n^2 \pi^2} \left[(t \wedge s) \cos \frac{cn\pi(t - s)}{l} + \frac{l}{2cn\pi} \sin \frac{cn\pi(t + s - 2(t \wedge s))}{l} \right.
$$

$$
\left. - \frac{l}{2cn\pi} \sin \frac{cn\pi(t + s)}{l} \right].
$$

In particular, for $t = s$, we get the variance

$$
\mathrm{Var}(u(x, t)) = \sum_{n=1}^{\infty} \frac{\epsilon^2 l^2 q_n}{2c^2 n^2 \pi^2} \left[t - \frac{l}{2cn\pi} \sin \frac{2cn\pi}{l} t \right]. \tag{4.45}
$$

Now we consider the energy evolution for the system. Define an energy functional

$$
E(t) = \frac{1}{2} \int_0^l [u_t^2 + c^2 u_x^2] dx. \tag{4.46}
$$

Taking the time derivative, we obtain, in differential form,

$$
dE(t) = \epsilon \int_0^l u_t(x, t) dW(x, t) dx, \tag{4.47}
$$

or, in integral form,

$$
E(t) = E(0) + \epsilon \int_0^l \int_0^t u_t(x, s) dW(x, s) dx.
$$

Thus,

$$\mathbb{E}E(t) = E(0), \text{ and} \tag{4.48}$$

$$\text{Var}(E(t)) = \epsilon^2 \mathbb{E}\left(\int_0^l \int_0^t u_t(x,s) dW(x,s) dx\right)^2, \tag{4.49}$$

where $W(x,t)$ is in the following form

$$W(x,t) = \sum_{n=1}^{\infty} \sqrt{q_n} W_n(t) e_n(x). \tag{4.50}$$

Note that $u_t(x,t)$ can be written as

$$u_t(x,t) = \sum \left\{ -A_n \frac{cn\pi}{l} \sin \frac{cn\pi t}{l} + B_n \frac{cn\pi}{l} \cos \frac{cn\pi t}{l} \right.$$
$$+ \epsilon \sqrt{q_n} \left[\int_0^t \sin \frac{cn\pi s}{l} dW_n(s) \sin \frac{cn\pi t}{l} \right.$$
$$\left. + \int_0^t \cos \frac{cn\pi s}{l} dW_n(s) \cos \frac{cn\pi t}{l} \right] \right\} e_n(x). \tag{4.51}$$

Set $\mu_n \triangleq \frac{cn\pi}{l}$ and rewrite u_t as

$$u_t(x,t) = \sum \left\{ F_n(t) + \epsilon \sqrt{q_n} \left[\int_0^t (\sin \mu_n s \cdot \sin \mu_n t \right.\right.$$
$$\left.\left. + \cos \mu_n s \cdot \cos \mu_n t) dW_n(s) \right] \right\} e_n(x)$$
$$= \sum \left\{ F_n(t) + \epsilon \sqrt{q_n} \int_0^t \cos \mu_n (t-s) dW_n(s) \right\} e_n(x),$$

where

$$F_n(t) \triangleq -A_n \mu_n \sin \mu_n t + B_n \mu_n \cos \mu_n t, \quad n = 1, 2, \cdots.$$

For simplicity of notation, define

$$G_n(t) \triangleq F_n(t) + \epsilon \sqrt{q_n} \int_0^t \cos \mu_n (t-s) dW_n(s), \quad n = 1, 2, \cdots.$$

Then we have

$$u_t(x,t) = \sum G_n(t) e_n(x).$$

Thus,

$$\mathbb{E}\left(\int_0^l \int_0^t u_t(x,s)dW(x,s)dx\right)^2$$

$$= \mathbb{E}\left[\int_0^l \sum_{n=1}^{\infty} \sqrt{q_n} e_n(x) \int_0^t u_t(s)dW_n(s)dx\right]^2$$

$$= \mathbb{E}\left[\sum_{n=1}^{\infty} \sqrt{q_n} \int_0^l \int_0^t u_t(s)e_n(x)dW_n(s)dx\right]^2$$

$$= \mathbb{E}\left[\sum_{n=1}^{\infty} \sqrt{q_n} \int_0^t \left(\int_0^l e_n(x) \sum_{j=1}^{\infty} G_j(s)e_j(x)dx\right) dW_n(s)\right]^2$$

$$= \mathbb{E}\left[\sum_{n=1}^{\infty} \sqrt{q_n} \int_0^t \left(\sum_{j=1}^{\infty} G_j(s) \int_0^l e_n(x)e_j(x)dx\right) dW_n(s)\right]^2$$

$$= \mathbb{E}\left[\sum_{n=1}^{\infty} \sqrt{q_n} \int_0^t G_n(s)dW_n(s)\right]^2 = \sum_{n=1}^{\infty} q_n \mathbb{E}\int_0^t G_n^2(s)ds.$$

Using the expression for G_n, we further get

$$\mathbb{E}\left(\int_0^l \int_0^t u_t(x,s)dW(x,s)dx\right)^2$$

$$= \sum_{n=1}^{\infty} q_n \mathbb{E}\int_0^t \left[F_n(s) + \epsilon\sqrt{q_n}\int_0^s \cos\mu_n(s-r)dW_n(r)\right]^2 ds$$

$$= \sum_{n=1}^{\infty} q_n \int_0^t F_n^2(s)ds + \mathbb{E}\sum_{n=1}^{\infty} \epsilon^2 q_n^2 \int_0^t \left[\int_0^s \cos\mu_n(s-r)dW_n(r)\right]^2 ds$$

$$= \sum_{n=1}^{\infty} q_n \int_0^t F_n^2(s)ds + \sum_{n=1}^{\infty} \epsilon^2 q_n^2 \int_0^t \left[\int_0^s \cos^2\mu_n(s-r)dr\right] ds$$

$$= \sum_{n=1}^{\infty} q_n \left[A_n^2 \mu_n^2 \left(\frac{t}{2} - \frac{1}{4\mu_n}\sin 2\mu_n t\right) + B_n^2 \mu_n^2 \left(\frac{t}{2} + \frac{1}{4\mu_n}\sin 2\mu_n t\right)\right.$$

$$\left. - \frac{1}{2}A_n B_n \mu_n(1-\cos 2\mu_n t)\right] + \sum_{n=1}^{\infty} \epsilon^2 q_n^2 \left[\frac{t^2}{4} + \frac{1}{8\mu_n^2}(1-\cos 2\mu_n t)\right].$$

Therefore,

$$\text{Var}(E(t)) = \sum_{n=1}^{\infty} \epsilon^2 q_n \left[A_n^2 \mu_n^2 \left(\frac{t}{2} - \frac{1}{4\mu_n}\sin 2\mu_n t\right) + B_n^2 \mu_n^2 \left(\frac{t}{2} + \frac{1}{4\mu_n}\sin 2\mu_n t\right)\right.$$

$$\left. - \frac{1}{2}A_n B_n \mu_n(1-\cos 2\mu_n t)\right] + \sum_{n=1}^{\infty} \epsilon^4 q_n^2 \left[\frac{t^2}{4} + \frac{1}{8\mu_n^2}(1-\cos 2\mu_n t)\right].$$

4.6.2   Heat Equation with Multiplicative Noise

Next we consider the heat equation for the temperature $u(x, t)$ of a rod at position x and time t, subject to multiplicative noise

$$u_t = u_{xx} + \epsilon u \dot{w}(t), \quad 0 < x < 1, \tag{4.52}$$

$$u(0, t) = u(1, t) = 0, \tag{4.53}$$

$$u(x, 0) = f(x), \tag{4.54}$$

where ϵ is a positive parameter measuring the noise intensity and $w(t)$ is a scalar Brownian motion. We take Hilbert space $H \triangleq L^2(0, 1)$ with an orthonormal basis $e_n = \sqrt{2} \sin(n\pi x), n = 1, 2, \cdots$. We again use the method of eigenfunction expansion. Substituting

$$u(x, t) = \sum u_n(t) e_n(x) \tag{4.55}$$

into the above SPDE (4.52), with $\lambda_n = -(n\pi)^2$, we get

$$\sum \dot{u}_n(t) e_n(x) = \sum \lambda_n u_n(t) e_n(x) + \epsilon \sum u_n(t) e_n(x) \dot{w}(t). \tag{4.56}$$

Furthermore, we obtain the following system of SDEs

$$du_n(t) = \lambda_n u_n(t) dt + \epsilon u_n(t) dw(t), \quad n = 1, 2, 3, \cdots. \tag{4.57}$$

Thus,

$$u_n(t) = u_n(0) \exp\left(\left(\lambda_n - \tfrac{1}{2}\epsilon^2\right) t + \epsilon w(t)\right), \quad n = 1, 2, 3, \cdots, \tag{4.58}$$

where $u_n(0) = \langle f(x), e_n(x) \rangle$. Therefore, the final solution is

$$u(x, t) = \sum a_n e_n(x) \exp\left(b_n t + \epsilon w(t)\right), \tag{4.59}$$

where $a_n = \langle f(x), e_n(x) \rangle$ and $b_n = \lambda_n - \tfrac{1}{2}\epsilon^2$.

Note that $\mathbb{E} \exp(b_n t + \epsilon w_t) = \exp(b_n t) \mathbb{E} \exp(\epsilon w(t)) = \exp(b_n t) \exp(\tfrac{1}{2}\epsilon^2 t) = \exp(\lambda_n t)$. Hence, the mean and variance of the solution are

$$\mathbb{E} u(x, t) = \sum a_n e_n(x) \exp(\lambda_n t), \tag{4.60}$$

and

$$\begin{aligned}
\text{Var}(u(x, t)) &= \mathbb{E} \langle u(x, t) - E(u(x, t)), u(x, t) - E(u(x, t)) \rangle \\
&= \sum a_n^2 \exp(2\lambda_n t)[\exp(\epsilon^2 t) - 1].
\end{aligned} \tag{4.61}$$

Moreover, for $\tau \leq t$, we have

$$\begin{aligned}
\mathbb{E} \exp\{\epsilon(w(t) + w(\tau))\} &= \mathbb{E} \exp\{\epsilon(w(t) - w(\tau)) + 2\epsilon w(\tau)\} \\
&= \mathbb{E} \exp\{\epsilon(w(t) - w(\tau))\} \cdot \mathbb{E} \exp\{2\epsilon w(\tau)\}
\end{aligned}$$

$$= \exp\left\{\tfrac{1}{2}\epsilon^2(t-\tau)\right\} \cdot \exp\{2\epsilon^2\tau\}$$
$$= \exp\left\{\tfrac{1}{2}\epsilon^2[(t+\tau)+2(t\wedge\tau)]\right\}.$$

Additionally, by direct calculation, we get covariance

$$\text{Cov}(u(x,t),u(x,\tau))$$
$$= \sum a_n^2 \left\{\exp\left(b_n(t+\tau)+\tfrac{1}{2}\epsilon^2((t+\tau)+2(t\wedge\tau))\right)+\exp\left(\lambda_n(t+\tau)\right)\right.$$
$$\left.- \exp\left(\lambda_n\tau+b_nt+\tfrac{1}{2}\epsilon^2t\right)-\exp\left(\lambda_nt+b_n\tau+\tfrac{1}{2}\epsilon^2\tau\right)\right\}$$
$$= \sum a_n^2 \exp\{\lambda_n(t+\tau)\}[\exp\{\epsilon^2(t\wedge\tau)\}-1],$$

and correlation

$$\text{Corr}(u(x,t),u(x,\tau))$$
$$= \frac{\text{Cov}(u(x,t),u(x,\tau))}{\sqrt{\text{Var}(u(x,t))}\sqrt{\text{Var}(u(x,\tau))}}$$
$$= \frac{\sum a_n^2 \exp\{\lambda_n(t+\tau)\}[\exp\{c^2(t\wedge\tau)\}-1]}{\sqrt{\sum a_n^2 \exp(2\lambda_nt)[\exp(\epsilon^2t)-1]}\sqrt{\sum a_n^2 \exp(2\lambda_n\tau)[\exp(\epsilon^2\tau)-1]}}.$$

4.7 Effects of Noise on Solution Paths

In this section, we consider the effects of noise on solutions of SPDEs with techniques that we have developed so far. To illustrate what we can do at this stage, we consider a specific SPDE, i.e., the stochastic Burgers' equation.

4.7.1 Stochastic Burgers' Equation

We now consider Burgers' equation with additive random forcing e.g., [45]

$$\partial_t u + u\partial_x u = \nu\partial_{xx}u + \sigma \dot{W}(t), \quad 0 < x < l, \tag{4.62}$$
$$u(\cdot,0) = 0, \quad u(\cdot,l) = 0, \quad u(x,0) = u_0(x), \tag{4.63}$$

where $W(t)$ is a Brownian motion, with covariance Q, taking values in the Hilbert space $H = L^2(0,l)$. We assume that $\text{Tr}(Q) < \infty$, i.e., Q is a trace class operator. So $\dot{W}(t)$ is noise, colored in space but white in time.

Taking $F(u) = \frac{1}{2} \int_0^l u^2 dx = \frac{1}{2} \langle u, u \rangle$ and applying Itô's formula, we obtain

$$\frac{1}{2} d \|u\|^2 = \langle u, \sigma dW(t) \rangle + \left[\langle u, \nu u_{xx} - u u_x \rangle + \frac{1}{2} \sigma^2 \operatorname{Tr}(Q) \right] dt. \tag{4.64}$$

Thus,

$$\frac{d}{dt} \mathbb{E} \|u\|^2 = 2 \mathbb{E} \langle u, \nu u_{xx} - u u_x \rangle + \sigma^2 \operatorname{Tr}(Q)$$

$$= -2\nu \mathbb{E} \|u_x\|^2 + \sigma^2 \operatorname{Tr}(Q). \tag{4.65}$$

By the Poincáre inequality, $\|u\|^2 \leq c \|u_x\|^2$, for some positive constant depending only on the interval $(0, l)$, we have

$$\frac{d}{dt} \mathbb{E} \|u\|^2 \leq -\frac{2\nu}{c} \mathbb{E} \|u\|^2 + \sigma^2 \operatorname{Tr}(Q). \tag{4.66}$$

Then, using the Gronwall inequality, we get

$$\mathbb{E} \|u\|^2 \leq \mathbb{E} \|u_0\|^2 e^{-\frac{2\nu}{c} t} + \frac{1}{2} c \sigma^2 \operatorname{Tr}(Q) [1 - e^{-\frac{2\nu}{c} t}]. \tag{4.67}$$

Note that the first term in this estimate involves the initial data, and the second term involves the noise intensity σ as well as the trace of the noise covariance.

Moreover, we consider Burgers' equation with multiplicative random forcing

$$\partial_t u + u \partial_x u = \nu \partial_{xx} u + \sigma u \dot{w}(t), \tag{4.68}$$

with the same boundary condition and initial condition as above except here $w(t)$ is a scalar Brownian motion.

By Itô's formula, we obtain

$$\frac{1}{2} d \|u\|^2 = \langle u, \sigma u \, dw(t) \rangle + \left[\langle u, \nu u_{xx} - u u_x \rangle + \frac{1}{2} \sigma^2 \|u\|^2 \right] dt. \tag{4.69}$$

Thus,

$$\frac{d}{dt} \mathbb{E} \|u\|^2 = 2 \mathbb{E} \langle u, \nu u_{xx} - u u_x \rangle + \sigma^2 \mathbb{E} \|u\|^2$$

$$= -2\nu \mathbb{E} \|u_x\|^2 + \sigma^2 \mathbb{E} \|u\|^2$$

$$\leq \left(\sigma^2 - \frac{2\nu}{c} \right) \mathbb{E} \|u\|^2. \tag{4.70}$$

Therefore,

$$\mathbb{E} \|u\|^2 \leq \mathbb{E} \|u_0\|^2 e^{(\sigma^2 - \frac{2\nu}{c}) t}. \tag{4.71}$$

Further applying Itô's formula to $\log \|u(t)\|^2$ and using (4.69), we have

$$d \log \|u(t)\|^2 \leq \left(-\frac{2\nu}{c} - \sigma^2 \right) dt + 2\sigma \, dw(t),$$

that is,

$$\|u(t)\|^2 \le \|u(0)\|^2 \exp\left\{-\left(\tfrac{2\nu}{c} + \sigma^2\right)t + 2\sigma w(t)\right\}.$$

Define the growth rate λ of the solution orbit $u(t)$ by

$$\lambda \triangleq \lim_{t\to\infty} \frac{1}{t} \log \|u(t)\|^2.$$

In the linear SPDE case, this λ, when it exists, is a Liapunov exponent. Thus, by the growth property of Brownian motion,

$$\lim_{t\to\infty} \frac{1}{t} w(t) = 0, \quad \text{a.s.,}$$

the growth rate λ of the orbit $u(t)$ satisfies

$$\lambda \le -\left(\tfrac{2\nu}{c} + \sigma^2\right) < 0.$$

When the noise is stronger (i.e., σ is bigger), λ is smaller, that is, the stochastic Burgers' equation is more stable. So, the multiplicative noise stabilizes the system. Note that the additive noise only affects the mean energy upper bound, but without stabilization. Liapunov exponents for stochastic linear systems are defined in the context of multiplicative ergodic theorems, as in [19, p. 114] or [213].

Remark 4.31. Here we should point out that noise not only affects stability, but it may also induce, for example, bifurcation e.g., [61] and spatial correlation e.g., [101].

4.7.2 Likelihood for Remaining Bounded

Using the Chebyshev inequality, we can estimate the likelihood of solution orbits remaining inside or outside of a bounded set in Hilbert space $H = L^2(0, l)$. Taking the bounded set as the ball centered at the origin with radius $\delta > 0$, for example, for the above Burgers' Equation (4.68) with multiplicative noise, we have

$$\begin{aligned}
\mathbb{P}\{\omega : \|u\| \ge \delta\} &\le \frac{1}{\delta^2} \mathbb{E}\|u\|^2 \\
&\le \frac{\mathbb{E}\|u_0\|^2}{\delta^2} e^{\left(\sigma^2 - \frac{2\nu}{c}\right)t}.
\end{aligned} \tag{4.72}$$

and

$$\begin{aligned}
\mathbb{P}\{\omega : \|u\| < \delta\} &= 1 - \mathbb{P}\{\omega : \|u\| \ge \delta\} \\
&\ge 1 - \frac{\mathbb{E}\|u_0\|^2}{\delta^2} e^{\left(\sigma^2 - \frac{2\nu}{c}\right)t}.
\end{aligned} \tag{4.73}$$

This provides a lower bound on the probability that u remains inside the ball centered at the origin with radius δ.

4.8   Large Deviations for SPDEs

The likelihood estimates for solution orbits of a SPDE remaining inside or outside of a bounded set may be very crude. For example, in (4.73), if δ is large, we have $\mathbb{P}\{\omega : \|u\| < \delta\} \geq 0$, which is trivially true. The orbits' remaining inside or outside of a bounded set is caused by the stochastic force. The large deviation principles (LDP) presented in this section provide a more effective tool to quantify the probability of the orbits inside or outside of some sets, when noise is sufficiently *small*.

Consider the following stochastic partial differential equation in a separable Hilbert space H:

$$du^\epsilon = [Au^\epsilon + f(u^\epsilon)]dt + \sqrt{\epsilon}\Phi(u^\epsilon)dW(t), \quad u(0) = u_0 \in H, \tag{4.74}$$

where $W(t)$ is a U-valued Wiener process. Here A, f, and Φ satisfy the assumptions in Section 4.1, that is, for any $T > 0$,

(A_1') $f : H \to H$ is $(\mathcal{B}(H), \mathcal{B}(H))$ measurable,
(A_2') $\Phi : H \to \mathcal{L}^2(U^0, H)$ is $(\mathcal{B}(H), \mathcal{B}(\mathcal{L}^2(U^0, H)))$ measurable.

Recall that $U^0 = Q^{1/2}U$. Furthermore, we assume that there is a Lipschitz constant $L > 0$ such that, for all t in $[0, T]$,

(A_3') $\|f(u) - f(v)\| \leq L\|u - v\|$, $\|f(t, u)\|^2 \leq L(1 + \|u\|^2)$, $u, v \in H$.
(A_4') $\|\Phi(t, u)\|^2_{\mathcal{L}^2(U^0,H)} \leq L$, $u \in H$.

Let u be the solution of the corresponding deterministic partial differential equation

$$u' = Au + f(u), \quad u(0) = u_0 \in H.$$

We recall the following basic result [94, Proposition 12.1].

Theorem 4.32.   *For every $T > 0$ and $\delta > 0$, the following limit holds:*

$$\lim_{\epsilon \to 0} \mathbb{P}\left\{ \sup_{0 \leq t \leq T} \|u^\epsilon(t) - u(t)\| \geq \delta \right\} = 0.$$

The above result shows that u^ϵ converges to u as $\epsilon \to 0$ in probability (uniformly on the time interval $[0, T]$) but does not provide the rate of convergence. The theory of LDP is about the exponential decay of the probability in the above theorem.

A definition of large deviation principle was introduced by Varadhan [291, Ch. 2]. Here we consider the LDP for random processes. For more general theory of LDP we refer to [112,125,282,291].

Let $\rho_{0T}(\cdot, \cdot)$ be the usual metric in the space $C(0, T; H)$. A family of random processes $\{u^\epsilon\}_\epsilon$ is said to satisfy the LDP with rate function I if

1. (Lower bound) For every $\varphi \in C(0, T; H)$ and $\delta, \gamma > 0$, there is an $\epsilon_0 > 0$ such that, for $0 < \epsilon < \epsilon_0$,

$$\mathbb{P}\left\{ \rho_{0T}(u^\epsilon, \varphi) \leq \delta \right\} \geq \exp\left\{ -\frac{I(\varphi)+\gamma}{\epsilon} \right\}.$$

2. (Upper bound) For every $r > 0$ and $\delta, \gamma > 0$, there is an $\epsilon_0 > 0$ such that, for $0 < \epsilon < \epsilon_0$,

$$\mathbb{P}\left\{\rho_{0T}(u^\epsilon, K_T(r)) \geq \delta\right\} \leq \exp\left\{-\frac{r-\gamma}{\epsilon}\right\}.$$

Define the level set $K_T(r) \triangleq \{\varphi \in C(0, T; H) : I(\varphi) \leq r\}$. If $K_T(r)$ is compact, then the rate function I is called a good rate function.

LDP theory is an important field on its own. For the purpose of this book, we simply recall a LDP result for SPDEs without proof. More details are in e.g., [94, Ch. 12.1].

The associated control equation (or skeleton equation) of Equation (4.74) is defined as follows: For every $h \in L^2(0, T; U)$,

$$\dot{\varphi} = A\varphi + f(\varphi) + \Phi(\varphi)Q^{1/2}h, \quad \varphi(0) = u_0 \in H. \tag{4.75}$$

Denote by $\varphi^{u_0,h}$ the solution of the above skeleton equation, and also define the level set

$$K_T^{u_0}(r) \triangleq \left\{\varphi \in C(0, T; H) : \varphi = \varphi^{u_0,h}, \quad \frac{1}{2}\int_0^T \|h(s)\|_U^2 \, ds \leq r\right\}.$$

Then, we have the following LDP result for (4.74).

Theorem 4.33. [94, Theorem 12.15] *Assume that* (A_1')–(A_4') *hold. Let* $R_0 > 0$, $r_0 > 0$, *and* $T > 0$ *such that all level sets* $K_T^{u_0}(r_0)$, *for* $\|u_0\| \leq R_0$, *are contained in* $C(0, T; H)$. *Then the following conclusions hold:*

1. *For every* $\delta > 0$ *and* $\gamma > 0$, *there is an* $\epsilon_0 > 0$ *such that for* $0 < \epsilon < \epsilon_0$, *for* $u_0 \in H$ *with* $\|u_0\| < R_0$ *and for* $0 < r < r_0$,

$$\mathbb{P}\left\{\rho_{0T}(u^\epsilon, K_T^{u_0}(r)) < \delta\right\} \geq 1 - \exp\left\{-\frac{r-\gamma}{\epsilon}\right\}.$$

2. *For every* $\delta > 0$ *and* $\gamma > 0$, *there is an* $\epsilon_0 > 0$ *such that for* $0 < \epsilon < \epsilon_0$, *for* $h \in L^2(0, T; U)$ *with*

$$\int_0^T \|h(s)\|_U^2 \, ds < r_0$$

and $u_0 \in H$ *with* $\|u_0\| < R_0$,

$$\mathbb{P}\left\{\rho_{0T}(u^\epsilon, \varphi^{u_0,h}) < \delta\right\} \geq \exp\left\{-\frac{1}{\epsilon}\left(\frac{1}{2}\int_0^T \|h(s)\|_U^2 \, ds + \gamma\right)\right\}.$$

By the above result, the rate function $I(\varphi)$ is characterized as

$$I(\varphi) \triangleq \inf_{h \in L^2(0,T;U)} \left\{\frac{1}{2}\int_0^T \|h(s)\|_U^2 \, ds : \varphi = \varphi^{u_0,h}\right\}$$

with inf $\emptyset = \infty$.

4.9   Infinite Dimensional Stochastic Dynamics

We discuss some dynamical systems concepts in a Hilbert or Banach space H, with norm $\| \cdot \|$ and distance $d(\cdot, \cdot)$.

4.9.1   Basic Concepts

A random dynamical system [19, Definition 1.1.1] consists of two components. The first component is a *driving flow* θ on a probability space $(\Omega, \mathcal{F}, \mathbb{P})$, which acts as a model for a noise. The driving flow θ, also denoted as $\theta_t(\omega)$ or $\theta(t, \omega)$, is a $(\mathcal{B}(\mathbb{R}) \otimes \mathcal{F}, \mathcal{F})$-measurable flow:

$$\theta_t : \Omega \to \Omega,$$
$$\theta_0 = \mathrm{id}, \quad \theta_{t+\tau} = \theta_t \circ \theta_\tau =: \theta_t \theta_\tau$$

for t, $\tau \in \mathbb{R}$. Motivated by a dynamical systems approach to stochastic evolutionary equations, the measure $\mathbb{P}$ is assumed to be ergodic with respect to θ. The second component of a random dynamical system is a $(\mathcal{B}(\mathbb{R}) \otimes \mathcal{F} \otimes \mathcal{B}(H), \mathcal{B}(H))$-measurable mapping φ satisfying the *cocycle* property

$$\varphi(t + \tau, \omega, x) = \varphi(t, \theta_\tau \omega, \varphi(\tau, \omega, x)), \quad \varphi(0, \omega, x) = x,$$

where the phase space (or better called state space) H is a Hilbert space (or Banach space) and x is in H. We usually call φ a random dynamical system (RDS) over the driving flow θ, or just say that φ is a random dynamical system. Sometimes we also say that φ is a cocycle.

In the following we also need a concept of *temperedness* for random variables. A random variable x, taking values in H, is called tempered if

$$t \to \|x(\theta_t \omega)\|$$

is subexponentially growing, i.e.,

$$\lim_{t \to \pm\infty} \sup \frac{\log^+ \|x(\theta_t \omega)\|}{|t|} = 0 \quad \text{a.s.},$$

where $\log^+(y) \triangleq \max\{\log y, 0\}$ is the nonnegative part of the natural logarithm function $\log y$ (sometimes denoted as $\ln y$). This technical condition is not a very strong restriction, because the only alternative is that the above lim sup is ∞, which is a degenerate case [19, p. 164].

4.9.2   More Dynamical Systems Concepts

We recall several concepts in dynamical systems. A *manifold M* is a set, which locally looks like a Euclidean space. Namely, a "patch" of the manifold M looks like a "patch" in $\mathbb{R}^n$. For example, curves are one-dimensional manifolds, whereas tori and spheres are two-dimensional manifolds in $\mathbb{R}^3$. A manifold arising from the study of invariant sets for dynamical systems in $\mathbb{R}^n$ can be very complicated. So we give a formal definition of manifolds. For more discussion of manifolds, see [1, Ch. 3].

Recall that a homeomorphism from A to B is a continuous one-to-one and onto mapping $h : A \to B$ such that $h^{-1} : B \to A$ is also continuous.

Definition 4.34 (Differentiable manifold and Lipschitz manifold). An n-dimensional differentiable manifold M is a connected metric space with an open covering $\{U_\alpha\}$, i.e, $M = \bigcup_\alpha U_\alpha$, such that

 (i) for all α, U_α is homeomorphic to the open unit ball in $\mathbb{R}^n$, $B = \{x \in \mathbb{R}^n : |x| < 1\}$, i.e., for all α there exists a homeomorphism of U_α onto B, $h_\alpha : U_\alpha \to B$, and
(ii) if $U_\alpha \cap U_\beta \neq \varnothing$ and $h_\alpha : U_\alpha \to B, h_\beta : U_\beta \to B$ are homeomorphisms, then $h_\alpha(U_\alpha \cap U_\beta)$ and $h_\beta(U_\alpha \cap U_\beta)$ are subsets of $\mathbb{R}^n$, such that the mapping

$$h = h_\alpha \circ h_\beta^{-1} : h_\beta(U_\alpha \cap U_\beta) \to h_\alpha(U_\alpha \cap U_\beta) \tag{4.76}$$

is differentiable, and for all $x \in h_\beta(U_\alpha \cap U_\beta)$, the Jacobian determinant $\det Dh(x) \neq 0$.

If the mapping (4.76) is C^k differentiable for some natural number k, we call M a C^k differentiable manifold. When $k = \infty$, M is called a smooth manifold. If the mapping (4.76) is only Lispchitz continuous, then we say M is a Lispchitz manifold.

Differentiable manifolds can sometimes be represented as graphs of differentiable mappings. For example, a differentiable curve in $\mathbb{R}^n$ is a differentiable manifold, whereas a Lipschitz continuous curve is a Lipschitz manifold.

Just as invariant sets are important building blocks for deterministic dynamical systems, invariant sets are basic geometric objects that aid in understanding stochastic dynamics [19, Ch. 7]. The following concepts are from [109].

Definition 4.35 (Random set). A collection $M = \{M(\omega)\}_{\omega \in \Omega}$, of nonempty closed sets $M(\omega)$, contained in H is called a random set if

$$\omega \mapsto \inf_{y \in M(\omega)} d(x, y)$$

is a scalar random variable for every $x \in H$.

A random dynamical system is called *dissipative* if there exists a random set M that is bounded for every ω and that is absorbing: for every random variable $x(\omega) \in H$ there exists a $t_x(\omega) > 0$ such that if $t \geq t_x(\omega)$, then

$$\varphi(t, \omega, x(\omega)) \in M(\theta_t \omega).$$

In the deterministic case (φ is independent of ω), the last relation coincides with the definition of an absorbing set. In the case of parabolic partial differential equations, due to the smoothing property, it is usually possible to prove that a dissipative system possesses compact invariant absorbing sets. For more details, see [287, Ch. 1].

Definition 4.36 (Random invariant set). A random set $M(\omega)$ is called a forward random invariant set for a random dynamical system φ if

$$\varphi(t, \omega, M(\omega)) \subset M(\theta_t \omega), \ t > 0 \quad \text{and} \quad \omega \in \Omega.$$

It is called a random invariant set if

$$\varphi(t, \omega, M(\omega)) \subset M(\theta_t \omega), \quad t \in \mathbb{R} \quad \text{and} \quad \omega \in \Omega.$$

Definition 4.37 (Stationary orbit).   A random variable $y(\omega)$ is called a stationary orbit (or random fixed point or random invariant orbit) for a random dynamical system φ if

$$\varphi(t, \omega, y(\omega)) = y(\theta_t \omega), \quad a.s.$$

for every t.

Definition 4.38 (Random invariant manifold).   If a random invariant set M can be represented by a graph of a Lipschitz mapping

$$\gamma^*(\omega, \cdot) : H^+ \to H^-, \text{ with direct sum decomposition } H^+ \oplus H^- = H,$$

such that

$$M(\omega) = \{x^+ + \gamma^*(\omega, x^+), x^+ \in H^+\},$$

then M is called a Lipschitz continuous invariant manifold. If the mapping $\gamma^*(\omega, \cdot)$ is C^k (for a natural number k), then the manifold is called a C^k invariant manifold.

4.10   Random Dynamical Systems Defined by SPDEs

SPDEs with linear multiplicative noises generate random dynamical systems (cocycles) under quite general conditions [128–131]. However, it is unclear whether SPDEs with nonlinear multiplicative noises also generate random dynamical systems. For linear infinite random dynamical systems, including those generated by linear stochastic partial differential equations, Lian and Lu [213] recently proved a multiplicative ergodic theorem, which is the foundation for dynamical systems approaches to nonlinear stochastic partial differential equations.

In this section, we first define the canonical sample space for stochastic partial differential equations, then briefly discuss perfection of cocycles and, finally, present a few examples.

4.10.1   Canonical Probability Space for SPDEs

A standard model for the driving flow θ_t is induced by a two-sided *Brownian motion* $W(t)$, taking values in a separable Hilbert space H. In §3.1, we know that when we toss a die, the canonical or natural sample space is $\{1, 2, 3, 4, 5, 6\}$, which is the collection of all possible outcomes. In the context of an SPDE with $W(t)$ taking values in H, we could image that we are tossing a Brownian motion and what we see (i.e., the outcomes) are continuous curves, which take value 0 at time 0. The set of these continuous curves is usually denoted by $C_0(\mathbb{R}, H)$.

Thus we consider the probability space [109,110]

$$(C_0(\mathbb{R}, H), \mathcal{B}(C_0(\mathbb{R}, H)), \mathbb{P}),$$

where $C_0(\mathbb{R}, H)$ is the space of continuous functions that are zero at zero, with the metric that induces uniform convergence on compact intervals (the so-called compact-open metric). The corresponding Borel σ-field is denoted as $\mathcal{B}(C_0(\mathbb{R}, H))$. We identify sample paths of $W(t)$ with elements in $C_0(\mathbb{R}, H)$:

$$W(t, \omega) \equiv \omega(t).$$

In this way, W is a random variable taking values in $C_0(\mathbb{R}, H)$, and $\mathbb{P}$ is taken as the law (i.e., probability distribution measure) of this random variable. This probability distribution measure is called the *Wiener measure*. Note that $\mathbb{P}$ is ergodic with respect to the *Wiener shift* θ_t:

$$\theta_t \omega(s) = \omega(s + t) - \omega(t), \quad \text{for } \omega \in C_0(\mathbb{R}, H), \tag{4.77}$$

or, equivalently,

$$W(s, \theta_t \omega) = W(s + t, \omega) - W(t, \omega). \tag{4.78}$$

It is clear that $\theta_0 = \text{Id}$ and $\theta_{t\,|\,s} = \theta_t \theta_s$.

For an SPDE with Brownian motion $W(t)$, $(C_0(\mathbb{R}, H), \mathcal{B}(C_0(\mathbb{R}, H)), \mathbb{P})$ is often taken as the canonical probability space. Sometimes we denote this canonical probability space as $(\Omega_0, \mathcal{F}_0, \mathbb{P})$, or just $(\Omega, \mathcal{F}, \mathbb{P})$.

4.10.2 Perfection of Cocycles

In the definition of random dynamical systems, the cocycle property needs to be satisfied for all ω in Ω, whereas for stochastic partial differential equations most properties usually hold only almost surely. Indeed, the solution mapping for an SPDE usually defines a crude cocycle and it does not yet define a random dynamical system. To remedy this, a perfection procedure [128] is needed to define an indistinguishable random dynamical system for which the cocycle property holds for all ω in Ω. Recall that two random dynamical systems, $\varphi_1(t, \omega)$ and $\varphi_2(t, \omega)$, are indistinguishable if

$$\mathbb{P}\{\omega : \varphi_1(t, \omega) \neq \varphi_2(t, \omega) \text{ for some } t \in \mathbb{R}\} = 0.$$

For more details on this perfection procedure for SPDEs with linear multiplicative noise, see [128]. Some ingredients of this procedure are in §6.2. This perfection procedure for SDEs was investigated in detail in [21].

For more general discussion on random dynamical systems generated by SPDEs, we refer to [56,60,71,90,106,109,110,127,235], among others.

4.10.3 Examples

We present two examples of SPDEs that generate random dynamical systems (cocycles).

Example 4.39 (Heat equation with multiplicative noise). Consider the stochastic heat equation with zero Dirichlet boundary conditions

$$u_t = u_{xx} + \epsilon u \dot{w}(t), \quad 0 < x < 1, \tag{4.79}$$

$$u(0, t) = u(1, t) = 0, \tag{4.80}$$

$$u(x, 0) = u_0(x), \tag{4.81}$$

where ϵ is a positive parameter and $w(t)$ is a scalar Brownian motion. Introduce the canonical sample space $C_0(\mathbb{R}, \mathbb{R}^1)$ with the corresponding *Wiener shift* θ_t:

$$\theta_t \omega(s) = \omega(s + t) - \omega(t), \quad \text{for } \omega \in C_0(\mathbb{R}, \mathbb{R}^1). \tag{4.82}$$

Take Hilbert space $H = L^2(0, 1)$ with an orthonormal basis $e_n = \sqrt{2} \sin(n\pi x), n = 1, 2, \cdots$. We find the solution mapping by the method of eigenfunction expansion

$$u(x, t) = \sum u_n(t) e_n(x). \tag{4.83}$$

Inserting this expansion into the above SPDE (4.79), with $\lambda_n = -(n\pi)^2$, we get

$$\sum \dot{u}_n(t) e_n(x) = \sum \lambda_n u_n(t) e_n(x) + \epsilon \sum u_n(t) e_n(x) \, \dot{w}(t). \tag{4.84}$$

Thus, we further obtain the following system of SODEs:

$$du_n(t) = \lambda_n u_n(t) dt + \epsilon u_n(t) dw(t), \quad n = 1, 2, \cdots. \tag{4.85}$$

Therefore,

$$u_n(t) = u_n(0) \exp\left(\left(\lambda_n - \tfrac{1}{2}\epsilon^2\right) t + \epsilon w(t)\right), \tag{4.86}$$

where $u_n(0) = \langle u_0(x), e_n(x) \rangle$. The final solution for (4.79) is

$$u(x, t) = \sum a_n \exp\left(b_n t + \epsilon w(t)\right) e_n(x), \tag{4.87}$$

with $a_n = \langle u_0(x), e_n(x) \rangle$ and $b_n = \lambda_n - \tfrac{1}{2}\epsilon^2$.

Define the solution mapping

$$\varphi(t, \omega, u_0) \triangleq \sum a_n e_n(x) \exp\left(b_n t + \epsilon w(t, \omega)\right). \tag{4.88}$$

Obviously,

$$\varphi(0, \omega, u_0) = u_0. \tag{4.89}$$

Moreover,

$$\varphi(t + s, \omega, u_0) = \sum a_n e_n(x) \exp\left(b_n(t + s) + \epsilon w(t + s, \omega)\right),$$

$$\varphi(t, \theta_s \omega, \varphi(s, \omega, u_0)) = \sum a_n \exp\left(b_n s + \epsilon W(s, \omega)\right) e_n(x) \exp\left(b_n t + \epsilon w(t, \theta_s \omega)\right)$$

$$= \sum a_n e_n(x) \exp\left(b_n(t + s) + \epsilon w(t + s, \omega)\right),$$

that is,

$$\varphi(t + s, \omega, u_0) = \varphi(t, \theta_s \omega, \varphi(s, \omega, u_0)). \tag{4.90}$$

So φ is a crude cocycle, and after the perfection procedure in the last subsection, φ is a random dynamical system.

Example 4.40 (Heat equation with additive noise). Consider the following stochastic heat equation on $(0, 1)$ with zero Dirichlet boundary conditions:

$$u_t = u_{xx} + \dot{W}(t), \quad 0 < x < 1 \tag{4.91}$$
$$u(0, t) = u(1, t) = 0, \tag{4.92}$$
$$u(x, 0) = u_0(x), \tag{4.93}$$

where $W(t)$ takes values in $H = L^2(0, 1)$, with a trace class covariance operator Q (i.e., $\text{Tr}(Q) < \infty$).

Introduce the canonical sample space $C_0(\mathbb{R}, H)$ with the corresponding *Wiener shift* θ_t:

$$\theta_t \omega(s) = \omega(s + t) - \omega(t), \quad \text{for } \omega \in C_0(\mathbb{R}, H). \tag{4.94}$$

Denote by $S(t)$ the semigroup generated by ∂_{xx} with zero Dirichlet boundary conditions on $(0, 1)$. By (2.17) in Chapter 2,

$$S(t)h = \sum_{n=1}^{\infty} \langle h(x), e_n(x) \rangle \exp(\lambda_n t) e_n(x), \quad h \in H, \tag{4.95}$$

where $e_n(x) = \sqrt{2} \sin n\pi x$.

Then the solution to (4.91) is

$$u(x, t) = S(t)u_0 + \int_0^t S(t - \tau) dW(\tau).$$

Define the solution mapping

$$\varphi(t, \omega, u_0) \triangleq S(t)u_0 + \int_0^t S(t - \tau) dW(\tau, \omega).$$

Clearly, $\varphi(0, \omega, u_0) = u_0$. Moreover,

$$\varphi(t + s, \omega, u_0) = S(t + s)u_0 + \int_0^{t+s} S(t + s - \tau) dW(\tau, \omega)$$

and

$$\varphi(t, \theta_s\omega, \varphi(s, \omega, u_0))$$

$$= S(t)\varphi(s, \omega, u_0) + \int_0^t S(t - \tau)dW(\tau, \theta_s\omega)$$

$$= S(t + s)u_0 + \int_0^s S(t + s - \tau)dW(\tau, \omega) + \int_0^t S(t - \tau)dW(\tau, \theta_s\omega)$$

$$= S(t + s)u_0 + \int_0^s S(t + s - \tau)dW(\tau, \omega) + \int_s^{t+s} S(t + s - \tau)dW(\tau - s, \theta_s\omega)$$

$$= S(t + s)u_0 + \int_0^{t+s} S(t + s - \tau)dW(\tau, \omega).$$

Hence,

$$\varphi(t + s, \omega, u_0) = \varphi(t, \theta_s\omega, \varphi(s, \omega, u_0)). \tag{4.96}$$

So, φ is a crude cocycle, and after the perfection procedure in the last subsection, φ is a random dynamical system.

Remark 4.41. Equation (4.91) is a Langevin equation in Hilbert space whose solution u is also called an Ornstein–Uhlenbeck process. More properties of u can be found in [94, Ch. 5].

4.11 Problems

4.1. Strong, weak, and mild solutions for SPDEs

(a) Assume that $W(t)$ is an $L^2(0, l)$-valued Q-Wiener process. Consider the following SPDE defined on $(0, l)$ with the zero Dirichlet boundary conditions

$$u_t = u_{xx} + \dot{W}, \quad u(0) = 0.$$

Denote by A the Laplace operator ∂_{xx} on $(0, l)$ with zero Dirichlet boundary conditions. Assume that Q commutes with A and $\text{Tr}(AQ^{1/2}) < \infty$. First, find the mild solution and then show that the mild solution is actually a weak solution and also a strong solution.

(b) Assume that $w(t)$ is a scalar Wiener process. First, find the mild solution of the following SPDE:

$$u_t = u_{xx} + u\dot{w}, \quad u(t_0) = u_0 \in L^2(0, l)$$

with zero Dirichlet boundary conditions on $(0, l)$ and $t_0 > 0$. Then show that the mild solution is actually a weak solution and also a strong solution.

(c) Assume that $W(t)$ is an $L^2(0, l)$-valued Q-Wiener process with $\text{Tr}(Q) < \infty$. First, find the mild solution of the following stochastic wave equation:

$$u_{tt} = c^2 u_{xx} + \sigma\dot{W}, \quad u(0) = u_0 \in H_0^1(0, l), \quad u_t(0) = u_1 \in L^2(0, l)$$

with zero Dirichlet boundary conditions on $(0, l)$. Here c and σ are positive parameters. Further, assume that Q commutes with A. Then show that the mild solution is also a weak solution.

4.2. Martingale solutions for SPDEs

Assume that $W(t)$ is an $H = L^2(0, l)$-valued Q-Wiener process with $\text{Tr}(Q) < \infty$. Find the Kolmogorov operator $\mathcal{L}$ for the following SPDE:

$$u_t = \nu u_{xx} + f(u) + \dot{W}(t), \quad u(0) = u_0 \in H,$$

with zero Dirichlet boundary conditions on $(0, l)$. Here ν and σ are positive parameters and $\| f(u) \|_H < C$ for every $u \in H$. Prove the existence and uniqueness of the solution to the martingale problem for $\mathcal{L}$ by showing the existence and uniqueness of the martingale solution.

4.3. Stochastic Ginzburg–Landau equation

Consider the following SPDE

$$du = (u_{xx} + u - u^3)dt + \epsilon \, dW(t), \quad u(0) = u_0, \ 0 < x < l, \ t > 0,$$

with zero Dirichlet boundary conditions $u(0, t) = u(l, t) = 0$, where the covariance operator Q for the Wiener process $W(t)$ has eigenvalues $q_n = \frac{1}{n^2}, n = 1, 2, \ldots$ and ϵ is a real positive parameter. Take the Hilbert space $H = L^2(0, l)$ with the usual scalar product $\langle u, v \rangle = \int_0^l uv dx$ and norm $\|u\| = \sqrt{\langle u, u \rangle}$. Examine the well-posedness for appropriate solutions.

4.4. Stochastic Burgers equation

Consider the SPDE

$$u_t = \nu u_{xx} - u u_x + g(u) \, \dot{w}(t), \quad 0 < x < l, \ t > 0,$$

with zero Dirichlet boundary conditions $u(0, t) = u(l, t) = 0$, where $w(t)$ is a scalar Brownian motion, l is a positive constant, and $g(\cdot)$ is a smooth deterministic function. Take the Hilbert space $H = L^2(0, l)$ with the usual scalar product $\langle u, v \rangle = \int_0^l uv \, dx$ and norm $\|u\| = \sqrt{\langle u, u \rangle}$. Examine the well-posedness for appropriate solutions.

4.5. Conversion between two types of SPDEs

(a) Convert the Stratonovich SPDE to an Itô SPDE:

$$u_t = \nu u_{xx} + u u_x + g(u) \circ \dot{w}(t), \quad 0 < x < l, \ t > 0,$$

where $w(t)$ is a scalar Brownian motion, and $g(u)$ is a known function.

(b) Convert the Itô SPDE to a Stratonovich SPDE:

$$u_t = \nu u_{xx} + u u_x + g(u) \dot{w}(t), \quad 0 < x < l, \ t > 0,$$

where $w(t)$ is a scalar Brownian motion, and $g(u)$ is a known function.

(c) If g in (a) and (b) is either a constant or does not depend on u, are the two types of SPDEs the same or not? Explain.

4.6. Stochastic heat equation

Consider the following SPDE:

$$u_t = \nu u_{xx} + \sigma \dot{W}(t), \quad 0 < x < l, \ t > 0,$$

with zero Neumann boundary conditions $u_x(0, t) = 0 = u_x(l, t)$ and initial condition $u(x, 0) = u_0(x)$. Here ν and σ are positive real parameters, and $W(t)$ is a Brownian motion in $L^2(0, l)$ with covariance operator Q commuting with ∂_{xx}.

(a) Find the solution in a Fourier series.
    Hint: What is the appropriate orthonormal basis in $L^2(0, l)$, in the case of zero Neumann boundary conditions? We would like to select a basis formed from the eigenfunctions of $A = \nu \partial_{xx}$ under these boundary conditions, facilitating the construction of the solution.

(b) Calculate the mean, variance, and covariance for the solution. What is the difference with the case of zero Dirichlet boundary conditions?

(c) Is the solution mapping a cocycle? Does it define a random dynamical system in an appropriate Hilbert space?

4.7. A second-order stochastic differential equation

Let w be a scalar Brownian motion. Assume that c, λ, q are constants with properties: $c \neq 0, \lambda < 0$, and $q > 0$. Solve the following second-order stochastic differential equation:

$$\ddot{u}(t) = c^2 \lambda u + \epsilon \sqrt{q} \dot{w}(t)$$

with initial conditions $u(0) = u_0, \dot{u}(0) = u_1$.

4.8. Stochastic wave equation

Consider the following SPDE:

$$u_{tt} = c^2 u_{xx} + \sigma \dot{W}(t), \quad 0 < x < l,$$

with zero Neumann boundary conditions $u_x(0, t) = 0 = u_x(l, t)$ and initial conditions $u(x, 0) = u_0(x)$ and $u_t(x, 0) = v_0(x)$. Here c and σ are positive real parameters, and $W(t)$ is a Brownian motion in $L^2(0, l)$ with covariance operator Q commuting with ∂_{xx}.

(a) Find the Fourier series solution.
    Hint: What is the appropriate orthonormal basis in $L^2(0, l)$, in the case of zero Neumann boundary conditions? We would like to select a basis formed from the eigenfunctions of $A = c^2 \partial_{xx}$ under these boundary conditions, facilitating the construction of the solution.

(b) Calculate the mean, variance, and covariance for the solution. What is the difference with the case of zero Dirichlet boundary conditions?

4.9. A nonlinear SPDE

Consider the following nonlinear SPDE:

$$u_t = u_{xx} + \sin(u) + \sigma u \, \dot{w}(t), \quad 0 < x < l, \, t > 0,$$

with zero Dirichlet boundary conditions $u(0, t) = u(l, t) = 0$ and initial condition $u(x, 0) = u_0(x)$. Here σ is a real positive parameter and $w(t)$ is a scalar Brownian motion.

(a) Estimate the mean energy, $\mathbb{E}\|u\|^2$, for this system.
Hint: You may need to use Young's, Cauchy–Schwarz, Poincáre, Agmon, or Gronwall inequalities.
(b) What is the impact of the multiplicative noise on the mean energy, $\mathbb{E}\|u\|^2$, of the solution?

4.10. Stochastic Fisher–Kolmogorov–Petrovsky–Piscunov equation

Consider the SPDE in Example 1.6:

$$\partial_t u = D u_{xx} + \gamma u(1 - u) + \varepsilon \sqrt{u(1 - u)} \, \dot{w}(x, t), \quad u(0, t) = u(l, t) = 0,$$

with $0 < x < l, \, t > 0$, where $u(x, t)$ is the population density for a certain species, D, γ, and ε are parameters, and w is a scalar Brownian motion. Take the Hilbert space $H = L^2(0, l)$ with the usual scalar product $\langle u, v \rangle = \int_0^l uv \, dx$ and norm $\|u\| = \sqrt{\langle u, u \rangle}$. Examine the well-posedness for appropriate solutions.

4.11. A System of Coupled SPDEs

Consider the SPDEs in Example 1.5:

$$\partial_t a = D_1 a_{xx} + k_1 a - k_2 ab - k_3 a^2 + \dot{W}_1(x, t),$$
$$\partial_t b = D_2 b_{xx} + k_1 b - k_2 ab - k_3 b^2 + \dot{W}_2(x, t),$$

where $0 < x < l$, $D's$ and $k's$ are parameters (with $D's$ being positive), and W_1 and W_2 are independent Brownian motions. When $D_1 \ll D_2$, this is a slow-fast system of SPDEs. Assume that a and b satisfy the zero Dirichlet boundary conditions $a(0, t) = a(l, t) = b(0, t) = b(l, t) = 0$. Suppose that the covariance operator Q_1 for $W_1(t)$ has eigenvalues $q_n^1 = \frac{1}{n^2}, n = 1, 2, \cdots$, and the covariance operator Q_2 for $W_2(t)$ has eigenvalues $q_n^2 = \frac{1}{n^3}, n = 1, 2, \cdots$. Take the Hilbert space $H = L^2(0, l)$ with the usual scalar product $\langle u, v \rangle = \int_0^l uv \, dx$ and norm $\|u\| = \sqrt{\langle u, u \rangle}$. Estimate the mean energy $\mathbb{E}(\|a\|^2 + \|b\|^2)$ for this system.

(b) Calculate the mean, variance, and covariance for the solution. What is the difference with the case of zero Dirichlet boundary conditions?

4.9. A nonlinear SPDE.

Consider the following equation:

$$u_{tt} = u_{xx} + \sin(u) + \dot{W}(t,x), \quad 0 \le x \le 1, \quad 0 \le t,$$

with zero Dirichlet boundary conditions $u(t, 0) = u(t, 1) = 0$ and initial condition $u(x, 0) = u_0(x)$. Here $\dot{W}$ is a real positive parameter and $u_0(x)$ is a scalar Brownian motion.

(c) Estimate the mean energy $E|u|^2$ for the system.

Hint. You may need to use Young's, Cauchy–Schwarz, Poincaré, Agmon, or Gronwall inequalities.

(f) What is the impact of the multiplicative noise on the mean energy $E|u|^2$ of the solution?

4.10. Stochastic Fisher–Kolmogorov–Petrovsky–Piscunov equation.

Consider the SPDE in Example 4.x.x.

$$du + [u u_x - u_{xx} + \chi u] dt = P_i[u(K(1-u))u_x + v_x \dot{W}] + v(0,T) = v_x(1,T) = 0,$$

with $u(x, 0) = u_0 \ge 0$, where $u(x, t)$ is the population density for a certain species, P_i, χ are parameters, and u is a scalar Brownian motion. Take the Hilbert space $H = L^2(0, 1)$ with the usual scalar product $\langle u, v \rangle = \int_0^1 uv \, dx$ and norm $\|u\| = \langle u, u \rangle^{1/2}$. Examine the well-posedness for an H-valued solution.

4.11. A System of Coupled SPDEs.

Consider the SPDEs in Example 4.x.x.

$$du = (D_u u_{xx} + k u - \zeta u u_x - k u^2 - u W) dt$$

$$dv = (D_v v_{xx} + k v_{xx} - k u v - k v^2 + u v) dt,$$

where $D_u, D_v, k, \zeta, \chi, W_u, W_v$, and k, v are parameters; given D's being positive, and W_u and W_v being white in time free functions. When $D_u < D_v$, this is a two-fast system of SPDEs. Assume that u and v satisfy the zero Dirichlet boundary condition as $u(0, t) = u(1, t) = v(0, t) = v(1, t) = 0$. Suppose that the covariance operator Q_i for the eigenfunctions $\phi_{i,n} = \sqrt{2} \sin n \pi x$, and the covariance operator Q_j for the eigenfunctions $q_j = \sqrt{2} \sin n \pi x$. Take the Hilbert space $H = L^2(0, 1)$ with the usual scalar product $\langle u, v \rangle = \int_0^1 uv \, dx$ and norm $\|u\| = \langle u, u \rangle^{1/2}$. Estimate the mean energy $E\|u\|^2 + E\|v\|^2$ for the system.

5 Stochastic Averaging Principles

Averaging techniques, averaging as approximation, large deviations, error estimates for approximations

We now investigate time-averaging techniques for stochastic partial differential equations with slow and fast time scales. Averaged systems (i.e., effective systems) are derived, with the approximation error estimated via normal deviation techniques. Large deviation techniques further provide information about the effectiveness of the approximation. Moreover, averaging principles for partial differential equations with time-dependent random coefficients that are time-recurrent (e.g., periodic or quasiperiodic) are also briefly considered.

This chapter is arranged as follows. After reviewing several classical averaging results in § 5.1, an averaging principle and a normal deviation principle for SPDEs with slow and fast time scales are established in § 5.2 and § 5.4, respectively. A specific example is considered in § 5.6. Then large deviation principles are presented in § 5.7. Moreover, an averaging principle for a partial differential equation with random coefficients is proved in § 5.8. Finally, some remarks about stochastic climate modeling and singularly perturbed stochastic systems are presented in § 5.9, and several open research issues are discussed in the final section.

5.1 Classical Results on Averaging

Highly oscillating components in a dynamical system may be "averaged out" under certain suitable conditions, from which an averaged, effective system emerges. Such an effective system is more amenable for analysis and simulation, and it governs the evolution of the original system over a long time scale.

5.1.1 Averaging in Finite Dimension

The idea of averaging appeared in the perturbation theory developed by Clairaut, Laplace, and Lagrange in the 18th century. Then various averaging schemes (Gauss, Fatou, Delone–Hill) were widely applied in celestial mechanics in the 19th century. These were mainly formal or ad hoc techniques. Krylov and Bogoliubov [200, Ch. 2] presented a rigorous averaging method for nonlinear oscillations that is now called the Krylov–Bogolyubov method. There are several versions of averaging principles, based on Krylov and Bogoliubov's work, for systems with oscillating periodic

Effective Dynamics of Stochastic Partial Differential Equations. http://dx.doi.org/10.1016/B978-0-12-800882-9.00005-6

forcings [22, Ch. 6], [48, Ch. 5], [149, Ch. 5], [273, Ch. 2], [294]. These averaging principles are applicable to the following system in $\mathbb{R}^n$:

$$(u^\epsilon)' = \epsilon F(u^\epsilon, \tau), \quad u(0) = u_0, \quad 0 \le \epsilon \ll 1, \tag{5.1}$$

where $(\ )' = \frac{d}{d\tau}(\ )$. The vector field F is of C^r, $r \ge 2$ in all variables and T_1-periodic in τ. By changing the time scale

$$t = \epsilon\tau,$$

we transform (5.1) to

$$\dot{u}^\epsilon = F(u^\epsilon, t/\epsilon), \quad u(0) = u_0, \tag{5.2}$$

with $(\dot{\ }) = \frac{d}{dt}(\ )$. Define

$$\bar{F}(u) = \frac{1}{T_1} \int_0^{T_1} F(u, t)\, dt$$

and consider the averaged equation

$$\dot{u} = \bar{F}(u), \quad \bar{u}(0) = u_0. \tag{5.3}$$

A basic result on averaging [48, Ch. 5], [273, Theorem 2.8.1] is the following theorem.

Theorem 5.1. *For any $u_0 \in \mathbb{R}^n$,*

$$\|u^\epsilon(t) - u(t)\|_{\mathbb{R}^n} = \mathcal{O}(\epsilon), \quad \epsilon \to 0, \quad 0 \le t \le T,$$

for every fixed T.

In a more general setting, when $F(u, \tau)$ is not periodic in τ, we make the assumption that the time average of the vector field F asymptotically converges to a mean vector field $\bar{F}$:

$$\frac{1}{T^*} \int_0^{T^*} F(u, t)\, dt \to \bar{F}(u), \quad T^* \to \infty, \tag{5.4}$$

which is uniform in u on every bounded set of $\mathbb{R}^n$. Such a vector field F is called a KBM (Krylov, Bogoliubov, and Mitropolsky) vector field [273, p. 69].

Note that (5.4) holds in the special case when $F(u, t)$ is periodic, quasiperiodic, or almost periodic in time t. Under the assumption (5.4), a similar averaging principle holds, but the convergence rate, which is related to the difference between $\bar{F}(u)$ and $F(u, t)$ on $[0, T]$, may not be $\mathcal{O}(\epsilon)$ any more [273, Ch. 4.3]. Here we give a proof of Theorem 5.1 under the assumption that F is a KBM vector field.

Proof.   Let $U^\epsilon(t) = u^\epsilon(t) - u(t)$. Then

$$U^\epsilon(t) = \int_0^t \left[F(u^\epsilon(s), s/\epsilon) - \bar{F}(u(s)) \right] ds,$$

and

$$\|U^\epsilon(t)\|_{\mathbb{R}^n} \leq \int_0^t \left\|F(u^\epsilon(s), s/\epsilon) - F(u(s), s/\epsilon)\right\|_{\mathbb{R}^n} ds$$

$$+ \left\|\int_0^t \left[F(u(s), s/\epsilon) - \bar{F}(u(s))\right] ds\right\|_{\mathbb{R}^n}.$$

By the assumption on F, for every $T > 0$ there is a positive constant C_T such that $\max_{0 \leq t \leq T} \|u^\epsilon(t)\|_{\mathbb{R}^n} \leq C_T$. Hence for every $T > 0$ there is a positive constant L_T such that

$$\|U^\epsilon(t)\|_{\mathbb{R}^n} \leq L_T \int_0^t \left\|u^\epsilon(s) - u(s)\right\|_{\mathbb{R}^n} ds$$

$$+ \epsilon \left\|\int_0^{t/\epsilon} \left[F(u(\epsilon s), s) - F(u(\epsilon s))\right] ds\right\|_{\mathbb{R}^n}.$$

Define

$$\delta(\epsilon) = \epsilon \sup_{\|u\| \leq C_T} \sup_{0 \leq t \leq T} \left\|\int_0^{t/\epsilon} [F(u, s) - \bar{F}(u)] ds\right\|_{\mathbb{R}^n}. \tag{5.5}$$

Therefore, by the Gronwall inequality, we obtain the result.   □

Remark 5.2.   The function $\delta(\epsilon)$, defined in the above proof, is called the *order function* of F. Note that $\delta(\epsilon) \to 0$ as $\epsilon \to 0$ [273, Remark 4.3.2].

In the above averaging method, an essential point is the convergence of the time average of the oscillating component in a certain sense (5.4). We will see this convergence in different forms, which are the key assumptions in various averaging methods, such as (5.8), (5.14), (5.26) and (5.51) below.

Now we consider an application of the above averaging principle in the following example.

Example 5.3.   Consider a scalar ordinary differential equation:

$$\dot{u}^\epsilon = F(u^\epsilon, t/\epsilon) = 2u^\epsilon \sin^2 \frac{t}{2\epsilon}, \quad u^\epsilon(0) = u_0.$$

Notice that here $F(u, t) = u(1 - \cos t)$. Then we have

$$\bar{F}(u) = \frac{1}{2\pi} \int_0^{2\pi} u(1 - \cos t) dt = u,$$

from which we have the following averaged equation:

$$\dot{u} = u, \quad u(0) = u_0.$$

Then, by Theorem 5.1, for every $T > 0$,

$$\sup_{0 \leq t \leq T} |u^\epsilon(t) - u(t)| = \mathcal{O}(\epsilon), \quad \epsilon \to 0.$$

In some systems, the vector field in Equation (5.2) has both slow and fast components, and both components are fully coupled. Such a system is described as follows:

$$\text{(slow)} \quad \dot{u}^{\epsilon} = f(u^{\epsilon}, v^{\epsilon}), \quad u^{\epsilon}(0) = u_0 \in \mathbb{R}^n, \tag{5.6}$$

$$\text{(fast)} \quad \epsilon \dot{v}^{\epsilon} = g(u^{\epsilon}, v^{\epsilon}), \quad v^{\epsilon}(0) = v_0 \in \mathbb{R}^m. \tag{5.7}$$

Various physical problems that combine fast and slow motions are in this form, such as a Hamiltonian system in action-angle coordinates with small perturbation [22, Ch. 6.2] and a simple climate-weather model [172]. The fast part v^{ϵ} evolves on time scale $\mathcal{O}(1/\epsilon)$, whereas the slow part u^{ϵ} evolves on time scale $\mathcal{O}(1)$.

Suppose that both vector fields f and g are Lipschitz continuous and the fast part v^{ϵ} is mixing sufficiently so that a mean vector field $\bar{f}$ emerges: For every fixed $u \in \mathbb{R}^n$,

$$\frac{1}{T^*} \int_0^{T^*} f(u, v^{\epsilon}(t)) dt \rightarrow \bar{f}(u), \quad T^* \rightarrow \infty, \tag{5.8}$$

which is uniform in u on every bounded set of $\mathbb{R}^n$. Then $\bar{f}$ is also Lipschitz continuous. The condition (5.8) is in fact the same as (5.4) if we let $F(u, t) = f(u, v(t))$, with $v(t)$ the solution of (5.7) with fixed $u \in \mathbb{R}^n$ and $\epsilon = 1$. A simple case is that for every fixed $u \in \mathbb{R}^n$, the fast part has a unique exponentially stable equilibrium. Define the following averaged equation:

$$\dot{u} = \bar{f}(u), \quad u(0) = u_0.$$

We then have the following averaging principle due to Anosov [15].

Theorem 5.4. *For every $u_0 \in \mathbb{R}^n$ and almost all $v_0 \in \mathbb{R}^m$,*

$$\lim_{\epsilon \rightarrow 0} \sup_{0 \leq t \leq T} \|u^{\epsilon}(t) - u(t)\|_{\mathbb{R}^n} = 0$$

for every fixed $T > 0$.

This theorem will be seen as a special case of an averaging principle for the stochastic slow-fast system (5.11)–(5.12) below. Moreover, under the assumption (5.8), the slow-fast system (5.6)–(5.7) is in the form of (5.2). In fact, let us assume that, for every fixed $u \in \mathbb{R}^n$, the flow $\Phi^{\epsilon}_{v_0, u}(t)$ defined by the fast part v^{ϵ} is mixing enough such that (5.8) holds. This is equivalent to that the flow $\Phi^{\epsilon}_{\cdot, u}(t)$ has a unique stable stationary measure μ_u. Then, by an invariant manifold reduction [143, Ch. 4] from the slow-fast system (5.6)–(5.7), we have

$$\dot{u}^{\epsilon} = f(u^{\epsilon}, \Phi^{\epsilon}_{v_0, u}(t)), \quad u^{\epsilon}(0) = u_0, \tag{5.9}$$

which describes long time behavior of the slow-fast system (5.6)–(5.7). Furthermore, for every $\epsilon > 0$ by a time scale change $t \rightarrow \epsilon t$, $\Phi^{\epsilon}_{v_0, u}(t)$ has the same distribution as that of $\Phi^1_{v_0, u}(t/\epsilon)$. Thus we have

$$\dot{u}^{\epsilon} = f(u^{\epsilon}, \Phi^1_{v_0, u}(t/\epsilon)) \triangleq F(u^{\epsilon}, t/\epsilon), \quad u^{\epsilon}(0) = u_0,$$

which is exactly the form (5.2). Due to the mixing property, $\Phi^1_{v_0,u}(t)$ is independent of v_0 as $\epsilon \to 0$ for fixed u. Then, by the same discussion as for Theorem 5.1, we have Theorem 5.4.

There are various results for averaging of such slow-fast systems (5.6)–(5.7) under the ergodic assumption or other weaker conditions; see e.g., [23,65,189] and [217, Ch. 5].

Remark 5.5.   We have just mentioned an invariant manifold reduction. In fact, for slow-fast systems, there is an interesting connection between averaging and slow manifold reduction, which will be discussed in Chapter 6.

Let us look at a simple example of slow-fast systems.

Example 5.6.   Consider the following system of ODEs:

$$\dot{u}^\epsilon = u^\epsilon - v^\epsilon, \quad u^\epsilon(0) = u_0 \in \mathbb{R}^1,$$
$$\dot{v}^\epsilon = -\frac{1}{\epsilon}[v^\epsilon - 2u^\epsilon], \quad v^\epsilon(0) = v_0 \in \mathbb{R}^1.$$

By Theorem 5.4, there is an averaged equation

$$\dot{u} = -u, \quad u(0) = u_0, \tag{5.10}$$

and for every $T > 0$

$$\sup_{0 \le t \le T} |u^\epsilon(t) - u(t)| \to 0, \quad \epsilon \to 0.$$

In fact, for fixed $u \in \mathbb{R}^1$, the solution of the fast part is

$$v^\epsilon(t) = e^{-\frac{1}{\epsilon}t}v_0 + 2u\frac{1}{\epsilon}\int_0^t e^{-\frac{1}{\epsilon}(t-s)}ds = e^{-\frac{1}{\epsilon}t}v_0 + 2u(1 - e^{-\frac{1}{\epsilon}t}).$$

Then, for every $t > 0$, as $\epsilon \to 0$, $v^\epsilon(t) \to 2u$, which is independent of v_0. Formally, substituting this into the u^ϵ equation, we obtain the averaged Equation (5.10).

Krylov and Bogoliubov type averaging is also developed for randomly perturbed nonlinear systems, described by stochastic *ordinary* differential equations with slow and fast time scales e.g., [134, Ch. 7.9], [181,186], [278, Ch. II.3], and [321, Ch. 3]:

$$du^\epsilon(t) = f(u^\epsilon(t), v^\epsilon(t))dt + \sigma_1(u^\epsilon(t), v^\epsilon(t))dw_1(t), \tag{5.11}$$
$$\epsilon dv^\epsilon(t) = g(u^\epsilon(t), v^\epsilon(t))dt + \sqrt{\epsilon}\sigma_2(u^\epsilon(t), v^\epsilon(t))dw_2(t), \tag{5.12}$$
$$u^\epsilon(0) = u \in \mathbb{R}^n, \quad v^\epsilon(0) = v \in \mathbb{R}^m, \tag{5.13}$$

where $w_1(t)$ and $w_2(t)$ are independent standard Brownian motions in $\mathbb{R}^k$ and $\mathbb{R}^l$, respectively. Suppose that the nonlinearity $f : \mathbb{R}^{n+m} \to \mathbb{R}^n$ and the matrix valued function $\sigma_1 : \mathbb{R}^{n+m} \to \mathbb{R}^{n \times k}$ are Lipschitz continuous and uniformly bounded in $v \in \mathbb{R}^m$. Furthermore, suppose that the nonlinearity $g : \mathbb{R}^{n+m} \to \mathbb{R}^m$ and the matrix valued function $\sigma_2 : \mathbb{R}^{n+m} \to \mathbb{R}^{m \times l}$ are also Lipschitz continuous in (u, v).

The strength of noise in the fast equation is chosen to be $\sqrt{\epsilon}$ to balance the stochastic force and the deterministic force. In fact, for the noise strength ϵ^α, there are three cases:

1. $\alpha > 1/2$: Deterministic force overwhelms stochastic force.
2. $\alpha < 1/2$: Stochastic force overwhelms deterministic force.
3. $\alpha = 1/2$: There is a balance between stochastic force and deterministic force, which is also the condition for the existence of an ϵ independent invariant measure. This can be seen from the generator (see (5.16) below) of the diffusion process defined by the fast Equation (5.12).

The above fact can also be seen by a time scale change, $\tau = t/\epsilon$. In fact, under the new time τ, the fast equation is

$$dv = g(u, v)d\tau + \epsilon^{\alpha - 1/2}\sigma_2(u, v)d\tilde{w}_2(\tau),$$

where $\tilde{w}_2(\tau) = w_2(\epsilon\tau)/\sqrt{\epsilon}$ has the same distribution as that of $w_2(\tau)$.

Now, we assume that the fast part (5.12) with frozen $u \in \mathbb{R}^n$ has a unique stationary solution $\bar{v}$ with distribution μ such that the following ergodic theorem [19, p. 538] holds: For $\bar{f}(u) \triangleq \int_{\mathbb{R}^m} f(u, v)\mu(dv) = \mathbb{E}f(u, \bar{v})$,

$$\lim_{T \to \infty} \frac{1}{T} \int_0^T f(u, v(t))dt = \bar{f}(u), \tag{5.14}$$

uniformly in u on bounded sets. Note that $\bar{f}$ is also Lipschtiz continuous. Then we have the following averaging principle for the stochastic slow-fast system (5.11)–(5.12). See [134, Ch. 7.9], [186], and [278, Ch. II.3] for more details.

Theorem 5.7. *For every $T > 0$, the slow motion u^ϵ converges in distribution in the space $C(0, T; \mathbb{R}^n)$ to an averaged effective motion $\bar{u}$, which solves*

$$d\bar{u}(t) = \bar{f}(\bar{u})dt + \bar{\sigma}_1(\bar{u})dw_1(t), \quad \bar{u}(0) = u_0,$$

where $\bar{\sigma}_1$ is the square root of the nonnegative matrix

$$\int_{\mathbb{R}^m} \sigma_1(u, v)\sigma_1^T(u, v)\mu(dv),$$

in which ^{T} denotes transpose of a matrix.

We present a heuristic argument for the proof of the above theorem via multiscale expansion [252, Ch. 17].

Outline of the Proof of Theorem 5.7

For a bounded continuous function φ on $\mathbb{R}^n$, let $p^\epsilon(t, u, v) \triangleq \mathbb{E}\varphi(u^\epsilon(t))$. Then p^ϵ satisfies the Fokker–Planker equation

$$\frac{\partial p^\epsilon}{\partial t} = \left[\frac{1}{\epsilon}\mathcal{L}_2 + \mathcal{L}_1 \right] p^\epsilon, \quad p^\epsilon(0, u, v) = \varphi(u), \tag{5.15}$$

where

$$\mathcal{L}_2 = \frac{1}{2}\sigma_2(u^\epsilon, v^\epsilon)\sigma_2^T(u^\epsilon, v^\epsilon)\partial_{vv} + g(u^\epsilon, v^\epsilon)\partial_v, \tag{5.16}$$

$$\mathcal{L}_1 = \frac{1}{2}\sigma_1(u^\epsilon, v^\epsilon)\sigma_1^T(u^\epsilon, v^\epsilon)\partial_{uu} + f(u^\epsilon, v^\epsilon)\partial_u. \tag{5.17}$$

By the assumption on fast part, for any fixed $u^\epsilon = u$ the operator $\mathcal{L}_2$ has a unique invariant measure denoted by μ_u. Define a projection onto null($\mathcal{L}_2$), the null space of $\mathcal{L}_2$, by

$$\mathbf{P}\varphi(u) \triangleq \int_{\mathbb{R}^m} \varphi(u, v)\mu_u(dv).$$

Formally, we assume that p^ϵ has the following expansion:

$$p^\epsilon = p_0 + \epsilon p_1 + \epsilon^2 p_2 + \cdots$$

Inserting this expansion into (5.15) and equating coefficients of equal powers of ϵ yields

$$\mathcal{L}_2 p_0 = 0, \quad \frac{\partial p_0}{\partial t} = \mathcal{L}_2 p_1 + \mathcal{L}_1 p_0, \quad \cdots$$

By the first equation and the fact that initial value φ is independent of v, we have $\mathbf{P}p_0 = p_0$. Moreover, from the second equation, we have the following orthogonality property

$$\left(\frac{\partial p_0}{\partial t} - \mathcal{L}_1 p_0\right) \perp \text{null}(\mathcal{L}_2).$$

That is, acting with $\mathbf{P}$ on both sides of the second equation yields

$$\frac{\partial p_0}{\partial t} = \mathbf{P}\mathcal{L}_1 p_0 = \overline{\mathcal{L}}_1 p_0, \quad p_0(0) = \varphi, \tag{5.18}$$

where

$$\overline{\mathcal{L}}_1 = \frac{1}{2}\overline{\sigma_1\sigma_1^T}(u)\partial_{uu} + \bar{f}(u)\partial_u,$$

with

$$\overline{\sigma_1\sigma_1^T}(u) \triangleq \int_{\mathbb{R}^m} \sigma_1(u, v)\sigma_1^T(u, v)\mu_u(dv), \quad \text{and} \quad \bar{f}(u) \triangleq \int_{\mathbb{R}^m} f(u, v)\mu_u(dv).$$

Let $q^\epsilon = p^\epsilon - p_0$. Then

$$\frac{\partial q^\epsilon}{\partial t} = \epsilon\mathcal{L}_1 p_1 + \text{high order term of } \epsilon.$$

Thus $p^\epsilon - p_0 = \mathcal{O}(\epsilon)$, provided that p_0 and p_1 are bounded. Notice that (5.18) is in fact the Fokker–Planker equation associated with the following stochastic differential equation

$$du(t) = \bar{f}(u)dt + \bar{\sigma}_1(u)dw_1(t).$$

This yields the result.   □

For a rigorous proof, we need some *a priori* estimates on p^ϵ, p_0, and p_1, which can be guaranteed by, for example, smoothness of f and some strong mixing property of the fast part [246, Sec. 3]. The multiscale expansion method is expected to be applicable to SPDEs; however, the corresponding Fokker–Planker Equation (5.15), which is a PDE defined in infinite dimensional space, is difficult to solve. Some classical methods for deterministic slow-fast PDEs may be generalized to SPDEs.

We apply Theorem 5.7 to a stochastic version of *Example* 5.6.

Example 5.8.   Consider the following slow-fast system of SODEs:

$$\dot{u}^\epsilon = u^\epsilon - v^\epsilon, \quad u^\epsilon(0) = u_0 \in \mathbb{R}^1,$$
$$dv^\epsilon = \frac{1}{\epsilon}[-v^\epsilon + 2u^\epsilon]dt + \frac{1}{\sqrt{\epsilon}}dW(t), \quad v^\epsilon(0) = v_0 \in \mathbb{R}^1,$$

where $W(t)$ is a standard scalar Wiener process. Then by Theorem 5.7 we have the following averaged equation:

$$\dot{u} = -u, \quad u(0) = u_0,$$

which is the same as that of the deterministic case (*Example* 5.6). However, u^ϵ indeed behaves randomly due to the stochastic fast part v^ϵ. So, the averaged equation is not a good approximation for small $\epsilon > 0$. The deviation $u^\epsilon - u$ needs to be examined for more effective approximations.

Remark 5.9.   We note that averaging for slow-fast SODEs is also related to that of differential equations with random oscillating coefficients in the following form:

$$\dot{u}^\epsilon = F(u^\epsilon, t/\epsilon), \quad u^\epsilon(0) = u_0. \tag{5.19}$$

In fact, the slow-fast SODE (5.11)–(5.12) with $\sigma_1 = 0$ can be reduced onto a random slow manifold under some appropriate assumptions. Similar results for SPDEs will be discussed in Section 6.5.

To obtain a more "effective" equation than the one provided by Theorem 5.7, it is necessary to take the error made in the averaging approach into account. That is, we consider the normalized deviation

$$\frac{1}{\sqrt{\epsilon}}\left(u^\epsilon - u\right),$$

which converges in distribution to a Gaussian process under appropriate conditions [181]. This is in fact a result of central limit theorem type, as seen in the following averaging result for random differential equations. This actually is a finite dimensional version of Theorem 5.34 for slow-fast SPDEs.

We now present an averaging principle for systems with random fast oscillating coefficients. This also provides a method to derive a normalized deviation estimate. We consider the following simple equation:

$$\dot{u}^\epsilon(t) = \frac{1}{\sqrt{\epsilon}}\eta(t/\epsilon), \quad u^\epsilon(0) = u_0 \in \mathbb{R}, \tag{5.20}$$

where $\eta(t)$ is a stationary process with every order of finite moment, $\mathbb{E}\eta(t) = 0$ and further satisfies the following strong mixing condition:

(**SM**) Let $\mathcal{F}_s^t = \sigma\{\eta(\tau) : s \leq \tau \leq t\}$ and

$$\alpha(t) = \sup_{s\geq 0} \sup_{A\in\mathcal{F}_0^s, B\in\mathcal{F}_{s+t}^\infty} |\mathbb{P}(AB) - \mathbb{P}(A)\mathbb{P}(B)|.$$

Then

$$\int_0^\infty \alpha^{1/4}(t)dt < \infty.$$

Remark 5.10. An example for $\eta(t/\epsilon)$ is the stationary solution of the following Langevin equation:

$$d\eta = -\lambda\eta\, dt + dw(t),$$

where $\lambda > 0$ and $w(t)$ is a standard scalar Brownian motion.

Then we have the following averaging principle.

Theorem 5.11. *Assume that the strong mixing condition* (**SM**) *holds. Then, the solution u^ϵ of random Equation (5.20) converges in distribution to u that solves the following stochastic differential equation:*

$$du = \sqrt{A}\, dW(t), \quad u(0) = u_0, \tag{5.21}$$

where $W(t)$ is a standard scalar Brownian motion and

$$A = 2\,\mathbb{E}\int_0^\infty \eta(0)\eta(t)dt.$$

Remark 5.12. The above result is classical [246, Sec. 3], [278, Ch. II.1]. The basic idea of proof will also be used to estimate the deviation between averaged equation and original equation.

To study the limit of u^ϵ, we first need the tightness of u^ϵ in $C[0, T]$ for every $T > 0$. For this, we recall the following results on mixing stationary processes [183, Lemma 1 and Lemma 2].

Lemma 5.13.   *Let $\xi_1(t/\epsilon)$ be a $\mathcal{F}_{t/\epsilon}^{\infty}$ measurable process such that $\mathbb{E}\xi_1(t/\epsilon) = 0$. Also let $\xi_2(s/\epsilon)$ be a $\mathcal{F}_0^{s/\epsilon}$ measurable process. Then, for $p, q, r \geq 1$ with $1/p + 1/q + 1/r = 1$,*

$$|\mathbb{E}\{\xi_1(t/\epsilon)\xi_2(s/\epsilon)\}| \leq C\left(\mathbb{E}|\xi_1(t/\epsilon)|^p\right)^{1/p}\left(\mathbb{E}|\xi_2(s/\epsilon)|^q\right)^{1/q}\alpha\left(\frac{t-s}{\epsilon}\right)^{1/r}.$$

Lemma 5.14.   *Let $\xi(s/\epsilon)$ be $\mathcal{F}_0^{s/\epsilon}$ measurable, $\xi_1(t/\epsilon)$ be $\mathcal{F}_{t/\epsilon}^{t/\epsilon}$ measurable, and $\xi_2(\tau/\epsilon)$ be $\mathcal{F}_{\tau/\epsilon}^{\tau/\epsilon}$ measurable, for $\tau > t > s$. Assume further that $\mathbb{E}\xi_2(\tau/\epsilon) = 0$, and set*

$$H(t/\epsilon, \tau/\epsilon) := \mathbb{E}\left[\xi_1(t/\epsilon)\xi_2(\tau/\epsilon)\right].$$

Then

$$\left|\mathbb{E}\left\{\xi(s/\epsilon)\left[\xi_1(t/\epsilon)\xi_2(\tau/\epsilon) - H(t/\epsilon, \tau/\epsilon)\right]\right\}\right|$$
$$\leq C(\mathbb{E}|\xi(s/\epsilon)|^8)^{1/8}(\mathbb{E}|\xi_1(t/\epsilon)|^8)^{1/8}(\mathbb{E}|\xi(\tau/\epsilon)|^8)^{1/8}$$
$$\times\left[\alpha\left(\frac{t-s}{\epsilon}\right)\alpha\left(\frac{\tau-t}{\epsilon}\right)\right]^{1/8}.$$

Proof of Theorem 5.11

For every $T > 0$, denote by μ^{ϵ} the distribution associated with u^{ϵ} in $C[0, T]$. We first show that $\{\mu^{\epsilon}\}_{0<\epsilon\leq 1}$ is tight. In fact,

$$u^{\epsilon}(t) = u_0 + \frac{1}{\sqrt{\epsilon}}\int_0^t \eta(s/\epsilon)ds, \quad 0 \leq t \leq T,$$

and we prove that u^{ϵ} is uniformly bounded in $C^{\gamma}[0, T]$ for some $0 < \gamma < 1$. Note that for $0 < \alpha < 1$,

$$\frac{1}{\sqrt{\epsilon}}\int_0^t \eta(s/\epsilon)ds = \frac{\sin\pi\alpha}{\pi\sqrt{\epsilon}}\int_0^t\int_s^t (t-s)^{\alpha-1}(\sigma-s)^{-\alpha}d\sigma\eta(s/\epsilon)ds. \quad (5.22)$$

Here we have used the following equality:

$$\int_s^t (t-s)^{\alpha-1}(\sigma-s)^{-\alpha}d\sigma = \frac{\pi}{\sin\pi\alpha},$$

which is related to the so-called factorization method. By the stochastic Fubini theorem ([94], p. 109) we rewrite (5.22) as

$$\frac{1}{\sqrt{\epsilon}}\int_0^t \eta(s/\epsilon)ds = \int_0^t (t-s)^{\alpha-1}Y^{\epsilon}(s)ds,$$

where

$$Y^{\epsilon}(s) := \frac{1}{\sqrt{\epsilon}}\int_0^s (s-r)^{-\alpha}\eta(r/\epsilon)dr.$$

Then, by the Hölder inequality, there is a positive constant $C_{T,1}$ such that

$$\sup_{0 \le t \le T} |u^{\epsilon}(t)|^2 \le C_{T,1} \int_0^T |Y^{\epsilon}(s)|^2 ds.$$

Notice that, by Lemma 5.13 and the assumption on η, there is a $C_{T,2} > 0$ such that, for $0 \le s \le T$,

$$\begin{aligned}
\mathbb{E}|Y^{\epsilon}(s)|^2 &= \frac{2}{\epsilon} \left| \mathbb{E} \int_0^s \int_{\tau}^s (s-r)^{-\alpha}(s-\tau)^{-\alpha} \eta(r/\epsilon)\eta(\tau/\epsilon) dr \, d\tau \right| \\
&\le \frac{C}{\epsilon} \int_0^s \int_{\tau}^s (s-r)^{-\alpha}(s-\tau)^{-\alpha} \alpha^{1/4}((r-\tau)/\epsilon) dr \, d\tau \\
&\le C_{T,2}.
\end{aligned}$$

Thus we have, for some positive constant C_T,

$$\mathbb{E} \sup_{0 \le t \le T} |u^{\epsilon}(t)|^2 \le C_T. \tag{5.23}$$

Moreover, for s, t satisfying $0 \le s < t \le T$,

$$\mathbb{E}|u^t(t) - u^{\epsilon}(s)|^2 = \frac{2}{\epsilon} \mathbb{E} \int_s^t \int_s^{\tau} \eta(\tau/\epsilon)\eta(\sigma/\epsilon) d\sigma \, d\tau,$$

by Lemma 5.13. Thus for some $0 < \gamma < 1$,

$$\mathbb{E}|u^{\epsilon}(t) - u^{\epsilon}(s)|^2 \le C_T |t-s|^{\gamma},$$

which implies the tightness of $\{\mu^{\epsilon}\}_{0 < \epsilon \le 1}$ with (5.23). Then there is a sequence $\{\epsilon_n\}_{n=1}^{\infty}$ such that $\epsilon_n \to 0$, $n \to \infty$, and u^{ϵ_n} converges in distribution to some u in $C[0, T]$. Denote by P the limit of μ^{ϵ_n}. We follow a martingale approach [183] to show that for every $\varphi \in C_0^{\infty}(\mathbb{R})$,

$$\varphi(u(t)) - \varphi(u_0) - \frac{1}{2} \int_0^t A\varphi(u(\tau)) d\tau, \quad 0 \le t \le T,$$

is a P-martingale on $C(0, T; \mathbb{R})$ with

$$A = 2\mathbb{E} \int_0^{\infty} \eta(0)\eta(t) dt.$$

To this end, we just prove that for each $\mathcal{F}_0^s$-measurable bounded continuous function $\Phi : \mathbb{R} \to \mathbb{R}$, the following equality holds:

$$\mathbb{E}[\varphi(u(t))\Phi] - \mathbb{E}[\varphi(u(s))\Phi] = \mathbb{E}\left[\frac{1}{2} \int_s^t A\varphi(u(\tau)) d\tau \Phi \right]. \tag{5.24}$$

In fact, for every $\varphi \in C_0^\infty(\mathbb{R})$,

$$\varphi(u^\epsilon(t)) - \varphi(u^\epsilon(s)) = \frac{1}{\sqrt{\epsilon}} \int_s^t \varphi'(u^\epsilon(\tau))\eta(\tau/\epsilon)d\tau$$

$$= \frac{1}{\sqrt{\epsilon}} \int_s^t \varphi'(u^\epsilon(s))\eta(\tau/\epsilon)d\tau$$

$$+ \frac{1}{\sqrt{\epsilon}} \int_s^t (\varphi'(u^\epsilon(\tau)) - \varphi'(u^\epsilon(s)))\eta(\tau/\epsilon)d\tau$$

$$= \frac{1}{\sqrt{\epsilon}} \int_s^t \varphi'(u^\epsilon(s))\eta(\tau/\epsilon)d\tau$$

$$+ \frac{1}{\epsilon} \int_s^t \int_s^\tau \varphi''(u^\epsilon(\sigma))\eta(\sigma/\epsilon)\eta(\tau/\epsilon)d\sigma \, d\tau.$$

Notice that $\eta(\tau/\epsilon)$ is $\mathcal{F}_{\tau/\epsilon}^\infty$-measurable and $\varphi'(u^\epsilon(s))\Phi$ is $\mathcal{F}_0^{s/\epsilon}$-measurable. By Lemma 5.13 with $r = 2$,

$$\frac{1}{\sqrt{\epsilon}}\mathbb{E}\int_s^t \varphi'(u^\epsilon(s))\eta(\tau/\epsilon)\Phi \, d\tau = \mathcal{O}(\epsilon).$$

Furthermore, define

$$H(\tau/\epsilon, \sigma/\epsilon) \triangleq \eta(\sigma/\epsilon)\eta(\tau/\epsilon),$$
$$\bar{H}(\tau/\epsilon, \sigma/\epsilon) \triangleq \mathbb{E}\eta(\sigma/\epsilon)\eta(\tau/\epsilon),$$

and

$$\tilde{H} \triangleq H - \bar{H}.$$

Then,

$$\frac{1}{\epsilon}\mathbb{E}\int_s^t \int_s^\tau \varphi''(u^\epsilon(\sigma))\eta(\sigma/\epsilon)\eta(\tau/\epsilon)\Phi \, d\sigma \, d\tau$$

$$= \frac{1}{\epsilon}\mathbb{E}\int_s^t \int_s^\tau \varphi''(u^\epsilon(\sigma))\bar{H}(\tau/\epsilon, \sigma/\epsilon)\Phi \, d\sigma \, d\tau$$

$$+ \frac{1}{\epsilon}\mathbb{E}\int_s^t \int_s^\tau \varphi''(u^\epsilon(\sigma))\tilde{H}(\tau/\epsilon, \sigma/\epsilon)\Phi \, d\sigma \, d\tau \triangleq I_1 + I_2.$$

By changing the order of integration and noticing the assumption on η, we have

$$\lim_{\epsilon \to 0} I_1 = \lim_{\epsilon \to 0} \frac{1}{\epsilon} \int_s^t \mathbb{E}[\varphi''(u^\epsilon(\sigma))\Phi] \int_\sigma^t \mathbb{E}\eta(\sigma/\epsilon)\eta(\lambda/\epsilon)d\lambda \, d\sigma$$

$$= \lim_{\epsilon \to 0} \int_s^t \mathbb{E}[\varphi''(u^\epsilon(\sigma))\Phi] \int_{\sigma/\epsilon}^{t/\epsilon} \mathbb{E}\eta(\sigma/\epsilon)\eta(\lambda)d\lambda \, d\sigma$$

$$= \lim_{\epsilon \to 0} \int_s^t \mathbb{E}[\varphi''(u^\epsilon(\sigma))\Phi] \left[\int_0^{\frac{t-\sigma}{\epsilon}} \mathbb{E}\eta(0)\eta(\lambda)d\lambda + \int_{\frac{t-\sigma}{\epsilon}}^\infty \mathbb{E}\eta(0)\eta(\lambda)d\lambda \right] d\sigma$$

$$= \int_s^t \mathbb{E}[\varphi''(u(\sigma))\Phi] \int_0^\infty \mathbb{E}\eta(0)\eta(\lambda)d\lambda$$

$$= \mathbb{E} \int_s^t \left[\frac{1}{2} A\varphi''(u(\sigma))\Phi \right].$$

By integration by parts and using Lemma 5.14, we conclude that

$$\lim_{\epsilon \to 0} I_2 = \lim_{\epsilon \to 0} \frac{1}{\epsilon} \mathbb{E} \int_s^t \int_s^\tau \varphi''(u^\epsilon(s))\tilde{H}(\tau/\epsilon, \sigma/\epsilon)\Phi \, d\sigma \, d\tau$$

$$+ \lim_{\epsilon \to 0} \frac{1}{\epsilon} \mathbb{E} \int_s^t \int_s^\tau \int_s^\sigma \varphi'''(u^\epsilon(\lambda))\tilde{H}(\tau/\epsilon, \sigma/\epsilon)\eta(\lambda/\epsilon)d\lambda \, d\sigma \, d\tau \Phi = 0.$$

Thus, combining the above discussions, we have (5.24). Finally, by the uniqueness of solutions and Theorem 4.26, the limit of ν^ϵ, still denoted by P, is unique and solves the martingale problem related to the SDE (5.21). This completes the proof. □

As an application of Theorem 5.11, we consider the following example.

Example 5.15. Suppose that $\eta(t)$ is defined as in Remark 5.10 with $\lambda = \frac{1}{2}$, that is, it is the stationary solution of the following SDE:

$$d\eta = -\eta \, dt + dw(t).$$

Then u^ϵ, the solution of

$$\dot{u}^\epsilon = \frac{1}{\sqrt{\epsilon}} \eta(t/\epsilon), \quad u^\epsilon(0) = u_0,$$

converges in distribution to u that solves

$$du = dW(t), \quad u(0) = u_0,$$

where $W(t)$ is a standard scalar Wiener process. This example also shows that

$$\frac{1}{\epsilon} \int_0^t \exp\left\{ \frac{t-s}{\epsilon} \right\} dw(s)$$

behaves like white noise in the sense of distribution for small $\epsilon > 0$.

5.1.2 Averaging in Infinite Dimension

Averaging for weakly nonlinear partial differential equations has been investigated by many authors e.g., [148]. So far, there is no complete theory in this field, and some significant problems are to be solved. Here we only consider an averaging problem for some PDEs with highly oscillating forcing terms.

Let $H = L^2(0, l)$, a separable Hilbert space with the usual norm $\| \cdot \|$ and scalar product $\langle \cdot, \cdot \rangle$. Consider the following parabolic partial differential equation on interval $(0, l)$ with zero Dirichlet boundary condition

$$u_t^\epsilon = u_{xx}^\epsilon + F(u^\epsilon, t/\epsilon), \quad u^\epsilon(0) = u_0 \in H, \tag{5.25}$$

where $F(u, t) : \mathbb{R} \times \mathbb{R} \to \mathbb{R}$ is continuous, with bounded Fréchet derivative, i.e., $|F_u'| \leq L_F$ for some positive constant L_F. Moreover, assume that

$$\lim_{T \to \infty} \frac{1}{T} \int_0^T F(u, t) dt = \bar{F}(u), \tag{5.26}$$

which is uniform for u in every bounded set of H for some function $\bar{F}$. By these assumptions, both F and $\bar{F}$ are Lipschitz continuous in u. Define the following averaged equation:

$$u_t = u_{xx} + \bar{F}(u), \quad u(0) = u_0. \tag{5.27}$$

Then we have the following averaging principle.

Theorem 5.16. *For every $T > 0$, the unique solution u^ϵ of PDE (5.25) converges in $C(0, T; H)$ to the unique solution u of PDE (5.27), as $\epsilon \to 0$.*

Proof. First, note that basic *a priori* estimates yield that $\{u^\epsilon\}$ is compact in $C(0, T; H)$ and there is positive constant C_T such that

$$\|u^\epsilon(t)\| \leq C_T, \quad 0 \leq t \leq T.$$

Then, in order to determine the limit of u^ϵ as $\epsilon \to 0$, we consider the limit of $\langle u^\epsilon, \varphi \rangle$ for every $\varphi \in C_0^\infty(0, L)$. That is, we examine the weak limit of u^ϵ. Note that

$$\langle u^\epsilon(t), \varphi \rangle = \langle u_0, \varphi \rangle + \int_0^t \langle u^\epsilon(s), \varphi_{xx} \rangle ds + \left\langle \int_0^t \bar{F}(u^\epsilon(s)), \varphi \right\rangle ds$$
$$+ \left\langle \int_0^t [F(u^\epsilon(s), s/\epsilon) - \bar{F}(u^\epsilon(s))] ds, \varphi \right\rangle.$$

Denote by $R^\epsilon(t)$ the final term of the above equation. We next prove that

$$\lim_{\epsilon \to 0} R^\epsilon(t) = 0, \quad 0 \leq t \leq T.$$

In fact,

$$R^\epsilon(t) = \left\langle \int_0^t [F(u^\epsilon(s), s/\epsilon) - \bar{F}(u^\epsilon(s))] ds, \varphi \right\rangle$$
$$= \left\langle \epsilon \int_0^{t/\epsilon} [F(u^\epsilon(\epsilon s), s) - \bar{F}(u^\epsilon(\epsilon s))] ds, \varphi \right\rangle$$
$$\leq \delta(\epsilon) \|\varphi\|,$$

where

$$\delta(\epsilon) = \sup_{\|u\| \leq C_T} \sup_{0 \leq t < T} \epsilon \left\| \int_0^{t/\epsilon} [F(u, s) - \bar{F}(u)] ds \right\|. \tag{5.28}$$

By the assumption (5.26) we have $\delta(\epsilon) \to 0$ as $\epsilon \to 0$. Thus, by the well-posedness of (5.27), we complete the proof. $\square$

In this averaging principle for an infinite dimensional system, we have used the compactness of the solutions u^ϵ in $C(0, T; H)$ and obtained the averaged equation by determining a weak limit of u^ϵ. This is different from that of averaging for finite dimensional systems in Theorem 5.1. This will also be an important step in averaging for SPDEs.

In Theorem 5.16 the convergence is of $o(1)$ as $\epsilon \to 0$. In fact, by the proof of Theorem 5.16, similar to that of Theorem 5.1, the convergence rate depends on $\delta(\epsilon)$, the difference between $F(u, t/\epsilon)$ and $\bar{F}$ in the sense of averaging.

Averaged equations also describe long time dynamics of the systems with fast oscillating forces. For example, consider the following two-dimensional nonautonomous Navier–Stokes equation with an oscillating force

$$u_t = \Delta u + u \cdot \nabla u + g_0(x) + g_1(x, t/\epsilon) \quad \text{on} \quad D,$$

with the zero Dirichlet boundary condition on a bounded domain D. Here $g_1(x, t)$ is T-periodic in t and smooth enough in x. A global attractor $\mathcal{A}^\epsilon$ is constructed in $L^2(D)$ and the limit is proved to be the global attractor $\mathcal{A}^0$ for the following averaged equation [74]:

$$u_t = \Delta u + u \cdot \nabla u + g_0(x) + \bar{g}_1(x) \quad \text{on} \quad D,$$

with $\bar{g}_1 = \frac{1}{T} \int_0^T g_1(x, t)dt$. Similar results also hold for more general force $g_1(x, t, t/\epsilon)$, as seen in, e.g., [74].

Remark 5.17. To derive an averaged, effective system for a system with highly oscillating forcing, an essential issue is the convergence rate in ϵ of

$$\sup_{0 \le t \le T} \epsilon \left[\int_0^{t/\epsilon} F(u, s)ds - \bar{F}(u) \right] \tag{5.29}$$

on bounded set of u. This is sometimes called a generalized order function of F (see (5.5) and (5.28)). Different forms of (5.29), for example, (5.8) and (5.14), are used for slow-fast ODEs and SODEs.

5.2 An Averaging Principle for Slow-Fast SPDEs

Now consider a stochastic system in Hilbert space H with both slow and fast parts, that is, the following system of SPDEs:

$$du^\epsilon = [Au^\epsilon + f(u^\epsilon, v^\epsilon)]dt + \sigma_1 dW_1(t), \tag{5.30}$$

$$\epsilon dv^\epsilon = [Av^\epsilon + g(u^\epsilon, v^\epsilon)]dt + \sqrt{\epsilon}\, \sigma_2 dW_2(t), \tag{5.31}$$

where A is a negative definite unbounded operator and W_1, W_2 are independent Wiener processes with trace class covariance operators Q_1 and Q_2, respectively. Such a system may model, for example, a thermoelastic phenomenon in a random medium [77] or vibrating strings connected in parallel with various boundary conditions [240].

Here $\sqrt{\epsilon}$ is chosen to balance the stochastic and deterministic forces. Now, for small $\epsilon > 0$, by the classical averaging result for SODEs (Theorem 5.7), the following averaged equation:

$$du = [Au + \bar{f}(u)]dt + \sigma_1 dW_1(t) \tag{5.32}$$

is expected. We call this the averaging principle for slow-fast SPDEs. Similar to averaging for deterministic systems in infinite dimensions, estimates for solutions are important in the averaging, and estimation on the generalized order function (5.29) is also a key step. To improve the averaging principle, we consider the deviation, which describes the fluctuation (or "error") of the original slow-fast system from the averaged system.

To demonstrate the method of averaging for slow-fast SPDEs, we consider a specific case where A is the second-order Laplace operator $\Delta = \partial_{xx}$ with zero Dirichlet boundary conditions on a bounded interval. In fact, let D be an open-bounded interval and let $H = L^2(D)$ be the Lebesgue space of square integrable functions on D. Denote by $0 < \lambda_1 \leq \lambda_2 \leq \cdots$ the eigenvalues of $-\partial_{xx} = -\Delta$ on D with zero Dirichlet boundary conditions, together with the corresponding eigenfunctions $\{e_i\}$, which form an orthonormal basis of H.

Consider the following specific slow-fast SPDE system:

$$du^{\epsilon} = \left[\Delta u^{\epsilon} + f(u^{\epsilon}, v^{\epsilon})\right]dt + \sigma_1 dW_1(t), \quad u^{\epsilon}(0) = u_0 \in L^2(D), \tag{5.33}$$

$$dv^{\epsilon} = \frac{1}{\epsilon}\left[\Delta v^{\epsilon} + g(u^{\epsilon}, v^{\epsilon})\right]dt + \frac{\sigma_2}{\sqrt{\epsilon}}dW_2(t), \quad v^{\epsilon}(0) = v_0 \in L^2(D), \tag{5.34}$$

with zero Dirichlet boundary conditions on D and ϵ a small positive parameter. Here, W_1 and W_2 are mutually independent Wiener processes, taking values in $H = L^2(D)$ and defined on a probability space $(\Omega, \mathcal{F}, \mathbb{P})$ with trace class covariance operators Q_1 and Q_2, respectively. Also, σ_1 and σ_2 may depend on u^{ϵ} and v^{ϵ}. Typically, we assume that:

(H$_1$) $f(u, v) : H \times H \to H$ is Lipschitz continuous in both variables u and v with Lipchitz constant C_f and $|f(u, v)| \leq C_f(|u| + |v| + 1)$.

(H$_2$) $g(u, v) : H \times H \to H$ is Lipschitz continuous in both variables u and v with Lipschitz constant C_g and $|g(u, v)| \leq C_g(|u| + |v| + 1)$.

(H$_3$) $C_g < \lambda_1$.

(H$_4$) W_1 and W_2 are H-valued Wiener processes with covariance operators Q_1 and Q_2, respectively. Moreover, $\mathrm{Tr}((-A)^{1/2}Q_1) < \infty$, and $\mathrm{Tr}(Q_2) < \infty$.

Remark 5.18. In the following, to be more specific, we only consider the case when σ_1 and σ_2 are constants. But the approach is valid for more general nonlinearity if some *a priori* estimates for solutions in a "good" Sobolev space, for example, $H_0^1(D)$, are available. This includes the case of, for example, polynomial nonlinearity:

$$f(u, v) = \sum_{k=0}^{2n-1} a_k u^k + bv, \tag{5.35}$$

with $a_{2n-1} < 0$ and σ_1, σ_2 being Lipchitz continuous in u^{ϵ} and v^{ϵ}. In fact, the nonlinearity f is locally Lipschitz continuous from H to $H^{-\beta}$ with $-1/2 \leq \beta \leq -1/4$. So, we should restrict u^ϵ to a bounded set of a smaller space, for example, $H_0^1(D)$, with a large probability such that f is bounded and Lipschitz continuous on the bounded set; see Section 5.4. A cubic nonlinearity case with additive noise has been discussed by Wang and Roberts [301]. The case of σ_1 dependent only on v^ϵ is considered by Cerrai [68]. Recent work by Fu and Duan [138] also provides an averaged result for SPDEs with two time scales. Ren et al. [263] have considered extracting effective dynamics for a coupled microscopic-macroscopic stochastic system. See also [67,69,164] for more results on related issues.

Remark 5.19.   The assumptions (H$_2$) and (H$_3$) ensure the existence of a unique stationary solution, which is strongly mixing with exponential decay rate, to the fast equation (5.34) with fixed u^ϵ. Another assumption is that there is a Lipschitz continuous mapping $\bar{f} : H \to H$ such that

$$\mathbb{E}\left\| \frac{1}{T} \int_t^{t+T} f(u, v^{\epsilon,u}(s))ds - \bar{f}(u) \right\| \leq \alpha(T)(1 + \|u\| + \|v_0\|),$$

with $\alpha(T) \to 0, T \to \infty$ and $v^{\epsilon,u}$ being the solution to the fast Equation (5.34) with fixed $u^\epsilon = u \in H$. Then, $\alpha(T/\epsilon)$ indeed describes the order function of f.

Denote by μ^u the unique stationary measure for the fast Equation (5.34) with fixed $u^\epsilon = u \in H$. Define the average

$$\bar{f}(u) \triangleq \int_H f(u, v)\mu^u(dv). \tag{5.36}$$

Then we consider the following averaged equation:

$$du = \left[\Delta u + \bar{f}(u) \right]dt + \sigma_1 dW_1(t), \tag{5.37}$$
$$u(0) = u_0 \quad \text{and} \quad u|_{\partial D} = 0, \tag{5.38}$$

and the following averaged principle will be established.

Theorem 5.20 (Averaging principle for slow-fast SPDEs).   *Assume that* (H$_1$)–(H$_4$) *hold. Given $T > 0$, for every $u_0 \in H$, solution $u^\epsilon(t, u_0)$ of (5.33) converges in probability to u in $C(0, T; H)$, which solves (5.37)–(5.38). Moreover, the rate of convergence is $1/2$. That is, for any $\kappa > 0$,*

$$\mathbb{P}\left\{ \sup_{0 \leq t \leq T} \|u^\epsilon(t) - u(t)\| \leq C_T^\kappa \sqrt{\epsilon} \right\} > 1 - \kappa,$$

for some positive constant $C_T^\kappa > 0$.

Remark 5.21.   The above averaging principle shows that the rate of convergence of u^ϵ to u in $C(0, T; H)$ is $\sqrt{\epsilon}$. In fact, u can be seen as the first term of asymptotic expansion of u^ϵ in ϵ, that is,

$$u^\epsilon = u + \sqrt{\epsilon} \cdot deviation + \mathcal{O}(\epsilon).$$

The study of *deviation* in Section 5.4 further confirms this expansion.

Before proving the above result, we first present some examples to illustrate the application of the above averaging principle.

Example 5.22. Consider a linear slow-fast stochastic system, where the fast part is one dimensional:

$$du^\epsilon = \left[u^\epsilon_{xx} + v^\epsilon\right]dt, \quad -l < x < l, \tag{5.39}$$

$$dv^\epsilon = -\frac{1}{\epsilon}v^\epsilon dt + \frac{1}{\sqrt{\epsilon}}dw, \tag{5.40}$$

where $l > 0$, $w(t)$ is a standard scalar Wiener process, and u^ϵ satisfies the zero Dirichlet boundary conditions. The system (5.39)–(5.40) is in the form of (5.33)–(5.34) with $f(u, v) = v, g(u, v) = 0, \sigma_1 = 0$ and $\sigma_2 = 1$.

The fast part (5.40) has a unique stationary distribution $\mu = \mathcal{N}\left(0, \frac{1}{2}\right)$ which is independent of slow part u^ϵ. Thus

$$\bar{f}(u) = 0,$$

and the averaged equation is

$$\partial_t u = \partial_{xx} u. \tag{5.41}$$

Let $\bar{u}^\epsilon \triangleq \mathbb{E}u^\epsilon$ and $\bar{v}^\epsilon \triangleq \mathbb{E}v^\epsilon$. Then

$$\partial_t \bar{u}^\epsilon = \partial_{xx}\bar{u}^\epsilon + \bar{v}^\epsilon,$$

$$\dot{\bar{v}}^\epsilon = -\frac{1}{\epsilon}\bar{v}^\epsilon.$$

A standard analysis shows that

$$\max_{0 \le t \le T} \|\bar{u}^\epsilon(t) - u(t)\| \to 0, \quad \epsilon \to 0.$$

The above result implies that u approximates the expectation of u^ϵ on a long time scale for small $\epsilon > 0$. However, the expectation is not necessarily a "good" approximation for a random process. In fact, for small ϵ, $\frac{1}{\sqrt{\epsilon}}\int_0^t v^\epsilon(s)ds$ behaves like a standard one-dimensional Wiener process $W(t)$ in the sense of distribution, so we can write out a more effective system than the above averaged system:

$$d\tilde{u}^\epsilon = \partial_{xx}\tilde{u}^\epsilon dt + \sqrt{\epsilon}\,dW(t).$$

A further approximation is discussed in Section 5.4.

Example 5.23. Consider the following stochastic slow-fast system with the zero Dirichlet boundary conditions on $(-l, l)$:

$$du^\epsilon = \left[u^\epsilon_{xx} + u^\epsilon - (u^\epsilon)^3 + v^\epsilon\right]dt, \tag{5.42}$$

$$dv^\epsilon = \frac{1}{\epsilon}\left[v^\epsilon_{xx} - v^\epsilon + u^\epsilon\right]dt + \frac{3}{\sqrt{\epsilon}}dW(t), \tag{5.43}$$

where $W(t)$ is an $L^2(-l, l)$-valued Wiener process with trace class covariance operator Q. Then (5.42)–(5.43) is in the form of (5.33)–(5.34) with $f(u, v) = u - u^3 + v$, $g(u, v) = -v + u$, $\sigma_1 = 0$ and $\sigma_2 = 3$.

Here, assumption (H_1) does not hold. But the solutions to (5.42)–(5.43) have uniform energy estimates. By the proof of Theorem 5.20 and Remark 5.28, we, in fact, still have the averaging result. Moreover, $g(u, v)$ is a linear function; for any fixed $u^\epsilon = u \in H$, the SPDE (5.43) has a unique stationary distribution

$$\mu_u = \mathcal{N}\left((I - \partial_{xx})^{-1} u, \frac{9(I - \partial_{xx})^{-1} Q}{2}\right).$$

Then:

$$\bar{f}(u) = u - u^3 + (I - \partial_{xx})^{-1} u,$$

and the averaged equation is a deterministic PDE:

$$\partial_t u = \partial_{xx} u + u - u^3 + (I - \partial_{xx})^{-1} u. \tag{5.44}$$

One can see that if the slow part and the fast part are linearly coupled, the effect of noise in the fast part does not appear in the averaged equation. The situation is different in a nonlinear coupling case.

Example 5.24. Consider the following stochastic slow-fast system, with the zero Dirichlet boundary conditions on $(-l, l)$,

$$du^\epsilon = \left[u^\epsilon_{xx} + u^\epsilon - (u^\epsilon + v^\epsilon)^3\right]dt, \tag{5.45}$$

$$dv^\epsilon = \frac{1}{\epsilon} v^\epsilon_{xx} dt + \frac{\sigma}{\sqrt{\epsilon}} dW(t). \tag{5.46}$$

There is nonlinear coupling between the slow and fast parts. In this case, $f(u, v) = -(u + v)^3$ is locally Lipschitz continuous and $g(u, v) = 0$. However, the uniform energy estimates for the solutions hold, and as mentioned in Example 5.23, we still have the averaged result. Notice that there is no input from the slow part to the fast part, and (5.46) has a unique stationary distribution

$$\mu = \mathcal{N}\left(0, \frac{\sigma^2 Q}{2\partial_{xx}}\right).$$

Hence,

$$\bar{f}(u) = u - u^3 - \frac{3\sigma^2}{2} \partial_{xx}^{-1} Q u,$$

and the averaged equation is

$$\partial_t u = \partial_{xx} u + u - u^3 - \frac{3\sigma^2}{2} \partial_{xx}^{-1} Q u. \tag{5.47}$$

Now, one can see that due to the nonlinear coupling between the slow and fast parts, if the strength of noise is large enough, that is if $|\sigma|$ is large enough, the stationary solution $u = 0$ of the effective Equation (5.47) is stable. This shows stabilization of the noise by nonlinear coupling.

5.3   Proof of the Averaging Principle Theorem 5.20

In order to derive the averaging principle, Theorem 5.20, the estimates on the generalized rate function of $f(u, v^{\epsilon, u}(t))$ (see Remark 5.2 and Remark 5.17) for every fixed $u \in H$ is a key step. To this end, we need some preliminary results.

5.3.1   Some a priori Estimates

Well-posedness of the stochastic slow-fast system (5.33)–(5.34) is standard [94, Theorem 7.4] due to Lipschitz continuous nonlinearity. We state this result below without proof.

Lemma 5.25.   *Assume* (H_1)–(H_4). *For every* $u_0 \in H^2 \cap H_0^1, v_0 \in H$ *and* $T > 0$, *there is a unique solution,* $u^{\epsilon}(t), v^{\epsilon}(t)$, *in* $L^2(\Omega, C(0, T; H) \cap L^2(0, T; H_0^1))$ *for (5.33)–(5.34). Furthermore,*

$$\mathbb{E}\|u^{\epsilon}(t) - u^{\epsilon}(s)\|^{2p} \leq C_p|t - s|^{\alpha p} \tag{5.48}$$

for some $\alpha \in (0, 1)$ *and some positive constant* C_p.

Global existence follows from the following estimates for $(u^{\epsilon}, v^{\epsilon})$.

Lemma 5.26.   *Assume* (H_1)–(H_4). *For every* $u_0 \in H_0^1, v_0 \in H_0^1$ *and* $T > 0$, *there is a positive constant* C_T *that is independent of* ϵ *such that*

$$\mathbb{E} \sup_{0 \leq t \leq T} \|u^{\epsilon}(t)\|_1^2 + \sup_{0 \leq t \leq T} \mathbb{E}\|v^{\epsilon}(t)\|^2 \leq C_T(\|u_0\|_1^2 + \|v_0\|^2). \tag{5.49}$$

Proof.   By applying Itô's formula to $\|u^{\epsilon}\|^2$ and $\|v^{\epsilon}\|^2$ and noticing assumptions (H_1)–(H_2), we obtain, for some constant $C > 0$,

$$\frac{1}{2}\frac{d}{dt}\|u^{\epsilon}\|^2 = -\|u^{\epsilon}\|_1^2 + \langle f(u^{\epsilon}, v^{\epsilon}), u^{\epsilon}\rangle + \langle \dot{W}_1, u^{\epsilon}\rangle + \frac{\sigma_1^2}{2}\mathrm{Tr}(Q_1)$$

$$\leq -\lambda_1\|u^{\epsilon}\|^2 + C\|u^{\epsilon}\|^2 + \frac{\lambda_1 - C_g}{4}\|v^{\epsilon}\|^2 + C$$

$$+ \langle \dot{W}_1, u^{\epsilon}\rangle + \frac{\sigma_1^2}{2}\mathrm{Tr}(Q_1),$$

and

$$\epsilon\frac{1}{2}\frac{d}{dt}\|v^{\epsilon}\|^2 = -\|v^{\epsilon}\|_1^2 + \langle g(u^{\epsilon}, v^{\epsilon}), v^{\epsilon}\rangle + \langle \sqrt{\epsilon}\dot{W}_1, v^{\epsilon}\rangle + \epsilon\frac{\sigma_2^2}{2}\mathrm{Tr}(Q_2)$$

$$\leq -\lambda_1\|v^{\epsilon}\|^2 + C_g\|v^{\epsilon}\|^2 + C\|u^{\epsilon}\|^2 + \frac{\lambda_1 - C_g}{4}\|v^{\epsilon}\|^2 + C$$

$$+ \langle \sqrt{\epsilon}\dot{W}_1, v^{\epsilon}\rangle + \epsilon\frac{\sigma_2^2}{2}\mathrm{Tr}(Q_2).$$

Then, there is a positive constant, still denoted by C, such that

$$\frac{d}{dt}\mathbb{E}(\|u^{\epsilon}\|^2 + \epsilon\|v^{\epsilon}\|^2) \leq C\mathbb{E}(\|u^{\epsilon}\|^2 + \epsilon\|v^{\epsilon}\|^2) + \sigma_1^2\mathrm{Tr}(Q_1) + \sigma_2^2\mathrm{Tr}(Q_2),$$

which yields

$$\mathbb{E}(\|u^\epsilon(t)\|^2 + \epsilon\|v^\epsilon(t)\|^2) \le C_T(\|u_0\|^2 + \|v^\epsilon\|^2), \quad 0 \le t \le T,$$

for some constant $C_T > 0$. Having this estimate together with Itô's formula applied to $\|v^\epsilon\|^2$, we have

$$\mathbb{E}\int_0^T \|v^\epsilon(t)\|_1^2 dt \le C_T$$

for some positive constant (still denoted as) C_T. Now applying Itô's formula to $\|u^\epsilon(t)\|_1^2$, we have

$$\begin{aligned}
\frac{1}{2}\|u^\epsilon\|_1^2 &= -\|Au^\epsilon\|^2 - \langle f(u^\epsilon, v^\epsilon), Au^\epsilon\rangle + \langle \sigma_1 \dot{W}_1, Au^\epsilon\rangle + \frac{\sigma_1^2}{2}\mathrm{Tr}((-A)^{1/2}Q_1)\\
&\le -\|Au^\epsilon\|^2 + 2C_f\|u^\epsilon\|_1^2 + C_f\|v^\epsilon\|_1^2 + \langle \sigma_1\dot{W}_1, Au^\epsilon\rangle\\
&\quad + \frac{\sigma_1^2}{2}\mathrm{Tr}((-A)^{1/2}Q_1).
\end{aligned}$$

Therefore, by the inequality in Lemma 3.24,

$$\mathbb{E}\sup_{0\le t\le T}\|u^\epsilon(t)\|_1^2 \le C_T(\|u_0\|_1^2 + \|v_0\|^2).$$

This completes the proof.   □

Classical averaging methods depend heavily on the periodicity of the fast part on the fast time scale [22, Ch. 6], [48, Ch. 5], [149, Ch. V.3], [273, Ch. 2], [294]. A weaker condition is that the fast part is sufficiently mixing [187, p. 386]. So the properties of the fast system (5.34) for fixed u is a key step to carry out the averaging approach for slow-fast SPDEs (5.33)–(5.34). We first have the following result for the fast system (5.34) with fixed $u \in H$.

Lemma 5.27. *Assume* (H_1)–(H_4). *For fixed* $u \in H$, *the fast system (5.34) has a unique stationary solution* $\eta_u^\epsilon(t)$, *with distribution* μ_u *independent of* ϵ, *and the stationary measure* μ_u *is strong mixing with exponential decay rate. Moreover, for every solution* v_u^ϵ *of (5.34) with initial value* v,

$$\mathbb{E}\|v_u^\epsilon(t) - \eta_u^\epsilon(t)\|^2 \le e^{-2(\lambda_1 - C_g)t/\epsilon}\mathbb{E}\|v - \eta_u^\epsilon(0)\|^2, \tag{5.50}$$

and, for $T \ge t \ge r \ge 0$ *and* $\delta > 0$, *there is a constant* $C_T > 0$ *such that*

$$\mathbb{E}\left\|\int_r^{r+\delta} e^{A(t-s)}[f(u, v_u^\epsilon(s)) - \bar{f}(u)]ds\right\| \le \sqrt{\epsilon\delta}\frac{C_T(1 + \|v\|^2)}{\lambda_1 - C_g}. \tag{5.51}$$

Proof. The first result follows from the assumptions (H_2) and (H_3). In fact, for $u \in H$, define the transition semigroup P_t^u associated with the fast equation by

$$P_t^u\varphi(v) \triangleq \mathbb{E}\varphi(v_u^\epsilon(t)), \quad t \ge 0, \quad v \in H \tag{5.52}$$

for any Lipschitz continuous function φ. For $f(u, \cdot) : H \to H$, define

$$P_t^u f(u, v) \triangleq \sum_{i=1}^{\infty} [P_t^u \langle f(u, v), e_i \rangle] e_i = \mathbb{E} f(u, v_u^{\epsilon}(s)) = \sum_{i=1}^{\infty} \langle P_t^u f(u, v), e_i \rangle e_i.$$

Then, by (5.50), we have

$$\| P_t^u f(u, v) - \bar{f}(u) \| \le C_f (\| v \| + \| \eta_u^{\epsilon}(0) \| + 1) e^{-(\lambda_1 - C_g) t / \epsilon}. \tag{5.53}$$

For the second result, for every $\psi \in H$, we estimate

$$\mathbb{E} \left\langle \int_r^{r+\delta} [f(u, v_u^{\epsilon}(s)) - \bar{f}(u)] ds, \psi \right\rangle^2$$

$$= \int_r^{r+\delta} \int_r^{r+\delta} \mathbb{E} \langle f(u, v_u^{\epsilon}(s)) - \bar{f}(u), \psi \rangle \langle f(u, v_u^{\epsilon}(\tau)) - \bar{f}(u), \psi \rangle \, d\tau \, ds$$

$$\le \int_r^{r+\delta} \int_r^{r+\delta} \left| \mathbb{E} \langle f(u, v_u^{\epsilon}(s)) - \bar{f}(u), \psi \rangle \langle f(u, v_u^{\epsilon}(\tau)) - \bar{f}(u), \psi \rangle \right| \, d\tau \, ds$$

$$= 2 \int_r^{r+\delta} \int_s^{r+\delta} \left| \mathbb{E} \langle f(u, v_u^{\epsilon}(s)) - \bar{f}(u), \psi \rangle \langle f(u, v_u^{\epsilon}(\tau)) - \bar{f}(u), \psi \rangle \right| \, d\tau \, ds.$$

By the properties of conditional expectation and Markovian property of the process v_u^{ϵ}, we conclude

$$|\mathbb{E} \langle f(u, v_u^{\epsilon}(s)) - \bar{f}(u), \psi \rangle \langle f(u, v_u^{\epsilon}(\tau)) - \bar{f}(u), \psi \rangle|$$

$$= |\mathbb{E} \{ \langle f(u, v_u^{\epsilon}(s)) - \bar{f}(u), \psi \rangle \mathbb{E} [\langle f(u, v_u^{\epsilon}(\tau)) - \bar{f}(u), \psi \rangle | \mathcal{F}_s] \}|$$

$$= |\mathbb{E} \{ \langle f(u, v_u^{\epsilon}(s)) - \bar{f}(u), \psi \rangle P_{\tau - s}^u \langle f(u, v_u^{\epsilon}(s)) - \bar{f}(u), \psi \rangle \}|$$

$$\le \{ \mathbb{E} [\langle f(u, v_u^{\epsilon}(s)) - \bar{f}(u), \psi \rangle]^2 \}^{1/2} \{ \mathbb{E} [P_{\tau - s}^u \langle f(u, v_u^{\epsilon}(s)) - \bar{f}(u), \psi \rangle]^2 \}^{1/2}$$

$$= \{ \mathbb{E} [\langle f(u, v_u^{\epsilon}(s)) - \bar{f}(u), \psi \rangle]^2 \}^{1/2} \{ \mathbb{E} [\langle P_{\tau - s}^u f(u, v_u^{\epsilon}(s)) - \bar{f}(u), \psi \rangle]^2 \}^{1/2}$$

$$\le C \mathbb{E} (\| u \|^2 + \| v_u^{\epsilon} \|^2 + 1) \| \psi \|^2 e^{-(\lambda_1 - C_g)(\tau - s) / \epsilon}.$$

Now, by Lemma 5.26, we have

$$\mathbb{E} \left\| \int_r^{r+\delta} [f(u, v_u^{\epsilon}(s)) - \bar{f}(u)] ds \right\| \le \sqrt{\epsilon} \delta \frac{C_T (1 + \| v \|^2 + \| u \|^2)}{\lambda_1 - C_g}.$$

Then, by the fact that $\| e^{At} \|_{\mathcal{L}(H,H)} \le 1$ for $t \ge 0$, we have the estimate (5.51). $\quad \square$

Remark 5.28.   We have used the Lipschitz continuity of f in (5.53). However, the above discussion is also applicable to some locally Lipschitz f. In fact, for example, we assume that $f(u, v)$ satisfies

$$|f(u, v)| \le C(|u|^{2p+1} + |v|^{2p+1}) \quad \text{for} \quad u, v \in \mathbb{R}$$

and

$$|f(u, v_1) - f(u, v_2)| \le [|v_1|^{2p} + |v_2|^{2p}] |v_1 - v_2| \quad \text{for} \quad u, v_1, v_2 \in \mathbb{R}.$$

Then,

$$\mathbb{E}\|f(u, v_u^\epsilon) - \bar{f}(u)\| \le \left[\mathbb{E}[\|v_u^\epsilon\|_1^{2p} + \|\eta_u^\epsilon\|_1^{2p}]^2\right]^{1/2} \left[\mathbb{E}\|v_u^\epsilon - \eta_u^\epsilon\|^2\right]^{1/2}.$$

Thus, we also have the result of Lemma 5.27, provided that we have estimates for $\mathbb{E}\|v_u^\epsilon\|_1^{4p}$ and $\mathbb{E}\|\eta_u^\epsilon\|_1^{4p}$.

We need to examine further properties of the fast system. To this end, we consider the fast system on a slow time scale. Making the time scale transformation $t \to \tau = \epsilon t$, the system (5.34) becomes

$$dv = [\Delta v + g(u, v)]d\tau + \sigma_2 d\tilde{W}_2(\tau), \quad v(0) = v_0, \tag{5.54}$$

where $\tilde{W}_2$ is the scaled version of W_2 (both have the same distribution). For every $t > 0$, the solution v_u^ϵ of (5.34) and the solution v_u of (5.54) have the same distribution. Indeed, SPDE (5.54) has a unique stationary solution η_u with the distribution μ_u.

By a version of the contraction mapping principle depending on a parameter [66, Appendix C], the solution $v_u(t)$ of (5.54) is differentiable with respect to u, with the Fréchet derivative $D_u v$, along the direction $h \in H$, satisfying

$$d\langle D_u v, h\rangle = [\langle \Delta D_u v, h\rangle + \langle g_u'(u, v) D_u v, h\rangle]dt.$$

Then we have

$$\sup_{u, v_0 \in H, 0 \le t < \infty} |D_u v_u|_{\mathcal{L}(H)} \le C, \tag{5.55}$$

for some deterministic constant $C > 0$. Moreover, $\bar{f}$ is Lipschitz continuous with Lipschitz constant C_f. Then, by a standard analysis [94, Theorem 7.4] for the averaged equation (5.37)–(5.38), we have the following well-posedness result for the averaged equation.

Lemma 5.29. *Assume* (H$_1$)–(H$_4$) *hold. For every* $u_0 \in H_0^1$ *and* $T > 0$, *the system* (5.37)–(5.38) *has a unique solution* $u \in L^2(\Omega, C(0, T; H) \cap L^2(0, T; H_0^1))$. *Moreover, there is a positive* $C_T > 0$ *such that for every integer* $m \ge 2$ *and every* $T > 0$,

$$\mathbb{E}\|u(t)\|^m \le C_T(1 + \|u_0\|^m). \tag{5.56}$$

5.3.2 Averaging as an Approximation

Having the above results, we can now prove Theorem 5.20. In order to derive the approximation in probability, we restrict the system to a smaller probability space that is arbitrarily close to the original probability space.

By Lemma 5.26, for every $\kappa > 0$, there is a compact set K_κ in $C(0, T; H)$ such that

$$\mathbb{P}\{u^\epsilon \in K_\kappa\} > 1 - \kappa/3.$$

Then there is a constant $C_T^\kappa > 0$ that is only dependent on T and κ, such that

$$\sup_{0 \le t \le T} \|u^\epsilon(t)\|^2 \le C_T^\kappa$$

for $u^\epsilon \in K_\kappa$. Here, K_κ is chosen as a family of decreasing sets with respect to κ. Furthermore, for $\kappa > 0$ and $u \in K_\kappa$, by the estimate (5.48), (5.51) and Markov inequality [39, p. 74], there is a positive constant, which we still denote by C_T^κ, such that for $T \geq t > s \geq 0$,

$$\mathbb{P}\{\|u^\epsilon(t) - u^\epsilon(s)\|^2 \leq C_T^\kappa |t - s|^\alpha\} \geq 1 - \kappa/3, \tag{5.57}$$

where $\alpha \in (0, 1)$ is chosen in (5.48), and for every $T \geq t \geq 0$,

$$\mathbb{P}\left\{\left\|\int_0^t e^{A(t-s)}[f(u, v^\epsilon(s)) - \bar{f}(u)]ds\right\| \leq \sqrt{\epsilon}C_T^\kappa(1 + \|u_0\|^2 + \|v_0\|^2)\right\}$$
$$\geq 1 - \kappa/3, \tag{5.58}$$

for $u \in K_\kappa$.

Now, for $\kappa > 0$, we introduce a new subprobability space $(\Omega_\kappa, \mathcal{F}_\kappa, \mathbb{P}_\kappa)$ defined by

$$\Omega_\kappa = \{\omega \in \Omega : u^\epsilon(\omega) \in K_\kappa, \text{ and the events of (5.57) and (5.58) hold}\},$$

and

$$\mathcal{F}_\kappa := \{S \cap \Omega_\kappa : S \in \mathcal{F}\}, \quad \mathbb{P}_\kappa(S) := \frac{\mathbb{P}(S \cap \Omega_\kappa)}{\mathbb{P}(\Omega_\kappa)} \quad \text{for} \quad S \in \mathcal{F}_\kappa.$$

Remark 5.30. The choice of such Ω_κ makes the nonlinearity $f(u(\omega), v(\omega))$ Lipschitz continuous for $\omega \in \Omega_\kappa$. This is even true for the nonlinearity of f with the form of (5.35).

Now we restrict $\omega \in \Omega_\kappa$ and introduce auxiliary processes. For $T > 0$, we partition the interval $[0, T]$ into subintervals of length $\delta = \epsilon^{1/\alpha}$, and we construct processes $(\tilde{u}^\epsilon, \tilde{v}^\epsilon)$, such that for $t \in [k\delta, (k + 1)\delta)$,

$$\tilde{u}^\epsilon(t) = e^{A(t-k\delta)}u^\epsilon(k\delta) + \int_{k\delta}^t e^{A(t-s)}f(u^\epsilon(k\delta), \tilde{v}^\epsilon(s))ds$$
$$+ \sigma_1 \int_{k\delta}^t e^{A(t-s)}dW_1(s), \quad \tilde{u}^\epsilon(0) = u_0, \tag{5.59}$$

$$d\tilde{v}^\epsilon(t) = \frac{1}{\epsilon}\left[A\tilde{v}^\epsilon(t) + g(u^\epsilon(k\delta), \tilde{v}^\epsilon(t))\right]dt + \frac{\sigma_2}{\sqrt{\epsilon}}dW_2(t),$$
$$\tilde{v}^\epsilon(k\delta) = v^\epsilon(k\delta). \tag{5.60}$$

Then, for $t \in [k\delta, (k + 1)\delta)$,

$$\frac{1}{2}\frac{d}{dt}\|v^\epsilon(t) - \tilde{v}^\epsilon(t)\|^2 \leq -\frac{1}{\epsilon}(\lambda_1 - C_g)\|v^\epsilon(t) - \tilde{v}^\epsilon(t)\|^2$$
$$+ \frac{1}{\epsilon}C_g\|v^\epsilon(t) - \tilde{v}^\epsilon(t)\|\|u^\epsilon(t) - u^\epsilon(k\delta)\|.$$

By the choice of Ω_κ, we have (5.57) for $t \in [k\delta, (k+1)\delta)$, and by the Gronwall inequality,

$$\|v^\epsilon(t) - \tilde{v}^\epsilon(t)\|^2 \le C_T^\kappa \delta^\alpha, \quad t \in [0, T]. \tag{5.61}$$

Moreover, by the choice of Ω_κ and the assumption of $f(\cdot, v)$, we see that $f(\cdot, v) : H \to H$ is Lipschitz continuous. Hence, for $t \in [k\delta, (k+1)\delta)$,

$$\|u^\epsilon(t) - \tilde{u}^\epsilon(t)\| \le C_f \int_{k\delta}^t \|v^\epsilon(s) - \tilde{v}^\epsilon(s)\| ds + C_f \int_{k\delta}^t \|u^\epsilon(k\delta) - u^\epsilon(s)\| ds.$$

So, by noticing (5.57), we imply

$$\|u^\epsilon(t) - \tilde{u}^\epsilon(t)\|^2 \le C_T^\kappa \delta^\alpha, \quad t \in [0, T]. \tag{5.62}$$

Notice that in the mild sense

$$u(t) = e^{At}u_0 + \int_0^t e^{A(t-s)}\bar{f}(u(s))ds + \sigma_1 \int_0^t e^{A(t-s)}dW_1(s).$$

Then, using $\lfloor z \rfloor$ to denote the largest integer less than or equal to z,

$$\sup_{0 \le s \le t} \|\tilde{u}^\epsilon(s) - u(s)\|$$

$$\le \left\| \int_0^t e^{A(t-s)}[f(u^\epsilon(\lfloor s/\delta \rfloor \delta), \tilde{v}^\epsilon(s)) - \bar{f}(u^\epsilon(\lfloor s/\delta \rfloor \delta))]ds \right\|$$

$$+ \int_0^t e^{A(t-s)} \left\| \bar{f}(u^\epsilon(\lfloor s/\delta \rfloor \delta)) - \bar{f}(u^\epsilon(s)) \right\| ds$$

$$+ \int_0^t e^{A(t-s)} \left\| \bar{f}(u^\epsilon(s)) - \bar{f}(u(s)) \right\| ds.$$

Noticing the definition of Ω_κ and the Lipschitz property of $\bar{f}$, by (5.58) and (5.36), we have for $t \in [0, T]$,

$$\sup_{0 \le s \le t} \|\tilde{u}^\epsilon(s) - u(s)\| \le C_T^\kappa \left[\sqrt{\epsilon} + \int_0^T \sup_{0 \le \tau \le s} \|u^\epsilon(\tau) - u(\tau)\| ds \right]. \tag{5.63}$$

Realizing that

$$\sup_{0 \le s \le t} \|u^\epsilon(s) - u(s)\| \le \sup_{0 \le s \le t} \|u^\epsilon(s) - \tilde{u}^\epsilon(s)\| + \sup_{0 \le s \le t} \|\tilde{u}^\epsilon(s) - u(s)\| \tag{5.64}$$

by the Gronwall inequality and (5.57), (5.62), and (5.63), we conclude that for $t \in [0, T]$,

$$\sup_{0 \le s \le t} \|u^\epsilon(s) - u(s)\| \le C_T^\kappa \sqrt{\epsilon} \quad \text{in} \quad \Omega_\kappa. \tag{5.65}$$

This estimate shows that the convergence of u^ϵ to u is of order $\sqrt{\epsilon}$ in probability. This completes the proof of Theorem 5.20. $\qquad\square$

Remark 5.31.   In the above averaging approach, the auxiliary process $\tilde{u}^\epsilon$ is defined as a time discretized approximation to u^ϵ on finite time interval $[0, T]$. The length of partition δ determines the approximation of $\tilde{u}^\epsilon$ to u^ϵ. So, by the relation (5.64), a better approximation of $\tilde{u}^\epsilon$ to u^ϵ leads to a better averaging approximation of u to u^ϵ. In fact, one can check that, for example, if $\delta = \epsilon^\gamma$, with $\gamma \geq \frac{1}{2\alpha}$ and $0 < \alpha < 1$, we always have the approximation rate $\sqrt{\epsilon}$ from (5.64). But for $0 < \gamma < \frac{1}{2\alpha}$, the approximation rate is ϵ^γ. This implies that the averaging approximation rate is $\sqrt{\epsilon}$.

Remark 5.32.   This averaging result also holds for slow-fast SPDEs with multiplicative noise. Cerrai [68] derived an averaged system for the following slow-fast SPDEs:

$$du^\epsilon = \left[\Delta u^\epsilon + f(u^\epsilon, v^\epsilon)\right]dt + \sigma_1(u^\epsilon, v^\epsilon)dW_1(t), \quad u^\epsilon(0) = u_0 \in L^2(D),$$

$$dv^\epsilon = \frac{1}{\epsilon}\left[\Delta v^\epsilon + g(u^\epsilon, v^\epsilon)\right]dt + \frac{\sigma_2(u^\epsilon, v^\epsilon)}{\sqrt{\epsilon}}dW_2(t), \quad v^\epsilon(0) = v_0 \in L^2(D),$$

where f, g, σ_1, and σ_2 are Lipschitz continuous.

Remark 5.33.   The above discussion is also valid for PDEs with a random stationary coefficient

$$u_t = \Delta u + f(u, z(t/\epsilon)),$$

where f is Lipschitz continuous in u and z. Here, z is a stationary random process that is sufficiently mixing. One typical example is that $z(t)$ solves the following stochastic differential equation:

$$dz = \Delta z\, dt + \sigma dW(t)$$

on some bounded regular domain with the zero Dirichlet boundary conditions. This is the case when there is no input from the slow part to the fast equation.

5.4   A Normal Deviation Principle for Slow-Fast SPDEs

Hasselmann [155] considered a coupled climate-weather system where the fast system is weather and the slow system describes climate, in order to establish a stochastic model that is simpler than the original model but more precise than the averaged model. One method is to study the normal deviation from the averaged equation [20,155]. This deviation describes more qualitative properties of the slow system than the averaged equation, and it in fact satisfies the central limit theorem. Related results in the context of ordinary differential equations with stationary random coefficients that are sufficiently mixing are discussed by Kifer [188].

Here we present a similar result for a system described by the following slow-fast SPDEs:

$$du^\epsilon = [Au^\epsilon + f(u^\epsilon, v^\epsilon)]dt + \sigma_1 dW_1(t), \tag{5.66}$$

$$\epsilon dv^\epsilon = [Av^\epsilon + g(u^\epsilon, v^\epsilon)]dt + \sqrt{\epsilon}\sigma_2 dW_2(t). \tag{5.67}$$

For simplicity, we assume $\sigma_1 = 0$, that is, there is no stochastic force on the slow part, and we replace assumption (H$_2$) in the previous section by the following assumption:

(**H$_2'$**) $g(u, v) = g(v) : H \rightarrow H$ is Lipschitz continuous in variable u with Lipschitz constant C_g, and furthermore, $|g(v)| \leq C_g(|v| + 1)$.

We consider the deviation process between u^ϵ and u. For this, we introduce

$$z^\epsilon(t) \triangleq \frac{1}{\sqrt{\epsilon}}(u^\epsilon(t) - u(t)). \tag{5.68}$$

Then we have the following normal deviation principle.

Theorem 5.34 (Normal deviation principle). *Assume that* (H$_1$), (H$_2'$), (H$_3$), *and* (H$_4$) *hold, and set* $\sigma_1 = 0$. *Then the deviation process* z^ϵ, *in* $C(0, T; H)$, *converges in distribution to* z, *which solves the following linear stochastic partial differential equation:*

$$dz(t) = [\Delta z(t) + \overline{f_u'}(u(t))z(t)]dt + d\widetilde{W}(t), \quad z|_{\partial D} = 0, \quad z(0) = 0, \tag{5.69}$$

where

$$\overline{f_u'}(u) \triangleq \int_H f_u'(u, v)\mu(dv), \tag{5.70}$$

and $\widetilde{W}(t)$ *is an* H-*valued Wiener process defined on a new probability space* $(\bar{\Omega}, \bar{\mathcal{F}}, \bar{\mathbb{P}})$ *with covariance operator*

$$B(u) \triangleq 2 \int_0^\infty \mathbb{E}\Big[(f(u, \eta(t)) - \bar{f}(u)) \otimes (f(u, \eta(0)) - \bar{f}(u))\Big]dt. \tag{5.71}$$

Moreover, $u(t)$ *solves the averaged equation* (5.37).

Before proving this normal deviation principle, we make some remarks and present a couple of examples.

Remark 5.35. In this theorem, $B(u)$, as the covariance operator of a Gaussian process $\widetilde{W}$, is nonnegative definite. Furthermore, by the expression (5.71), $B(u)$ is in fact a Hilbert–Schmidt operator. Then, by the decomposition theorem [262, Ch. VIII.9] for positive linear operators, we have the square root of $B(u)$, denoted by $\sqrt{B(u)}$. Thus, Equation (5.69) can be rewritten as

$$dz(t) = [\Delta z(t) + \overline{f_u'}(u(t))z(t)]dt + \sqrt{B(u)}d\overline{W}(t), \quad z|_{\partial D} = 0, \quad z(0) = 0, \tag{5.72}$$

where $\overline{W}(t)$ is an H-valued cylindrical Wiener process with the identity covariance operator Id_H.

Remark 5.36. By adding the deviation to the averaged equation, we formally have an approximation equation

$$d\bar{u}^\epsilon = \big[A\bar{u}^\epsilon + \bar{f}(\bar{u}^\epsilon)\big]dt + \sqrt{\epsilon}\,d\widetilde{W}(t), \quad \bar{u}^\epsilon(0) = u_0, \tag{5.73}$$

where $\widetilde{W}(t)$ is an H-valued Wiener process, defined on a new probability space $(\bar{\Omega}, \bar{\mathcal{F}}, \bar{\mathbb{P}})$, with covariance operator $B(u)$. This equation in fact gives a much better approximation than the averaged Equation (5.37). We will not give a rigorous justification here, but a random slow manifold reduction via computer algebra [267] on a long time scale, as in Section 5.6, also gives the same result.

By the above normal deviation principle, we draw the following approximation result.

Corollary 5.37. *Assume that* (H$_1$), (H$_2'$), (H$_3$), *and* (H$_4$) *hold and set* $\sigma_1 = 0$. *There exists a new probability space* $(\bar{\Omega}, \bar{\mathcal{F}}, \bar{\mathbb{P}})$, *such that the following error estimate holds, for every* $T > 0$ *and* $\kappa > 0$,

$$\bar{\mathbb{P}}\left\{ \sup_{0 \le t \le T} \|u^\epsilon(t) - u(t) - \sqrt{\epsilon}z(t)\| \le C_T^\kappa \epsilon \right\} > 1 - \kappa,$$

with some positive constant C_T^κ.

Before giving the proof of Theorem 5.34, we illustrate the normal deviation result in two examples.

Example 5.38. Consider equations (5.39)–(5.40) in *Example* 5.22. By (5.71), we have

$$B(u) = 1,$$

which is a scalar function. Then z, the limit of the deviation z^ϵ, satisfies the following linear SPDE:

$$dz = z_{xx}dt + d\bar{w}(t), \quad z(0) = 0,$$

where $\bar{w}(t)$ is a new scalar Brownian motion defined on some probability space $(\bar{\Omega}, \bar{\mathcal{F}}, \bar{\mathbb{P}})$. Similar to $u^\epsilon = u + \sqrt{\epsilon}z^\epsilon$, we define $\bar{u}^\epsilon := u + \sqrt{\epsilon}z$ and we have the following equation:

$$d\bar{u}^\epsilon = \bar{u}^\epsilon_{xx}dt + \sqrt{\epsilon}d\bar{w}(t).$$

Then, on this new probability space, for every $T > 0$ and $\kappa > 0$, there is $C_T^\kappa > 0$ such that

$$\bar{\mathbb{P}}\left\{ \sup_{0 \le t \le T} \|u^\epsilon(t) - \bar{u}^\epsilon(t)\| \le C_T^\kappa \epsilon \right\} > 1 - \kappa.$$

This example shows that if the fast part has no coupling from a slow part, (5.73) is a better effective approximation model. But, for the nonlinear case, this is not a trivial problem, even when the coupling is linear [219].

Example 5.39. Consider equations (5.42)–(5.43) in *Example* 5.23. By (5.71), we have

$$B(u) = 9(I - \partial_{xx})^{-1}Q,$$

which is a Hilbert–Schmidt operator. Then z satisfies the following linear SPDE:

$$dz(t) = [z_{xx}(t) + (1 - 3u^2(t))z(t)]dt + 3\sqrt{(I - \partial_{xx})^{-1}}\, d\overline{W}(t), \quad z(0) = 0,$$

where $\overline{W}(t)$ is a Wiener process with covariance Q, defined on a new probability space $(\overline{\Omega}, \overline{\mathcal{F}}, \overline{\mathbb{P}})$. The above equation shows that z depends on u, which solves the averaged Equation (5.44). Then, for every $T > 0$ and $\kappa > 0$,

$$\overline{\mathbb{P}}\left\{ \sup_{0 \le t \le T} \|u^\epsilon(t) - u(t) - \sqrt{\epsilon}z(t)\| \le C_T^\kappa \epsilon \right\} > 1 - \kappa,$$

for some positive constant C_T^κ.

5.5 Proof of the Normal Deviation Principle Theorem 5.34

We apply a martingale approach to prove the normal deviation principle, Theorem 5.34. Notice that z^ϵ satisfies

$$\dot{z}^\epsilon = Az^\epsilon + \frac{1}{\sqrt{\epsilon}}[f(u^\epsilon, v^\epsilon) - \bar{f}(u)] \tag{5.74}$$

with $z^\epsilon(0) = 0$. To determine the limit of z^ϵ, we study one convergent subsequence of z^ϵ and determine the limit of this subsequence. To this end, we first show the tightness of the distributions of z^ϵ in $C(0, T; H)$. We need the following uniform estimate on $v^\epsilon(t)$ for every $t > 0$.

Lemma 5.40. *Assume that (H'_2), (H_3), and (H_4) hold. For every $v_0 \in H$ and $m > 0$, there is $C_m > 0$ such that the solution $v^\epsilon(t)$ to Equation (5.34), with $v^\epsilon(0) = v_0$, satisfies*

$$\mathbb{E}\|v^\epsilon(t)\|^m \le C_m, \tag{5.75}$$

for $t \ge 0$.

Proof. This is a direct application of Itô's formula to $\|v^\epsilon(t)\|^2$ and then to $\|v^\epsilon(t)\|^m$ with $m \ge 2$. The details are omitted here. $\qquad \square$

By (5.74), we have

$$\|z^\epsilon(t)\| = \left\| \frac{1}{\sqrt{\epsilon}} \int_0^t e^{A(t-s)}[f(u^\epsilon(s), v^\epsilon(s)) - f(u(s), v^\epsilon(s))]ds \right.$$
$$+ \frac{1}{\sqrt{\epsilon}} \int_0^t e^{A(t-s)}[f(u(s), v^\epsilon(s)) - f(u(s), \eta^\epsilon(s))]ds$$
$$\left. + \frac{1}{\sqrt{\epsilon}} \int_0^t e^{A(t-s)}[f(u(s), \eta^\epsilon(s)) - \bar{f}(u(s))]ds \right\|$$
$$\le C_f \int_0^t \|z^\epsilon(s)\|ds + \frac{1}{\sqrt{\epsilon}} \int_0^t \|v^\epsilon(s) - \eta^\epsilon(s)\|ds$$
$$+ \frac{1}{\sqrt{\epsilon}} \left\| \int_0^t e^{A(t-s)}[f(u(s), \eta^\epsilon(s)) - \bar{f}(u(s))]ds \right\|.$$

Thus,

$$\mathbb{E} \sup_{0 \le s \le t} \|z^\epsilon(s)\| \le C_f \int_0^t \mathbb{E} \sup_{0 \le \tau \le s} \|z^\epsilon(\tau)\| ds + \frac{1}{\sqrt{\epsilon}} \int_0^T \mathbb{E} \|v^\epsilon(s) - \eta^\epsilon(s)\| ds$$

$$+ \frac{1}{\sqrt{\epsilon}} \mathbb{E} \sup_{0 \le t \le T} \left\| \int_0^t e^{A(t-s)} [f(u(s), \eta^\epsilon(s)) - \bar{f}(u(s))] ds \right\|.$$

Also, by the same discussion for (5.22), we obtain, for some $\alpha \in (0, 1)$,

$$\frac{1}{\sqrt{\epsilon}} \int_0^t e^{A(t-s)} [f(u(s), \eta^\epsilon(s)) - \bar{f}(u(s))] ds$$

$$= \frac{\sin \pi \alpha}{\alpha} \int_0^t (t-s)^{\alpha-1} e^{A(t-s)} Y^\epsilon(s) ds,$$

where

$$Y^\epsilon(s) = \frac{1}{\sqrt{\epsilon}} \int_0^s (s-r)^{-\alpha} e^{A(s-r)} [f(u(r), \eta^\epsilon(r)) - \bar{f}(u(r))] dr$$

$$= \frac{1}{\sqrt{\epsilon}} \sum_{k=1}^\infty \int_0^s (s-r)^{-\alpha} e^{-\lambda_k(s-r)} \langle f(u(r), \eta^\epsilon(r)) - \bar{f}(u(r)), e_k \rangle e_k dr.$$

$$(5.76)$$

Here, $\{\lambda_k\}$ are the eigenvalues of $-A$ with the associated eigenfunctions $\{e_k\}$, which form an orthonormal basis. By Lemma 5.13, Lemma 5.40, and the same discussion as in the proof of Theorem 5.11, we have

$$\mathbb{E} \|Y^\epsilon(s)\|^2 \le C_T, \quad 0 \le s \le T,$$

for some positive constant (still denoted by) C_T. Then, by the Gronwall inequality, we have

$$\mathbb{E} \sup_{0 \le t \le T} \|z^\epsilon(t)\| \le C_T, \tag{5.77}$$

for some positive constant (still denoted by) C_T. Furthermore, for $\theta > 0$,

$$\|(-A)^\theta z^\epsilon(t)\| \le \left\| \frac{1}{\sqrt{\epsilon}} \int_0^t (-A)^\theta e^{A(t-s)} [f(u^\epsilon(s), v^\epsilon(s)) - f(u(s), v^\epsilon(s))] ds \right\|$$

$$+ \left\| \frac{1}{\sqrt{\epsilon}} \int_0^t (-A)^\theta e^{A(t-s)} [f(u(s), v^\epsilon(s)) - f(u(s), \eta^\epsilon(s))] ds \right\|$$

$$+ \left\| \frac{1}{\sqrt{\epsilon}} \int_0^t (-A)^\theta e^{A(t-s)} [f(u(s), \eta^\epsilon(s)) - \bar{f}(u(s))] ds \right\|$$

$$\triangleq J_1 + J_2 + J_3.$$

Notice that for $0 < \theta < 1/2$,

$$\frac{1}{\sqrt{\epsilon}} \|(-A)^\theta e^{A(t-s)} [f(u^\epsilon, v^\epsilon) - f(u, v^\epsilon)]\| \le C \left(1 + \frac{1}{\sqrt{s}}\right) C_f \|z^\epsilon(s)\|,$$

and

$$\frac{1}{\sqrt{\epsilon}}\|(-A)^{\theta}e^{A(t-s)}[f(u, v^{\epsilon}) - f(u, \eta^{\epsilon})]\| \leq C\left(1 + \frac{1}{\sqrt{s}}\right)\frac{C_f}{\sqrt{\epsilon}}\|v^{\epsilon} - \eta^{\epsilon}\|,$$

for some constant $C > 0$. Then, by estimates (5.77) and (5.50),

$$\mathbb{E} \sup_{0 \leq t \leq T} J_1 \leq C_T, \quad \mathbb{E} \sup_{0 \leq t \leq T} J_2 \leq C_T,$$

for some constant $C_T > 0$. Now we consider J_3. By the factorization method, we have

$$J_3 = \frac{\sin \pi \alpha}{\alpha} \int_0^t (t - s)^{\alpha - 1} e^{A(t-s)}(-A)^{\theta} Y^{\epsilon}(s)ds,$$

where Y^{ϵ} is defined by (5.76). Notice that

$$\|(-A)^{\theta} Y^{\epsilon}(s)\|^2$$
$$= \frac{1}{\epsilon}\sum_{k=1}^{\infty}\lambda_k^{2\theta}\left|\int_0^s (s - r)^{\alpha} e^{\lambda_k(s-r)}\langle f(u(r), \eta^{\epsilon}(r)) - \bar{f}(u(r)), e_k\rangle dr\right|^2.$$

Hence using Lemma 5.13 and choosing positive α and θ such that $0 < \alpha + \theta < 1/2$, we get

$$\mathbb{E}\|(-A)^{\theta} Y^{\epsilon}(s)\|^2 \leq C_T, \quad 0 \leq s \leq T,$$

for some constant $C_T > 0$. This yields

$$\mathbb{E} \sup_{0 \leq t \leq T} J_3 < C_T.$$

So, for $\theta > 0$

$$\mathbb{E}\|z^{\epsilon}\|_{C(0,T;H^{\theta/2})} \leq C_T, \tag{5.78}$$

for some constant $C_T > 0$. Next, we show a Hölder property in time t for $z^{\epsilon}(t)$ which, with estimate (5.78), implies the tightness of the distribution of $\{z^{\epsilon}\}$ in $C(0, T; H)$. In fact, for s, t with $0 \leq s < t \leq T$, we obtain

$$\|z^{\epsilon}(t) - z^{\epsilon}(s)\|^2$$
$$\leq \frac{2}{\sqrt{\epsilon}}\left\|\int_s^t e^{A(t-r)}[f(u^{\epsilon}(r), v^{\epsilon}(r)) - \bar{f}(u(r))]dr\right\|^2$$
$$+ \frac{2}{\sqrt{\epsilon}}\left\|(I - e^{A(t-s)})\int_0^s e^{A(s-r)}[f(u^{\epsilon}(r), v^{\epsilon}(r)) - \bar{f}(u(r))]dr\right\|^2.$$

By the same discussion as for the estimate of (5.77), and noticing the strong continuity of e^{At}, we conclude that for $0 < \gamma < 1$,

$$\mathbb{E}\|z^{\epsilon}(t) - z^{\epsilon}(s)\|^2 \leq C_T|t - s|^{\gamma}. \tag{5.79}$$

This implies the tightness of the distributions of z^{ϵ} in $C(0, T; H)$.

Now, decompose z^ϵ into z_1^ϵ and z_2^ϵ, which solve

$$\dot{z}_1^\epsilon = Az_1^\epsilon + \frac{1}{\sqrt{\epsilon}}[f(u, \eta^\epsilon) - \bar{f}(u)], \quad z_1(0)^\epsilon = 0, \tag{5.80}$$

and

$$\dot{z}_2^\epsilon = Az_2^\epsilon + \frac{1}{\sqrt{\epsilon}}[f(u^\epsilon, v^\epsilon) - f(u, \eta^\epsilon)], \quad z_2^\epsilon(0) = 0, \tag{5.81}$$

respectively. Similar to the discussion for z^ϵ above, we see that the probability distribution measures of z_1^ϵ and z_2^ϵ are both tight in $C(0, T; H)$.

Now we consider z_1^ϵ and z_2^ϵ separately. Denote by ν_1^ϵ the probability distribution measure (or law) of z_1^ϵ induced on $C(0, T; H)$. By the tightness of ν_1^ϵ, there is a sequence $\epsilon_n \to 0$, $n \to \infty$, such that $z_1^{\epsilon_n}$ converges in distribution to some z_1 in $C(0, T; H)$. Denote by P^0 the distribution of z_1 in $C(0, T; H)$. In fact, P^0 is the weak limit of $\nu_1^{\epsilon_n}$. Next we show that P^0 uniquely solves a martingale problem.

For $\gamma > 0$, denote by $UC^\gamma(H, \mathbb{R})$ the space of functions from H to $\mathbb{R}^1$, which, with their Fréchet derivatives up to the order γ, are all uniformly continuous. For $h \in UC^\gamma(H, \mathbb{R})$, denote by h' and h'' the first- and second-order Fréchet derivatives, respectively. Then we have the following important lemma.

Lemma 5.41. *Assume that* (H_1), (H'_2), (H_3), *and* (H_4) *hold, and set* $\sigma_1 = 0$. *Then* P^0 *solves the following martingale problem on* $C(0, T; H) : P^0\{z_1(0) = 0\} = 1$, *and*

$$h(z_1(t)) - h(z_1(0)) - \int_0^t \langle h'(z_1(\tau)), Az_1(\tau)\rangle d\tau - \frac{1}{2}\int_0^t \text{Tr}\big[h''(z_1(\tau))B(u)\big]d\tau$$

is a P^0 *martingale for every* $h \in UC^2(H, \mathbb{R})$. *Here*

$$B(u) := 2\int_0^\infty \mathbb{E}\big[(f(u, \eta(t)) - \bar{f}(u)) \otimes (f(u, \eta(0)) - \bar{f}(u))\big]dt,$$

and $h' = h_z$, $h'' = h_{zz}$ *are Fréchet derivatives as defined in* § *3.2.*

Proof. We adapt the approach used in the proof of Theorem 5.11 in order to prove this lemma.

For $0 < s \leq t < \infty$ and $h \in UC^\infty(H)$, we compute, via integration by parts,

$$h(z_1^\epsilon(t)) - h(z_1^\epsilon(s)) = \int_s^t \left\langle h'(z_1^\epsilon(\tau)), \frac{dz_1^\epsilon}{d\tau}\right\rangle d\tau$$

$$= \int_s^t \langle h'(z_1^\epsilon(\tau)), Az_1^\epsilon(\tau)\rangle d\tau + \frac{1}{\sqrt{\epsilon}}\int_s^t \langle h'(z_1^\epsilon(s)), f(u(s), \eta^\epsilon(\tau)) - \bar{f}(u(s))\rangle d\tau$$

$$+ \frac{1}{\sqrt{\epsilon}}\int_s^t \int_s^\tau h''(z_1^\epsilon(\delta))\Big(f(u(\delta), \eta^\epsilon(\tau)) - \bar{f}(u(\delta)), Az_1^\epsilon(\delta)\Big)d\delta\, d\tau$$

$$+ \frac{1}{\epsilon} \int_s^t \int_s^\tau h''(z_1^\epsilon(\delta))\Big(f(u(\delta), \eta^\epsilon(\tau)) - \bar{f}(u(\delta)), f(u(\delta), \eta^\epsilon(\delta))$$

$$- \bar{f}(u(\delta))\Big)d\delta\, d\tau$$

$$+ \frac{1}{\sqrt{\epsilon}} \int_s^t \int_s^\tau \langle h'(z_1^\epsilon(\delta)), (f_u'(u(\delta), \eta^\epsilon(\tau)) - \bar{f}'(u(\delta)))u'(\delta)\rangle d\delta\, d\tau$$

$$= \int_s^t \langle h'(z_1^\epsilon(\tau)), Az_1^\epsilon(\tau)\rangle d\tau + L_1 + L_2 + L_3 + L_4.$$

Let $\{e_i\}_{i=1}^\infty$ be an orthonormal basis of H. Then

$$h''(z_1^\epsilon(\delta))\Big((f(u(\delta), \eta^\epsilon(\tau)) - \bar{f}(u(\delta))), f(u(\delta), \eta^\epsilon(\delta)) - \bar{f}(u(\delta))\Big)$$

$$= \sum_{i,j=1}^\infty \partial_{ij} h(z_1^\epsilon(\delta))\langle (f(u(\delta), \eta^\epsilon(\tau)) - \bar{f}(u(\delta)))$$

$$\otimes (f(u(\delta), \eta^\epsilon(\delta)) - \bar{f}(u(\delta))), e_i \otimes e_j\rangle,$$

where $\partial_{ij} = \partial_{e_i} \partial_{e_j}$, and ∂_{e_i} is the directional derivative operator in the direction e_i.
Define

$$A_{ij}^\epsilon(\delta, \tau, u) := \langle (f(u, \eta^\epsilon(\tau)) - \bar{f}(u)) \otimes (f(u, \eta^\epsilon(\delta)) - \bar{f}(u)), e_i \otimes e_j\rangle.$$

Then:

$$L_3 = \frac{1}{\epsilon} \sum_{ij} \int_s^t \int_s^\tau \partial_{ij} h(z_1^\epsilon(\delta)) A_{ij}^\epsilon(\delta, \tau, u(\delta)) d\delta\, d\tau$$

$$= \frac{1}{\epsilon} \sum_{ij} \int_s^t \int_s^\tau \int_s^\delta \langle \partial_{ij} h'(z_1^\epsilon(\lambda)),$$

$$\Big[Az_1^\epsilon(\lambda) + \frac{1}{\sqrt{\epsilon}}(f(u(\lambda), \eta^\epsilon(\lambda)) - \bar{f}(u(\lambda)))\Big]\rangle \tilde{A}_{ij}^\epsilon(\delta, \tau, u(\lambda))d\lambda\, d\delta\, d\tau$$

$$+ \frac{1}{\epsilon} \sum_{ij} \int_s^t \int_s^\tau \int_s^\delta \partial_{ij} h(z_1^\epsilon(\lambda)) \partial_u \tilde{A}_{ij}^\epsilon(\delta, \tau, u(\lambda)) u'(\lambda) d\lambda\, d\delta\, d\tau$$

$$+ \frac{1}{\epsilon} \sum_{ij} \int_s^t \int_s^\tau \partial_{ij} h(z_1^\epsilon(s)) \tilde{A}_{ij}^\epsilon(\delta, \tau, u(s)) d\delta\, d\tau$$

$$+ \frac{1}{\epsilon} \sum_{ij} \int_s^t \int_s^\tau \partial_{ij} h(z_1^\epsilon(\delta)) \mathbb{E}[A_{ij}^\epsilon(\delta, \tau, u(\delta))] d\delta\, d\tau$$

$$\triangleq L_{31} + L_{32} + L_{33} + L_{34},$$

where $\tilde{A}_{ij}^\epsilon(\delta, \tau, u) \triangleq A_{ij}^\epsilon(\delta, \tau, u) - \mathbb{E}[A_{ij}^\epsilon(\delta, \tau, u)]$. For every bounded continuous function Φ on $C(0, s; H)$, define $\Phi(\cdot, \omega) \triangleq \Phi(z_1^\epsilon(\cdot, \omega))$. Then, by the estimates on u

(Lemma 5.29 with $\sigma_1 = 0$), Lemma 5.14 and the same discussion as in the proof of Theorem 5.11, we have

$$|\mathbb{E}[(L_{31} + L_{32} + L_{33})\Phi]| \to 0 \quad \text{as} \quad \epsilon \to 0.$$

Now we determine the limit of $\int_s^\tau \mathbb{E}A_{ij}^\epsilon(\delta, \tau, u)d\delta$ as $\epsilon \to 0$. Note that $\eta(t)$ is stationary and correlated. Set

$$b^{ij}(\tau - \delta, u) \triangleq \mathbb{E}[\langle (f(u, \eta(\tau)) - \bar{f}(u)) \otimes (f(u, \eta(\delta)) - \bar{f}(u)), e_i \otimes e_j \rangle].$$

Hence,

$$\mathbb{E}\left[A_{ij}^\epsilon(\delta, \tau, u)\right] = b^{ij}\left(\frac{\tau - \delta}{\epsilon}, u\right).$$

Due to the property of strong mixing with exponential rate, we pass the limit $\epsilon \to 0$ and get

$$\int_0^{(\tau - \delta)/\epsilon} b^{ij}(\lambda, u)\, d\lambda \to \int_0^\infty b^{ij}(\lambda, u)\, d\lambda \triangleq \frac{1}{2}B_{ij}(u), \quad \epsilon \to 0.$$

The following step is similar to the estimation for I_1 in the proof of Theorem 5.11. If $\epsilon_n \to 0$ as $n \to \infty$, then $\nu^{\epsilon_n} \to P^0$ and

$$\lim_{n\to\infty} \mathbb{E}[L_3 \Phi] = \frac{1}{2}\int_s^t \mathbb{E}^{P^0}\left(\text{Tr}\left[h''(z_1(\tau))B(u(\tau))\right]\Phi\right)d\tau,$$

where

$$B(u) \triangleq \sum_{i,j} B_{ij}(u)(e_i \otimes e_j).$$

Moreover, by the assumption on f and Lemma 5.29, $B(u) : H \to H$ is a Hilbert–Schmidt operator.

Similarly, by Lemma 5.13 and the same discussion for the estimate of L_3,

$$\mathbb{E}[L_1 \Phi + L_2 \Phi] \to 0 \text{ as } \epsilon \to 0.$$

By the tightness of z^ϵ in $C(0, T; H)$, the sequence $z_1^{\epsilon_n}$ has a limit process, denoted by z_1, in the weak sense. Therefore,

$$\lim_{n\to\infty} \mathbb{E}\left[\int_s^t \langle h'(z_1^{\epsilon_n}(\tau)), Az_1^{\epsilon_n}(\tau)\rangle \Phi\, d\tau\right] = \mathbb{E}\left[\int_s^t \langle h'(z_1(\tau)), Az_1(\tau)\rangle \Phi\, d\tau\right],$$

and

$$\lim_{n\to\infty} \mathbb{E}\left[(h(z_1^{\epsilon_n}(t)) - h(z_1^{\epsilon_n}(s)))\Phi\right] = \mathbb{E}\left[(h(z_1(t)) - h(z_1(s)))\Phi\right].$$

Finally, we have

$$\mathbb{E}^{P^0}\big[\big(h(z_1)(t) - h(z_1(s))\big)\Phi\big]$$
$$= \mathbb{E}^{P^0}\bigg[\int_s^t \langle h'(z_1(\tau)), Az_1(\tau)\rangle \Phi \, d\tau\bigg]$$
$$+ \frac{1}{2}\mathbb{E}^{P^0}\bigg\{\int_s^t \mathrm{Tr}[h''(z_1(\tau))B(u(\tau))]\Phi \, d\tau\bigg\}. \tag{5.82}$$

By an approximation argument, we know that (5.82) actually holds for all $h \in UC^2(H)$. This completes the proof of this lemma. $\qquad\square$

Back to the proof of Theorem 5.34. By the uniqueness of solution and Theorem 4.26, the limit of v_1^ϵ, denoted by P^0 as above is unique and solves the martingale problem related to the following stochastic partial differential equation:

$$dz_1 = Az_1 \, dt + d\widetilde{W}, \tag{5.83}$$

where $\widetilde{W}(t)$ is an H-valued Wiener process, defined on a new probability space $(\bar{\Omega}, \bar{\mathcal{F}}, \bar{\mathbb{P}})$, with covariance operator $B(u)$, such that z_1^ϵ converges in probability $\bar{\mathbb{P}}$ to z_1 in $C(0, T; H)$.

Moreover, recall that the distribution (or law) of z_2^ϵ on $C(0, T; H)$ is also tight. Let z_2 be one weak limit point of z_2^ϵ in $C(0, T; H)$. Then z_2 solves the following equation:

$$\dot{z}_2 = Az_2 + \overline{f_u'}(u)z, \quad z_2(0) = 0. \tag{5.84}$$

By the well-posedness of the above problem, z^ϵ uniquely converges in distribution to z, which solves

$$dz = [Az + \bar{f}_u'(u)z]dt + d\widetilde{W}. \tag{5.85}$$

This finishes the proof of the normal deviation result, Theorem 5.34. $\qquad\square$

Remark 5.42. By the martingale representation theorem, the approximation of deviation is in the sense of distribution rather than probability.

In the case when the fast part has nonlinear coupling with the slow part, we assume that $g = g(u, v) : H \times H \to H$ is Lipschitz continuous in (u, v). Then the stationary solution $\eta_u(t)$ of the fast system $\eta(t)$ depends on u. Now, in the proof of the deviation result, $\overline{A}_{ij}^\epsilon(\delta, \tau)$ should be defined as

$$\overline{A}_{ij}^\epsilon(\delta, \tau, u) \triangleq \big\langle (f(u, \eta_u^\epsilon(\tau)) - \bar{f}(u)) \otimes (f(u, \eta_u^\epsilon(\delta)) - \bar{f}(u)), e_i \otimes e_j\big\rangle.$$

The above proof for the deviation result still holds, since $u(t)$ is deterministic. Thus, we also have the deviation result under assumptions (H$_1$)–(H$_4$).

Remark 5.43. By applying a direct verification of the Guassian property, Cerrai [67] has also obtained the limit of z^ϵ. Moreover, the covariance operator of the driven process for z is given in the following weak sense:

$$\Phi(u)(h, k) \triangleq \int_0^\infty \bigg\{\int_H \big[\langle f(u, \cdot), h\rangle\langle P_t f(u, \cdot), k\rangle$$
$$+ \langle f(u, \cdot), k\rangle\langle P_t f(u, \cdot), h\rangle\big] d\mu - 2\langle \bar{f}(u), h\rangle\langle \bar{f}(u), k\rangle\bigg\} dt,$$

for $h, k \in H$. Here, P_t is the semigroup defined by (5.52) and it is independent of u. Notice that μ is the unique stationary measure of P_t. By direct calculation, we can verify that

$$\Phi(u)(h, k) = B(u)(h, k), \quad \text{for} \quad h, k \in H.$$

5.6   Macroscopic Reduction for Stochastic Systems

We apply the above averaging and normal deviation principles to a stochastic partial differential equation with non-Lipschitz nonlinearity and derive a macroscopic reduction model. This result is applied to macroscopically reduce stochastic reaction-diffusion equations by artificially separating the system into two distinct slow and fast time parts [303].

Consider the following stochastic partial differential equation on the interval $(0, \pi)$:

$$\partial_t w^\epsilon = \partial_{xx} w^\epsilon + (1 + \epsilon\gamma)w^\epsilon - (w^\epsilon)^3 + \sigma\sqrt{\epsilon}\partial_t W \tag{5.86}$$

$$w^\epsilon(0, t) = w^\epsilon(\pi, t) = 0 \tag{5.87}$$

with parameter $\gamma \in \mathbb{R}$. We assume that only the second spatial mode is forced by white noise, that is, $W(x, t) = w_2(t)\sin 2x$, where $w_2(t)$ is a standard scalar Wiener process. Let $H = L^2(0, \pi)$ and $A = \partial_{xx}$ with the zero Dirichlet boundary conditions. Then A has eigenvalues $\lambda_k = -k^2$ with corresponding eigenfunctions $e_k(x) = \sin(kx)$, $k = 1, 2, \ldots$

Notice that the system is dissipative. Consider the stochastic system (5.86) on long time scales of order ϵ^{-1}. Decompose the field $w^\epsilon(t) = \sqrt{\epsilon}u^\epsilon(t') + \sqrt{\epsilon}v^\epsilon(t')$ in the slow time $t' = \epsilon t$. Then we have the following slow-fast system:

$$\partial_{t'} u^\epsilon = \gamma u^\epsilon - \mathcal{P}_1(w^\epsilon)^3, \tag{5.88}$$

$$\partial_{t'} v^\epsilon = \frac{1}{\epsilon}\left[\partial_{xx} + 1 + \epsilon\gamma\right]v^\epsilon - \mathcal{Q}_1(w^\epsilon)^3 + \frac{\sigma}{\sqrt{\epsilon}}\mathcal{Q}_1\partial_{t'} W', \tag{5.89}$$

where $W'(t')$ is the scaled Wiener process $\sqrt{\epsilon}W(\epsilon^{-1}t')$, which has the same distribution as $W(t)$; $\mathcal{P}_1$ is the projection from $H = L^2(0, \pi)$ to the subspace $H_1 = \{a\sin x : a \in \mathbb{R}\}$; and $\mathcal{Q}_1 = Id_H - \mathcal{P}_1$. We denote $H_1^\perp = \mathcal{Q}_1 H$.

To apply Theorem 5.20 and Theorem 5.34 to the above slow-fast system whose nonlinearity is non-Lipschitz, we need some estimates for the solutions.

First, we still denote $w^\epsilon = (u^\epsilon, v^\epsilon)$. Thus,

$$\partial_{t'} w^\epsilon = A_\epsilon w^\epsilon + f(w^\epsilon) + \Sigma_\epsilon \partial_{t'} W', \quad \text{on} \quad 0 < x < \pi, \tag{5.90}$$

where $A_\epsilon = \text{diag}(\gamma, \epsilon^{-1}(\partial_{xx} + 1 + \epsilon\gamma))$ with zero Dirichlet boundary conditions, $f(w) = -w^3$, and $\Sigma_\epsilon = \text{diag}(0, \sigma Q/\sqrt{\epsilon})$ with $Q = \sin(2x)$. We also introduce the following linear system

$$\partial_{t'} z^\epsilon = A_\epsilon z^\epsilon + \Sigma_\epsilon \partial_{t'} W, \quad z^\epsilon(0) = 0.$$

By the assumption on W,

$$z^\epsilon(t') = \frac{\sigma}{\sqrt{\epsilon}} \int_0^{t'} e^{-(3-\epsilon\gamma)(t'-s)/\epsilon} dw_2(s) \sin(2x).$$

Then we have the following result.

Lemma 5.44. *For every $T > 0$ and $q \geq 2$,*

$$\sup_{0 \leq t' \leq T} \mathbb{E}\|z^\epsilon(t')\|^q + \mathbb{E}\int_0^T |z_{xx}^\epsilon(t')|_{L^q}^q \, dt' \leq C_q T. \tag{5.91}$$

Next, we present an estimate on w^ϵ. Define $\tilde{w}^\epsilon \triangleq w^\epsilon - z^\epsilon$. By (5.90),

$$\partial_{t'}\tilde{w}^\epsilon = A_\epsilon \tilde{w}^\epsilon + f(w^\epsilon), \tag{5.92}$$

which is equivalent to

$$\partial_{t'}\tilde{u}^\epsilon = \gamma\tilde{u}^\epsilon + \mathcal{P}_1 f(w^\epsilon),$$

$$\partial_{t'}\tilde{v}^\epsilon = \epsilon^{-1}[\partial_{xx}\tilde{v}^\epsilon + \tilde{v}^\epsilon] + \gamma\tilde{v}^\epsilon + \mathcal{Q}_1 f(w^\epsilon),$$

with zero Dirichlet boundary conditions on $(0, \pi)$ and $\tilde{w}^\epsilon = (\tilde{u}^\epsilon, \tilde{v}^\epsilon)$.
We estimate

$$\frac{1}{2}\frac{d}{dt}\|\tilde{w}^\epsilon\|^2 = -\gamma\|\tilde{u}^\epsilon\|_1^2 - \epsilon^{-1}(\|\tilde{v}^\epsilon\|_1^2 + \|\tilde{v}^\epsilon\|^2) + \gamma\|v^\epsilon\|^2 + \langle f^\epsilon(w^\epsilon), w^\epsilon\rangle$$

$$- \langle f^\epsilon(w^\epsilon), z^\epsilon\rangle$$

$$\leq -\gamma\|\tilde{w}^\epsilon\|_1^2 - |w^\epsilon|_{L^4(I)}^4 + |w^\epsilon|_{L^4(I)}^3 |z^\epsilon|_{L^4(I)}$$

$$\leq -\|\tilde{w}^\epsilon\|^2 - c_1|w^\epsilon|_{L^4(I)}^4 + c_2|z^\epsilon|_{L^4(I)}^4,$$

for some constants $c_1, c_2 > 0$. Integrating with respect to time yields

$$\sup_{0 \leq t' \leq T} \|\tilde{w}^\epsilon(t)\|^2 + 2\int_0^T \|\tilde{w}^\epsilon(s)\|^2 ds + 2c_1\int_0^T |w^\epsilon(s)|_{L^4}^4 ds$$

$$\leq \|w_0\|^2 + 2c_2\int_0^T |z^\epsilon(s)|_{L^4}^4 ds. \tag{5.93}$$

Note also that there are constants $c_3, c_4 > 0$ such that for small $\epsilon > 0$

$$\frac{1}{2}\frac{d}{dt'}\|\tilde{w}^\epsilon\|_1^2 = -\langle A_\epsilon\tilde{w}^\epsilon, -\tilde{w}_{xx}^\epsilon\rangle - \langle f(w^\epsilon), w_{xx}^\epsilon + z_{xx}^\epsilon\rangle$$

$$\leq -\|\tilde{w}_{xx}^\epsilon\|^2 + |w^\epsilon|_{L^3}^3 |z_{xx}^\epsilon|_{L^4}$$

$$\leq -\|\tilde{w}^\epsilon\|_1^2 + c_3|w^\epsilon|_{L^4}^4 + c_4|z_{xx}^\epsilon|_{L^4}^4.$$

Again, integrating with respect to time yields

$$\sup_{0 \leq s \leq t'} \|\tilde{w}^\epsilon(s)\|_1^2 \leq \|w_0\|^2 - \int_0^{t'} \sup_{0 \leq \tau \leq s} \|\tilde{w}^\epsilon(\tau)\|^2 ds + 2c_3\int_0^T |w^\epsilon(s)|_{L^4}^4 \, ds$$

$$+ 2c_4\int_0^T |z_{xx}^\epsilon(s)|_{L^4}^4 ds.$$

Using the Gronwall inequality and noticing (5.93), we obtain, for some constant $C > 0$,

$$\sup_{0 \le t' \le T} \|\tilde{w}^\epsilon(t')\|_1^2 \le C \left(1 + \|w_0\|^2 + \int_0^T |z_{xx}^\epsilon(s)|_{L^4}^4 \, ds \right). \tag{5.94}$$

Notice that for a fixed slow mode, the fast mode v^ϵ has a unique stationary solution whose distribution is difficult to write out. However, for small $\epsilon > 0$, the fast mode can be approximated by a simple Ornstein–Uhlenbeck process η, which solves the following linear equation for small $\epsilon > 0$,

$$\partial_{t'} \eta^\epsilon = \frac{1}{\epsilon} \left[\partial_{xx} + 1 \right] \eta^\epsilon + \frac{\sigma}{\sqrt{\epsilon}} \mathcal{Q}_1 \partial_{t'} W, \quad \eta^\epsilon(0) = v^\epsilon(0). \tag{5.95}$$

In fact, let $V^\epsilon \triangleq v^\epsilon - \eta^\epsilon$. We see that

$$\partial_{t'} V^\epsilon = \frac{1}{\epsilon} (\partial_{xx} + 1) V^\epsilon + \gamma v^\epsilon + \mathcal{Q}_1 f(u^\epsilon + v^\epsilon), \quad V^\epsilon(0) = 0.$$

Using the estimates on w^ϵ, we conclude that for every $T > 0$, there exists a constant $C_T > 0$ such that

$$\mathbb{E} \sup_{0 \le t' \le T} \|v^\epsilon(t') - \eta^\epsilon(t')\| \le \epsilon C_T (1 + \|w_0\|_1^4). \tag{5.96}$$

By the above simple approximation for high modes, we now approximate the slow mode u^ϵ by an averaged equation. The above linear Equation (5.95) has a unique stationary solution $\bar{\eta}^\epsilon$ with distribution μ that is independent of ϵ. Decomposing $\bar{\eta}^\epsilon = \sum_i \bar{\eta}_i e_i, \bar{\eta}_i = 0$ for $i \ne 2$, the scalar stationary process $\bar{\eta}_2$ satisfies the following stochastic ordinary differential equation:

$$d\eta_2 = -\frac{3}{\epsilon} \eta_2 dt' + \sigma \sqrt{\frac{1}{\epsilon}} dw_2',$$

where $w_2'(t')$ is the scaled Wiener process $\sqrt{\epsilon} w_2(\epsilon^{-1} t)$, which has the same distribution as $w_2(t)$. Note that the distribution of $\bar{\eta}_2$ is the one-dimensional normal distribution $\mathcal{N}\left(0, \frac{1}{6}\sigma^2\right)$. Define $\overline{\mathcal{P}_1 f}(u) \triangleq \int_{H_1^\perp} \mathcal{P}_1 f(u, v) \mu(dv)$, and introduce the following averaged equation for u^ϵ:

$$\partial_{t'} u = \gamma u + \overline{\mathcal{P}_1 f}(u) = \gamma u - \mathcal{P}_1 u^3 - \frac{3\sigma^2}{6} \mathcal{P}_1 (u \sin^2 2x). \tag{5.97}$$

In order to apply Theorem 5.20 to u^ϵ, first we need a similar result like (5.51), which follows from the discussion in Remark 5.28. Furthermore, notice that

$$\|u_1^3 - u_2^3\| \le 2[\|u_1\|_1^2 + \|u_2\|_1^2] \|u_1 - u_2\|.$$

This means $u^3 : H_0^1 \to H$ is Lipschitz on bounded sets of H_0^1. By the estimate (5.94), the discussion of the proof of Theorem 5.20 is applicable to w^ϵ. Hence, we arrive at the following theorem.

Theorem 5.45. *For every $T > 0$, u^ϵ for the slow-fast system (5.88)–(5.89) converges in probability to u for the averaged system (5.97), in $C(0, T; H_1)$. Moreover, the rate of convergence is $\frac{1}{2}$, that is, for every $\kappa > 0$,*

$$\mathbb{P}\left\{ \sup_{0 \leq t' \leq T} \|u^\epsilon(t') - u(t')\| \leq C_T^\kappa \sqrt{\epsilon} \right\} > 1 - \kappa,$$

for some constant $C_T^\kappa > 0$.

By the averaged Equation (5.97), we see a stabilization effect of the noise. In fact, if we write $u = A(t') \sin x$, then the amplitude A satisfies the Landau equation

$$\frac{dA}{dt'} = \left(\gamma - \frac{\sigma^2}{4}\right)A - \frac{3}{4}A^3. \tag{5.98}$$

Therefore, if $\sigma > 2\sqrt{\lambda}$, the system is stabilized.

Recently Blömker [43,44] also considered the amplitude equation for SPDEs by asymptotic methods. The same amplitude Equation (5.98) is obtained. Moreover, a deviation estimate provides a higher-order approximation in the sense of distribution.

Still, by the estimates (5.94) and the Lipschitz property of $u^3 : H_0^1 \to H$ on bounded sets of H_0^1, a proof similar to that of Lemma 5.41 is also applicable to w^ϵ. Now we calculate the deviation. Noticing that $\eta = \eta_2 \sin 2x$ and by the deviation result, we have

$$
\begin{aligned}
B(u) &= 2\mathbb{E} \int_0^\infty \left[\mathcal{P}_1(u + \eta(s))^3 - \overline{\mathcal{P}_1(u + \eta)^3} \right] \\
&\quad \otimes \left[\mathcal{P}_1(u + \eta(0))^3 - \overline{\mathcal{P}_1(u + \eta)^3} \right] ds \\
&= 18A^2 \int_0^\infty \mathbb{E}\left[(\eta_2^2(s) - \mathbb{E}\eta_2^2)(\eta_2^2(0) - \mathbb{E}\eta_2^2) \right] \langle e_1^2, e_2^2 \rangle^2 ds \, e_1(x) \otimes e_1(x) \\
&= \frac{\sigma^4}{24} A^2 e_1(x) \otimes e_1(x).
\end{aligned}
$$

Then, writing $\lim_{\epsilon \to 0} (u^\epsilon - u)/\sqrt{\epsilon} \triangleq \rho_1 \sin x$ and noticing that $3\mathcal{P}_1[(\mathbb{E}\eta^2)\rho_1] = \sigma^2 \rho_1/4$, the deviation ρ_1 solves the Ornstein–Uhlenbeck-like SDE

$$d\rho_1 = \left(\gamma - \frac{\sigma^2}{4} - \frac{9}{4}A^2\right)\rho_1 \, dt' + \frac{\sigma^2}{2\sqrt{6}} A \, dw, \quad \rho_1(0) = 0, \tag{5.99}$$

where w is a standard scalar Brownian motion.

In order to derive a macroscopic reduced model, we formally add the deviation to the averaged equation, which yields, on time scale t',

$$d\bar{A}^\epsilon = \left[\left(\gamma - \frac{\sigma^2}{4}\right)\bar{A}^\epsilon - \frac{3}{4}(\bar{A}^\epsilon)^3\right]dt' + \frac{\sqrt{\epsilon}\sigma^2}{2\sqrt{6}} \bar{A}^\epsilon dw(t'). \tag{5.100}$$

Here, we omit the higher-order terms. The above equation is not rigorously verified, but we adapt the construction of stochastic slow manifolds for dissipative SPDEs [266] to SPDE (5.86). Recall that we chose $W = w_2(t) \sin 2x$, and $\lambda_2 = 1$. In terms of the amplitude $a(t)$ of the fundamental mode $\sin x$, computer algebra readily derives that the stochastic slow manifold of the SPDE (5.86) is

$$
\begin{aligned}
w = a \sin x &+ \frac{1}{32} a^3 \sin 3x + \sqrt{\epsilon}\sigma \sin 2x \, e^{-3t} \star \dot{w}_2 \\
&+ \epsilon^{3/2} \gamma \sigma \sin 2x \, e^{-3t} \star e^{-3t} \star \dot{w}_2 + \cdots
\end{aligned}
\tag{5.101}
$$

The history convolution of the noise, $e^{-3t} \star \dot{w}_2 = \int_{-\infty}^{t} e^{-3(t-s)} dw_2(s)$, that appears in the shape of this stochastic slow manifold helps us eliminate such history integrals in the evolution *except* in the nonlinear interactions between noises. The amplitude equation is

$$
\dot{a} = \epsilon \gamma a - \frac{3}{4} a^3 - \frac{1}{2} \epsilon \sigma^2 a \big(\dot{w}_2 e^{-3t} \star \dot{w}_2 \big) + \cdots
$$

Analogous to the averaging and deviation theorems, an analysis of Fokker–Planck equations [266,310] then asserts that the canonical quadratic noise interaction term in this equation should be replaced by the sum of a mean drift and an effectively new independent noise process. Thus, the evolution on this stochastic slow manifold is

$$
da \approx \left[\epsilon \left(\gamma - \frac{1}{4}\sigma^2 \right) a - \frac{3}{4} a^3 \right] dt + \frac{1}{2\sqrt{6}} \epsilon \sigma^2 a \, d\tilde{w} + \cdots
\tag{5.102}
$$

where $\tilde{w}$ is a real valued standard Brownian motion. The stochastic model (5.102) is exactly consistent with the macroscopic model (5.100) on time scale t' with $a(t) \sin x = \sqrt{\epsilon} \bar{A}^{\epsilon}(\epsilon t)$.

Remark 5.46.   The above stochastic slow manifold discussion implies the effectiveness of Equation (5.73).

5.7   Large Deviation Principles for the Averaging Approximation

The averaged equation does not capture all qualitative properties of u^{ϵ}. Formally we can write out Equation (5.73), but rigorous verification is difficult. For Equation (5.73), $\bar{u}^{\epsilon}$ is perturbed by a small Gaussian white noise. A large deviation principle (LDP) may hold under some conditions, yielding occurrence of a small probability event such as the exit of $\bar{u}^{\epsilon}$ from a fixed neighborhood of a stable steady state of the averaged equation. This probability of exit is controlled by a rate function. So, if u^{ϵ} also has the LDP with the same rate function as $\bar{u}^{\epsilon}$, this will show the effectiveness of "the averaged equation plus (or corrected by) the deviation."

Such large deviations (LD) from the averaged equation were initiated by Freidlin and Wentzell [134, Ch. 7] for SDEs. In this section we present the LDP result for slow-fast SPDEs and show that the rate function is exactly that of (5.73), which further shows the effectiveness of "the averaged equation plus deviation," that is, Equation (5.73).

5.7.1   Large Deviations for Slow-Fast SPDEs

We consider the SPDEs (5.33)–(5.34) with $\sigma_1 = 0$. Additionally, the following assumption is made:

(H5) There are constants $c_0, c_1 > 0$ such that

$$\langle B(\varphi)h, h \rangle \geq c_0 \|h\|^2 \quad \text{and} \quad \langle DB(\varphi)h, h \rangle < c_1 \|h\|^2, \quad \text{for all} \quad \varphi, h \in H,$$

where B is defined in (5.71) and DB is the Fréchet derivative of B.

For the slow-fast stochastic system (5.33)–(5.34), we define the following skeleton equation:

$$\dot{\varphi} = A\varphi + \bar{f}(\varphi) + \sqrt{B(\varphi)}h, \quad \varphi(0) = u_0, \tag{5.103}$$

where $\sqrt{B(\varphi)}$ is the square root of positive operator $B(u)$, which is well defined by the decomposition theorem [262, Ch. VIII.9]. We also define the rate function

$$I_u(\varphi) \triangleq \inf_{h \in L^2(0,T;H)} \left\{ \frac{1}{2} \int_0^T \|h(s)\|^2 ds : \varphi = \varphi^h \right\}, \tag{5.104}$$

where φ^h solves (5.103) and $\inf \emptyset = +\infty$. We have the following result.

Theorem 5.47. *Assume that (H1)–(H5) hold and $\sigma_1 = 0$. Then, for every $T > 0$, $\{u^\epsilon\}$ satisfies the LDP with a good rate function I_u in $C(0, T; H)$.*

Remark 5.48. By an LDP result for SPDEs ([94, Theorem 12.15] or Section 4.8), the function $I_u(\varphi)$ is indeed the rate function of the following system:

$$d\tilde{u}^\epsilon = \left[A\tilde{u}^\epsilon + \bar{f}(\tilde{u}^\epsilon) \right]dt + \sqrt{\epsilon}\sqrt{B(\tilde{u}^\epsilon)}\,d\overline{W}(t), \tag{5.105}$$

$$\tilde{u}^\epsilon(0) = u_0, \tag{5.106}$$

which is exactly the averaged equation plus the deviation. This fact confirms the effectiveness of the averaged equation plus the fluctuating deviation to slow-fast SPDEs.

The theorem can be proved by adapting the approach for slow-fast stochastic *ordinary* differential equations [134, Ch. 7.4 and Ch. 7.5] to slow-fast SPDEs, although some different estimates are needed [304]. We omit the detailed proof [304] but present the following example.

Example 5.49. We consider the following slow-fast SPDEs on the interval $(0, l)$ with the zero Dirichlet boundary conditions:

$$\partial_t u^\epsilon = \partial_{xx} u^\epsilon + \lambda \sin u^\epsilon - v^\epsilon, \quad u^\epsilon(0) = u_0, \tag{5.107}$$

$$\epsilon \partial_t v^\epsilon = \partial_{xx} v^\epsilon - v^\epsilon + u^\epsilon + \sqrt{\epsilon}\sigma \partial_t W(t), \quad v^\epsilon(0) = v_0, \tag{5.108}$$

where $W(t)$ is a $L^2(0, l)$-valued Q-Wiener process with $\text{Tr}(Q) < \infty$, and λ, σ are positive constants. As usual, the small parameter ϵ measures the separation of time scales between the fast mode v and the slow mode u.

Note that the nonlinear interaction function, $f(u, v) = \lambda \sin u - v$, is Lipschitz continuous. Introduce the operator $A = \partial_{xx}$ with the zero boundary conditions on $(0, l)$. Now, for a fixed u, the fast system (5.108) has a unique stationary solution $\eta^{\epsilon, u}$ with distribution

$$\mu_u = \mathcal{N}\left((I - A)^{-1}u, \sigma^2 \frac{(I - A)^{-1}Q}{2} \right).$$

Then:

$$\bar{f}(u) = \lambda \sin u - (I - A)^{-1}u.$$

Let η^u be the stationary solution of

$$\partial_t v = \partial_{xx} v - v + u + \sigma \partial_t W(t)$$

for fixed $u \in L^2(0, l)$. Hence, η^u and μ_u have the same distribution, and

$$B(u) = 2\mathbb{E} \int_0^\infty \left[\eta^u(t) - (I - A)^{-1}u \right] \otimes \left[\eta^u(0) - (I - A)^{-1}u \right] dt.$$

Noticing that

$$\mathbb{E}\eta^u(t) \otimes \eta^u(0) = \sigma^2 \exp\{-(I - A)t\} \left[\frac{(I - A)^{-1}Q}{2} \right] + (I - A)^{-2}u \otimes u,$$

we then have

$$\sqrt{B(u)} = (I - A)^{-1}\sigma\sqrt{Q},$$

which is independent of u and satisfies assumption ($\mathbf{H_5}$). By Theorem 5.47, $\{u^\epsilon\}$ satisfies LDP on $C(0, T; H)$ with a good rate function

$$I_u(\varphi) \triangleq \inf_{h \in L^2(0,T;H)} \left\{ \frac{1}{2} \int_0^T \|h(s)\|^2 ds : \varphi = \varphi^h \right\},$$

where φ^h solves

$$\dot{\varphi} = A\varphi + \lambda \sin \varphi - (I - A)^{-1}\varphi + (I - A)^{-1}\sigma\sqrt{Q}h, \quad \varphi(0) = u_0.$$

Furthermore, the rate function is

$$I_u(\varphi) \triangleq \frac{1}{2} \int_0^T \left\| \frac{I - A}{\sqrt{Q}} \left[\dot{\varphi}(s) - A\varphi(s) - \lambda \sin \varphi + (I - A)^{-1}\varphi \right] \right\|^2 ds,$$

$$(5.109)$$

for absolutely continuous φ. Otherwise $I_u(\varphi) = \infty$. Here, the operator $(I - A)/\sqrt{Q}$ is well defined as $\text{Tr}(Q) < \infty$.

Now we write out the averaged equation plus the deviation for (5.107)–(5.108) as

$$d\bar{u}^\epsilon = [A\bar{u}^\epsilon + \lambda \sin \bar{u}^\epsilon - (I - A)^{-1}\bar{u}^\epsilon]dt + \sqrt{\epsilon}\sigma(I - A)^{-1}\sqrt{Q}d\overline{W}(t). \quad (5.110)$$

This is an effective model for the slow-fast SPDE system (5.107)–(5.108).

By the LDP for stochastic evolutionary equation [94, Ch. 12], $\{\bar{u}^\epsilon\}$ satisfies LDP with rate function $I_u(\varphi)$ defined by (5.109). For $\epsilon = 0$ and a large enough parameter λ, the model (5.110) is a deterministic system with two stable states near zero. For small $\epsilon > 0$, the noise causes orbits of the stochastic system (5.110) near one stable state to the position near the other one [123], which shows the metastability of the system (5.110). The original slow-fast stochastic system (5.107)–(5.108) also has this metastability, which shows that the averaged equation plus deviation, (5.110), indeed describes the metastability of $\{u^\epsilon\}$ solving (5.107)–(5.108) for small $\epsilon > 0$. In other words, the effective model (5.110) is a good approximation of the original slow-fast stochastic system (5.107)–(5.108).

5.8 PDEs with Random Coefficients

We present an averaging principle for some PDEs with random oscillating coefficients (called *random PDEs*), which generalizes the result of Theorem 5.16. This is not in the scope of SPDEs but can be derived from a slow-fast SPDE in the form of (5.30) (5.31) with $\sigma_1 = 0$. This result will also be used in a random slow manifold reduction in Chapter 6.

We consider the following PDE with random oscillating coefficients on a bounded interval $(0, l)$:

$$u_t^\epsilon = u_{xx}^\epsilon + f(u^\epsilon, t/\epsilon, \omega), \quad u^\epsilon(0) = u_0 \in H = L^2(0, l), \quad (5.111)$$

with zero Dirichlet boundary conditions. Here we assume that

(H) For every t, $f(t, \cdot)$ is Lipschitz continuous in u with Lipschitz constant L_f and $f(t, 0) = 0$. For every $u \in L^2(0, l)$, $f(\cdot, u)$ is a $L^2(0, l)$-valued stationary random process and is strongly mixing with exponential decay rate $\gamma > 0$, i.e.,

$$\sup_{s \geq 0} \sup_{A \in \mathcal{F}_0^s, B \in \mathcal{F}_{s+t}^\infty} |\mathbb{P}(A \cap B) - \mathbb{P}(A)\mathbb{P}(B)| \leq e^{-\gamma t}, \quad t \geq 0.$$

Here, $0 \leq s \leq t \leq \infty$, and $\mathcal{F}_s^t = \sigma\{f(\tau, u) : s \leq \tau \leq t\}$ is the σ-field generated by $\{f(\tau, u) : s \leq \tau \leq t\}$.

We introduce the notation $\varphi(t)$ to quantify the mixing as

$$\varphi(t) \triangleq \sup_{s \geq 0} \sup_{A \in \mathcal{F}_0^s, B \in \mathcal{F}_{s+t}^\infty} |\mathbb{P}(A \cap B) - \mathbb{P}(A)\mathbb{P}(B)|.$$

By the above assumption, for every $\alpha > 0$,

$$\int_0^\infty \varphi^\alpha(t) \, dt < \infty.$$

Remark 5.50. One special case is when $f(t, u) = F(\eta(t), y)$, where $F(\cdot, \cdot)$ is Lipschitz continuous and $\eta(t)$ is a stationary random process that solves the following stochastic differential equation:

$$dv = v_{xx}dt + dW,$$

with W an $L^2(0, l)$-valued Q-Wiener process. This $f(t, u)$ is strongly mixing and satisfies assumption (**H**).

In fact, for small $\epsilon > 0$, the random PDE (5.111) with $f(t, u) = F(\eta(t), u)$ can be seen as an effective approximation equation for the following slow-fast SPDE system:

$$u_t = u_{xx} + f(u, v),$$
$$\epsilon dv = v_{xx}dt + \sqrt{\epsilon}dW,$$

by a random slow manifold reduction; see Chapter 6.

For the random oscillating PDE (5.111), we have an averaging principle as well. Introduce the following averaged equation:

$$u_t = u_{xx} + \bar{f}(u), \quad u(0) = u_0, \tag{5.112}$$

where $\bar{f}(u) = \mathbb{E}f(t, u, \omega)$, and define the deviation process

$$z^{\epsilon}(t) \triangleq \frac{1}{\sqrt{\epsilon}}(u^{\epsilon}(t) - u(t)), \tag{5.113}$$

then the following averaging principle is established.

Theorem 5.51. *Assume* (**H**) *and let positive T be given. Then, for every $u_0 \in H$, the solution $u^{\epsilon}(t, u_0)$ of (5.111) converges in mean to the solution u of (5.112) in $C(0, T; H)$. Moreover, the rate of convergence is $\sqrt{\epsilon}$, that is,*

$$\mathbb{E} \sup_{0 \leq t \leq T} \|u^{\epsilon}(t) - u(t)\| \leq C_T \sqrt{\epsilon} \tag{5.114}$$

for some constant $C_T > 0$.

Furthermore, the deviation process z^{ϵ} converges in distribution in $C(0, T; H)$ to z, which solves the following linear SPDE:

$$dz(t) = [z_{xx}(t) + \overline{f'(u(t))}z(t)]dt + d\widetilde{W}, \quad z(0) = z(l) = 0, \tag{5.115}$$

where $\widetilde{W}(t)$ is an H-valued Wiener process, defined on a new probability space $(\bar{\Omega}, \bar{\mathcal{F}}, \bar{\mathbb{P}})$ and having covariance operator

$$B(u) \triangleq 2 \int_0^{\infty} \mathbb{E}\left[(f(u, t) - \bar{f}(u)) \otimes (f(u, 0) - \bar{f}(u))\right] dt.$$

Proof. The proof is similar to the proof of Theorem 5.20 and 5.34 for the slow-fast SPDEs.

First, by the assumption of Lipschitz property on f in (**H**), standard energy estimates yield that for $T > 0$,

$$\sup_{0 \le t \le T} \|u^\epsilon(t)\|^2 \le C_T \tag{5.116}$$

and

$$\|u^\epsilon(t) - u^\epsilon(s)\| \le C_T |t - s| \tag{5.117}$$

with some constant $C_T > 0$. Furthermore, for $T \ge t \ge 0$ and $\varphi \in L^2(0, l)$,

$$\left(\int_0^t e^{A(t-s)} [f(u, t/\epsilon) - \bar{f}(u)] ds, \varphi \right)^2$$

$$= \int_0^t \int_0^t \langle e^{A(t-s)} [f(u, s/\epsilon) - \bar{f}(u)], \varphi \rangle \langle e^{A(t-\tau)} [f(u, \tau/\epsilon) - \bar{f}(u)], \varphi \rangle \, ds \, d\tau,$$

where A denotes the Laplacian on $(0, l)$ with zero Dirichlet boundary conditions. Notice that for $s \le t$

$$\|e^{A(t-s)} [f(u, s/\epsilon) - \bar{f}(u)]\| \le L_f \|u\|$$

and

$$\mathbb{E} e^{A(t-s)} [f(u, s/c) - \bar{f}(u)] = 0.$$

Then, by the strong mixing assumption on $f(t, u)$ in (**H**) and Lemma 5.13 we have for some constant $C_T > 0$

$$\mathbb{E} \left\| \int_0^t e^{A(t-s)} [f(u, s/\epsilon) - \bar{f}(u)] ds \right\| \le \sqrt{\epsilon} \frac{C_T (1 + \|u\|^2)}{\gamma}, \quad 0 \le t \le T. \tag{5.118}$$

Now we prove the averaging approximation. We still partition the interval $[0, T]$ into subintervals of length δ with $\delta > 0$ and construct process $\tilde{u}^\epsilon$ such that, for $t \in [k\delta, (k+1)\delta)$,

$$\tilde{u}^\epsilon(t) = e^{A(t-k\delta)} u^\epsilon(k\delta) + \int_{k\delta}^t e^{A(t-s)} f(u^\epsilon(k\delta), s/\epsilon) ds, \quad u^\epsilon(0) = u_0.$$

Then, by (5.117),

$$\sup_{0 \le t \le T} \|u^\epsilon(t) - \tilde{u}^\epsilon(t)\| \le C_T \delta, \tag{5.119}$$

for some constant $C_T > 0$. Moreover,

$$u(t) = e^{At} u_0 + \int_0^t e^{A(t-s)} \bar{f}(u(s)) ds.$$

Then, using $\lfloor z \rfloor$ to denote the largest integer less than or equal to z,

$$
\sup_{0 \leq s \leq t} \| \tilde{u}^\epsilon(s) - u(s) \|
$$

$$
\leq \left\| \int_0^t e^{A(t-s)} [f(u^\epsilon(\lfloor s/\delta \rfloor \delta), s/\epsilon) - \bar{f}(u^\epsilon(\lfloor s/\delta \rfloor \delta))] \, ds \right\|
$$

$$
+ \int_0^t e^{A(t-s)} \| \bar{f}(u^\epsilon(\lfloor s/\delta \rfloor \delta)) - \bar{f}(u^\epsilon(s)) \| \, ds
$$

$$
+ \int_0^t e^{A(t-s)} \| \bar{f}(u^\epsilon(s)) - \bar{f}(u(s)) \| \, ds.
$$

Now, from (5.116)–(5.118), we obtain

$$
\mathbb{E} \sup_{0 \leq s \leq t} \| \tilde{u}^\epsilon(s) - u(s) \| \leq C_T \left[\sqrt{\epsilon} + \int_0^t \mathbb{E} \sup_{0 \leq \tau \leq s} \| u^\epsilon(\tau) - u(\tau) \| \, ds \right], \quad (5.120)
$$

for some constant $C_T > 0$. Notice that $\| u^\epsilon(t) - u(t) \| \leq \| u^\epsilon(t) - \tilde{u}^\epsilon(t) \| + \| \tilde{u}^\epsilon(t) - u(t) \|$. By the Gronwall lemma and (5.119)–(5.120), we have

$$
\mathbb{E} \sup_{0 \leq t \leq T} \| u^\epsilon(t) - u(t) \| \leq C_T \sqrt{\epsilon}. \quad (5.121)
$$

We next consider the deviation. Recall the definition of z^ϵ:

$$
\dot{z}^\epsilon = z_{xx}^\epsilon + \frac{1}{\sqrt{\epsilon}} [f(u^\epsilon, t/\epsilon) - \bar{f}(u)], \quad z^\epsilon(0) = 0
$$

with zero Dirichlet boundary conditions. Then, by exactly the same discussion for (5.77), (5.78), and (5.79), we have the tightness of $\{z^\epsilon\}$ in $C(0, T; H)$.

Now, decompose $z^\epsilon = z_1^\epsilon + z_2^\epsilon$ with

$$
\dot{z}_1^\epsilon = A z_1^\epsilon + \frac{1}{\sqrt{\epsilon}} [f(u, t/\epsilon) - \bar{f}(u)], \quad z_1^\epsilon(0) = 0
$$

and

$$
\dot{z}_2^\epsilon = A z_2^\epsilon + \frac{1}{\sqrt{\epsilon}} [f(u^\epsilon, t/\epsilon) - \bar{f}(u, t/\epsilon)], \quad z_2^\epsilon(0) = 0.
$$

Notice that $f(u, t/\epsilon)$ has the same property in u and t as that of $f(u, \eta^\epsilon(t))$ in (5.80). Hence, Lemma 5.41 holds for z_1^ϵ here, that is, z_1^ϵ converges in distribution to z_1, which solves

$$
dz_1 = A z_1 + d\widetilde{W}, \quad z_1(0) = 0 \quad (5.122)
$$

where $\widetilde{W}$ is an H-valued Wiener process, defined on a new probability space $(\bar{\Omega}, \bar{\mathcal{F}}, \bar{\mathbb{P}})$ and having the covariance operator $B(u)$. Furthermore, z_2^ϵ converges in distribution to z_2, which solves

$$
\dot{z}_2 = A z_2 + \bar{f}'(u) z, \quad z_2(0) = 0.
$$

Therefore z^ϵ converges in distribution to z with z solving (5.115). The proof is complete. $\qquad \square$

Remark 5.52. A similar result also holds for the system

$$u_t^\epsilon = u_{xx}^\epsilon + g(u) + f(u^\epsilon, t/\epsilon, \omega), \quad u^\epsilon(0) = u_0 \in H = L^2(0, l), \qquad (5.123)$$

where $g(u)$ is a (local) Lipschitz nonlinearity.

Remark 5.53. The assumption on the strong mixing property in (H) can be weakened as

$$\int_0^\infty \varphi^\alpha(t)dt < \infty$$

for some $\alpha > 0$. Under the above assumption we also have Theorem 5.51. See [205,305] for more details.

5.9   Further Remarks

In this section, we discuss the applicability of SPDEs' averaging principles to stochastic climate modeling and singularly perturbed stochastic systems.

5.9.1   Stochastic Climate Modeling

In climate-weather simulations, some variabilities or scales are not explicitly resolved. The parameterization by stochastic processes of these unresolved variabilities was suggested by Hasselmann [155] in 1976. Now stochastic climate modeling has attracted a lot of attention (e.g., [20,111,172,221]). It also has demonstrated considerable success in the study of, for example, the dynamics of El Niño and the Southern Oscillation [296].

An idealized formulation of general circulation models (GCM), under Hasselmann's idea [155], involves at least two scales, typically a fast one (the weather scale) and a slow one (the climate scale). Mathematically, this idealized GCM system is written as

$$\dot{U}^\epsilon = F(U^\epsilon, V^\epsilon), \qquad (5.124)$$

$$\dot{V}^\epsilon = \frac{1}{\epsilon}G(U^\epsilon, V^\epsilon), \qquad (5.125)$$

for a small parameter $\epsilon > 0$ and two nonlinear vector fields F and G. For a small-scale parameter $\epsilon > 0$, averaging out the fast variables gives a passage to statistical dynamical models (SDM). That is, define $u \triangleq \int U \, d\mu$ where μ is an invariant measure of the system for fixed U, and $f(u) \triangleq \int F(u, V) d\mu$. Then the corresponding SDM is

$$\dot{u} = f(u).$$

Thus, Hasselmann's idea may be viewed as the postulation that there is a connection between GCMs and SDMs provided by stochastic climate models, which are SODEs or SPDEs, for the error in averaging [172], that is, the deviation of U^ϵ from the average u. In contrast to Khasminskii's deviation [181], which is described by a linear SODE, Hasselmann's deviation leads to a nonlinear SODE. Kifer [186] gave a rigorous verification of Hasselmann's deviation for some fully coupled slow-fast SODEs.

Majda et al. [221] suggested that in stochastic climate modeling, the fast variable (on weather scale), which is not resolved in detail, is represented by a stochastic model such as

$$dV^\epsilon = \frac{1}{\epsilon} g(U^\epsilon, V^\epsilon)\, dt + \frac{\sigma}{\sqrt{\epsilon}} dW(t).$$

The random fluctuation on the weather scale is proved to affect the climate variable as a stochastic force [221].

The above stochastic climate models are SODEs instead of SPDEs [20,172,221]. The averaging results for SPDEs in this chapter in fact give a method to derive a stochastic climate model for spatially dependent complex system described by SPDEs. Recent works mentioned above applied a finite dimensional truncation to study a stochastic climate model described by SPDEs, replacing the fast unresolved part by a fast stochastic oscillating process. The averaging results of this chapter show that the full system also has an averaged model. The deviation principles in Section 5.4 imply Hasselmann's deviation [301,302].

5.9.2   Singularly Perturbed Stochastic Systems

We consider the applicability of the results in this chapter to a singularly perturbed stochastic system. In particular, we discuss the following stochastically damped non-linear wave equation with singular perturbation and zero Dirichlet boundary conditions:

$$\epsilon u_{tt}^\epsilon + u_t^\epsilon = \Delta u^\epsilon + f(u^\epsilon) + \sqrt{\epsilon}\eta(t), \quad 0 < x < l,$$

where the random force $\eta(t)$ is the formal derivative of a scalar Wiener process $W(t)$. In fact, the above system has the following slow-fast formulation:

$$u_t^\epsilon = v^\epsilon, \quad u^\epsilon(0) = u_0 \in H_0^1(0, l) \cap H^2(0, l), \tag{5.126}$$

$$\epsilon v_t^\epsilon = -v^\epsilon + \Delta u^\epsilon + f(u^\epsilon) + \sqrt{\epsilon}\eta(t), \quad v^\epsilon(0) = u_1 \in L^2(0, l). \tag{5.127}$$

Notice that for a fixed $u \in H_0^1(0, l) \cap H^2(0, l)$, if Equation (5.127) has a unique stationary solution that is strongly mixing with exponential decay rate, then, formally, we have the following averaged equation:

$$u_t = \Delta u + f(u), \quad u(0) = u_0,$$

and deviation equation

$$dz = d\overline{W}(t), \quad z(0) = 0,$$

where $\overline{W}(t)$ is a new Wiener process, defined on a new probability space $(\bar{\Omega}, \bar{\mathcal{F}}, \bar{\mathbb{P}})$ but with the same distribution as $W(t)$. By the deviation equation we have $z(t) = \overline{W}(t)$. Thus, we have the following expansion:

$$u^\epsilon(t) = u(t) + \sqrt{\epsilon}\tilde{W}(t) + \text{higher-order terms}$$

with some Wiener process $\tilde{W}(t)$. This shows that for small $\epsilon > 0$, damped nonlinear stochastic wave equations with singular perturbation can be viewed as a deterministic nonlinear heat equation with white noise fluctuation. In fact, in the sense of distribution, a rigorous discussion [219] implies the following effective approximation system, up to the order of error $\mathcal{O}(\epsilon)$:

$$d\bar{u}^{\epsilon} = [\Delta\bar{u}^{\epsilon} + f(\bar{u}^{\epsilon})]dt + \sqrt{\epsilon}\, d\tilde{W}(t), \quad \bar{u}^{\epsilon}(0) = u_0, \quad u(x,t)|_{x=0,L} = 0,$$

where $\tilde{W}(t)$ is a Wiener process that has the same distribution as $W(t)$.

5.10 Looking Forward

In this chapter, we have investigated time-averaging techniques for stochastic partial differential equations with slow and fast time scales. Averaged systems (i.e., effective systems) are derived and the approximation errors are estimated via normal deviation techniques. Large deviation principles further demonstrate the effectiveness of the approximation.

Some issues could be examined further, including averaging for the fully coupled slow-fast SPDEs systems, such as

$$du^{\epsilon} = [\Delta u^{\epsilon} + f^{\epsilon}(u^{\epsilon}, v^{\epsilon})]dt + \sigma_1(u^{\epsilon}, v^{\epsilon})dW_1(t) \tag{5.128}$$

$$dv^{\epsilon} = \frac{1}{\epsilon}[\Delta v^{\epsilon} + g^{\epsilon}(u^{\epsilon}, v^{\epsilon})]dt + \frac{\sigma_2(u^{\epsilon}, v^{\epsilon})}{\sqrt{\epsilon}}dW_2(t). \tag{5.129}$$

Cerrai [68] derived an averaged system for $\sigma_1(u, v) = \sigma_1(v)$. Normal deviations and large deviations are both interesting problems for (5.128)–(5.129). Another interesting case is to consider a large potential, that is,

$$f^{\epsilon}(u^{\epsilon}, v^{\epsilon}) = f_0(u^{\epsilon}, v^{\epsilon}) + \frac{1}{\sqrt{\epsilon}}f_1(u^{\epsilon}, v^{\epsilon}) \tag{5.130}$$

with f_1 satisfying suitable conditions.

It is also worthwhile to study the approximation of a stationary measure, if it exists, for u^{ϵ} by the stationary measure of the averaged equation. Similarly, approximation of the random attractor, if it exists, of (5.128)–(5.129) is also an interesting issue. An example is the singularly perturbed nonlinear wave equation in § 5.9.2, which is a special slow-fast system that has been studied in, e.g., [220]. Moreover, it should be interesting to examine how attractors and inertial manifolds evolve under the averaging process [81].

5.11 Problems

We assume that the Wiener process $W(t)$ takes values in Hilbert space $H = L^2(0, l)$ and that it has a trace class covariance operator Q that shares the same eigenfunctions with the Laplace operator ∂_{xx} under appropriate boundary conditions. Moreover, ϵ is a small positive parameter.

5.1. Stationary solution

Derive the explicit expression of the distribution for the unique stationary solution of the following SPDE on $(0, l)$:

$$d\eta = [\eta_{xx} + a(x)]dt + \sigma dW, \quad \eta(0, t) = \eta(l, t) = 0,$$

where $a(x) \in L^2(0, l)$. Prove the strong mixing property with exponential decay rate of the stationary distribution. Compute

$$\mathbb{E}(\bar{\eta}(x, t) \otimes \bar{\eta}(x, s)),$$

where $\bar{\eta}$ is the unique stationary solution for the above equation.

5.2. Averaging for a SPDE-ODE system with singular random force

Consider the following slow-fast SPDE on $(0, l)$

$$du^\epsilon = \left[u^\epsilon_{xx} + \frac{a(x)}{\sqrt{\epsilon}} \eta^\epsilon(t) \right] dt, \quad u(0) = u_0 \in H,$$

$$u(0, t) = u(l, t) = 0$$

$$d\eta^\epsilon = -\frac{1}{\epsilon} \eta^\epsilon dt + \frac{1}{\sqrt{\epsilon}} dW(t), \quad \eta(0) = 0 \in \mathbb{R}^1,$$

where $W(t)$ is a scalar standard Wiener process and $a(x)$ is a bounded smooth function. What is the averaged equation?

5.3. Averaging for a PDE-SPDE system

Consider the following slow-fast SPDEs on $(0, l)$

$$du^\epsilon = [u^\epsilon_{xx} + v^\epsilon u^\epsilon]dt, \quad u(0) = u_0 \in H = L^2(0, l),$$

$$dv^\epsilon = \frac{1}{\epsilon}[v^\epsilon_{xx} + a(x)]dt + \frac{1}{\sqrt{\epsilon}} dW(t), \quad v(0) = v_0 \in H,$$

with zero Dirichlet boundary conditions. What is the averaged equation? Derive the equation satisfied by the limit of the deviation process.

5.4. Averaging for a random PDE

Consider the following random PDE on $(0, l)$:

$$u^\epsilon_t = u^\epsilon_{xx} + \eta^\epsilon(t)u^\epsilon, \quad u^\epsilon(0) = u_0,$$

with zero Dirichlet boundary conditions. Here $\eta^\epsilon(t)$ is a stationary process that solves the following linear SPDE:

$$d\eta^\epsilon = \frac{1}{\epsilon}[\eta^\epsilon_{xx} + a(x)] dt + \frac{1}{\sqrt{\epsilon}} dW(t).$$

Write down the averaged equation and compare the result with Problem 5.3.

5.5. Averaging for a stochastic wave equation

Consider the averaged equation for the following slow-fast stochastic differential equations:

$$d\eta^\epsilon = \xi^\epsilon dt, \quad \eta^\epsilon(0) = \eta_0,$$
$$d\xi^\epsilon = -\frac{1}{\epsilon}[\xi^\epsilon - \lambda\eta^\epsilon]dt + \frac{\sigma}{\sqrt{\epsilon}}dw(t), \quad \xi^\epsilon(0) = \xi_0,$$

where $\lambda < 0$ and $w(t)$ is scalar standard Brownian motion. Show that the averaged equation is

$$d\eta = \lambda\eta, \quad \eta(0) = \eta_0.$$

Furthermore, consider the following singularly perturbed stochastic wave equation on $(0, l)$:

$$\epsilon d u_t^\epsilon + u_t^\epsilon dt = u_{xx}^\epsilon dt + \sqrt{\epsilon}dW(t),$$
$$u^\epsilon(0) = u_0 \in H^2(0, l) \cap H_0^1(0, l), \quad u_t^\epsilon(0) = u_1 \in H^1(0, l),$$

with zero Dirichlet boundary conditions. Show that the averaged equation is the following heat equation:

$$u_t = u_{xx}, \quad u(0) = u_0,$$

on $(0, l)$ with the zero Dirichelt boundary conditions.

5.6. Averaging for SDEs with cubic nonlinearity

Consider the following slow-fast stochastic differential equations:

$$du^\epsilon = \left[-u^\epsilon - (u^\epsilon + v^\epsilon)^3 \right] dt,$$
$$dv^\epsilon = \frac{1}{\epsilon}v^\epsilon dt + \frac{\sqrt{2}}{\sqrt{\epsilon}} dw,$$

where w is a standard Browian motion. By (5.71), find the expression of $B(u)$. Then, formally write down the averaged equation corrected by deviation.

6 Slow Manifold Reduction

Random invariant manifolds; random slow manifolds; random center manifolds; macroscopic reduction

In this chapter we first consider random center manifold reduction for a system of SPDEs and then random slow manifold reduction for SPDEs with slow and fast time scales. The effective dynamics is described by a lower-dimensional, reduced system on a random center manifold or on a random slow manifold, respectively. This provides a dimension reduction method for infinite dimensional or very high dimensional complex systems under uncertainty.

This chapter is organized as follows. After a background review in §6.1, random invariant manifolds are constructed for a class of stochastic partial differential equations in §6.2, then a reduction principle on random center-unstable manifolds is established for a class of stochastic partial differential equations with global or local Lipschitz nonlinearities in §6.3 and 6.4, respectively. A special case is reduction on a random center manifold. This random invariant manifold reduction principle is further applied to a slow-fast system of stochastic partial differential equations in §6.5. Finally, stochastic amplitude evolution via random slow manifold reduction is briefly discussed in §6.6, and some open problems are commented in §6.7.

6.1 Background

Invariant manifolds are special invariant sets, represented by graphs in state space, in which solution processes of a stochastic dynamical system live.

The theory of invariant manifolds for deterministic dynamical systems has a long history. Two main approaches to construct invariant manifolds had been developed: the Hadamard graph transform method [124,151], which is a geometric approach, and the Lyapunov–Perron method [78,182,256], which is analytic in nature. There are numerous works on stable, unstable, center, center-stable, and center-unstable manifolds for infinite dimensional, deterministic, autonomous dynamical systems; see, e.g., [28,29,72,78,82,89,108,203,234,259,269,276,287,307].

When an attracting invariant manifold exists, a reduction for a dynamical system is possible by restricting the dynamics to this manifold [78,222]. The reduced, effective system, defined on the attracting invariant manifold, has a lower dimension, and it captures long time dynamics of the original dynamical system.

Effective Dynamics of Stochastic Partial Differential Equations. http://dx.doi.org/10.1016/B978-0-12-800882-9.00006-8

In this section, we review an analytic method, the classical Lyapunov–Perron method, to construct an invariant manifold for a deterministic evolutionary system on a separable Hilbert space. This method will be adapted to stochastic evolutionary systems in the next section.

Let H be a separable Hilbert space, with norm $\|\cdot\|$ and scalar product $\langle\cdot,\cdot\rangle$. Consider the following deterministic evolutionary equation:

$$\dot{u}(t) = Au(t) + F(u(t)), \quad u(0) = u_0 \in H, \tag{6.1}$$

where $A : D(A) \subset H \to H$ is an unbounded linear operator generating a strongly continuous semigroup $\{e^{At}\}_{t\geq 0}$ on H, and nonlinearity $F : H \to H$ is Lipschitz continuous. Additionally, we assume $F(0) = 0$ for convenience.

We construct an invariant manifold for the dynamical system defined by Equation (6.1). In both the Hadamard graph transform method and the Lyapunov–Perron method (deterministic or stochastic), a spectral gap condition of the linear part of the dynamical system is a key ingredient. We make the following assumptions on the linear part, and they will be used throughout this chapter.

6.1.1   Spectrum of the Linear Operator

Assume that the spectrum, $\sigma(A)$, of linear operator A consists of a countable number of eigenvalues only, and it splits as

$$\sigma(A) = \{\lambda_k, k \in \mathbb{N}\} = \sigma_c \cup \sigma_s, \tag{6.2}$$

with both σ_c and σ_s nonempty, and

$$\sigma_c \subset \{z \in \mathbb{C} : \operatorname{Re} z \geq 0\} \quad \text{and} \quad \sigma_s \subset \{z \in \mathbb{C} : \operatorname{Re} z < 0\},$$

where $\mathbb{C}$ denotes the set of complex numbers and $\sigma_c = \{\lambda_1, \cdots, \lambda_N\}$ for some $N > 0$. Denote the corresponding eigenvectors for $\{\lambda_k, k \in \mathbb{N}\}$ by $\{e_1, \ldots, e_N, e_{N+1}, \cdots\}$. Assume also that the eigenvectors form an orthonormal basis of H. Thus there is an invariant orthogonal decomposition $H = H_c \oplus H_s$ with $\dim H_c = N$, such that for the restrictions $A_c = A|_{H_c}, A_s = A|_{H_s}$, one has $\sigma_c = \{z : z \in \sigma(A_c)\}$ and $\sigma_s = \{z : z \in \sigma(A_s)\}$. Moreover, $\{e^{tA_c}\}$ is a group of linear operators on H_c, and there exist projections Π_c and Π_s, such that $\Pi_c + \Pi_s = Id_H$, $A_c = \Pi_c A$ and $A_s = \Pi_s A$. Furthermore, we assume that the projections Π_c and Π_s commute with A. Additionally, suppose that there are constants $0 \leq \alpha < \beta$ such that

$$\|e^{tA_c}x\| \leq e^{\alpha t}\|x\|, \ t \leq 0, \tag{6.3}$$

$$\|e^{tA_s}x\| \leq e^{-\beta t}\|x\|, \ t \geq 0. \tag{6.4}$$

We call H_c the center-unstable subspace and H_s the stable subspace of semigroup e^{At} or of the linear dynamical system $\dot{u}(t) = Au(t)$. When σ_c only contains eigenvalues with zero real part, H_c is then called the center subspace. By the assumption on spectrum, $\alpha \geq 0$ and $-\beta < 0$, we have spectral gap $\alpha + \beta > 0$.

Remark 6.1.   The above property (6.3) and (6.4) is in fact the dichotomy property of the linear operator A. Sometimes this is also called the dichotomy property of the corresponding semigroup e^{At}, $t \geq 0$.

6.1.2   Construction of an Invariant Manifold: The Lyapunov–Perron Method

Denote by $\Phi(t)$, $t \geq 0$, the dynamical system (or semiflow) defined by Equation (6.1). We construct an N-dimensional invariant manifold for $\Phi(t)$ by the Lyapunov–Perron method, see, e.g., [78] and [156, Ch. 9].

First we consider the linear evolutionary equation (i.e., when $F(u)$ is absent),

$$\dot{u}(t) = Au(t), \quad u(0) \in H. \tag{6.5}$$

The dynamical system defined by this linear equation is $\Phi(t) = e^{At}$. Splitting the solution $u = u_c + u_s \in H_c \oplus H_s$, we get

$$\|(u_c(t), u_s(t)) - (u_c(t), 0)\| \to 0$$

with exponential rate $-\beta$ as $t \to \infty$. Hence, the center-unstable subspace $H_c = \{(u_c, 0)\}$ is an attracting invariant space for the linear system (6.5). Thus, we have the following reduced system:

$$\dot{u}_c(t) = A_c u_c(t),$$

which captures the long time dynamics of the original linear system (6.5). Invariant manifolds for linear systems are linear subspaces, e.g., the center-unstable subspace $H_c = \{(u_c, 0)\}$, and their structure is simple, but for a nonlinear system (6.1), the linear structure for invariant manifolds is destroyed due to the nonlinearity F. That is, the orbits of (6.5) in the invariant space H_c are now deformed by the nonlinearity. The set consisting of these orbits is expected to be an invariant manifold of the nonlinear system (6.1). However, in order to construct an invariant manifold, the nonlinearity of F should not change these orbits too dramatically. In fact, we need a gap condition: Assume that $F : H \to H$ is Lipschitz continuous with the Lipschitz constant $L_F > 0$, and for some η with $-\beta < \eta < 0$,

$$L_F \left(\frac{1}{\eta + \beta} + \frac{1}{\alpha - \eta} \right) < 1. \tag{6.6}$$

Under this gap condition, these orbits will not go too far away from H_c. The basic idea of the Lyapunov–Perron method is to find all orbits that stay close to H_c under the semiflow $\Phi(t)$.

Based on the spectral properties of A, for $-\beta < \eta < 0$, we define the following Banach space:

$$C_\eta^- = \left\{ u : (-\infty, 0] \to H : u \text{ is continuous and } \sup_{t \leq 0} e^{-\eta t} \|u(t)\| < \infty \right\}, \tag{6.7}$$

with norm

$$|u|_{C_\eta^-} = \sup_{t \leq 0} e^{-\eta t} \|u(t)\|. \tag{6.8}$$

We now construct an invariant manifold for $\Phi(t)$ by seeking a solution to (6.1) in C_η^-. This means that orbits that stay close to H_c under $\Phi(t)$ lie in C_η^-. The following result implies this fact.

Lemma 6.2.   *The function $u \in C_\eta^-$ is the solution of Equation (6.1) on $(-\infty, 0]$ if and only if $u(t)$ solves the following integral equation:*

$$u(t) = e^{A_c t} \Pi_c u_0 + \int_0^t e^{A_c(t-\tau)} \Pi_c F(u(\tau)) \, d\tau + \int_{-\infty}^t e^{A_s(t-\tau)} \Pi_s F(u(\tau)) \, d\tau.$$

$$(6.9)$$

Before the proof, we recall the variation of constants formula for an inhomogeneous evolutionary equation:

$$\dot{u} = Au + f(t),$$

where A is a matrix or a linear operator, and f is a "forcing" term. The solution is then

$$u(t) = e^{At} u(0) + \int_0^t e^{A(t-s)} f(s) ds.$$

$$(6.10)$$

Proof.   On one hand, if $u(t) = u_c(t) + u_s(t)$ in $H_c \oplus H_s$ is the solution to (6.1), and if $u \in C_\eta^-$, then, by the variation of constants formula (6.10), we have

$$u_c(t) = e^{A_c t} \Pi_c u_0 + \int_0^t e^{A_c(t-\tau)} \Pi_c F(u(\tau)) \, d\tau,$$

$$(6.11)$$

$$u_s(t) = e^{A_s(t-t_0)} \Pi_s u(t_0) + \int_{t_0}^t e^{A_c(t-\tau)} \Pi_s F(u(\tau)) \, d\tau.$$

$$(6.12)$$

For $u \in C_\eta^-$ and noticing $\beta + \eta > 0$, we have

$$\| e^{A_s(t-t_0)} \Pi_s u(t_0) \| \leq e^{-\beta t} e^{(\beta+\eta)t_0} |u|_{C_\eta^-} \to 0,$$

as $t_0 \to -\infty$. Now, letting $t_0 \to -\infty$ in (6.12), we conclude that

$$u_s(t) = \int_{-\infty}^t e^{A_s(t-\tau)} \Pi_s F(u(\tau)) \, d\tau.$$

Thus we see that (6.9) holds.

On the other hand, if u is of the form (6.9), then direct computation yields that u is a solution of (6.1) on $(-\infty, 0]$. The proof is complete. $\square$

By the above lemma, for $t \leq 0$,

$$\| u_s(t) \| \leq L_F \int_{-\infty}^t e^{-(\beta+\eta)(t-\tau)} \, d\tau |u|_{C_\eta^-} \leq \frac{L_F}{\beta+\eta} |u|_{C_\eta^-},$$

which implies that u_s is bounded on $(-\infty, 0]$ and u stays close to H_c. So, we define

$$\mathcal{M} = \{u(0) \in H : u(\cdot) \in C_\eta^-\}, \tag{6.13}$$

and, by uniqueness of the solution to Equation (6.1), $\mathcal{M}$ is invariant under $\Phi(t)$. We recall the following center-unstable manifold theorem.

Theorem 6.3 (Center-unstable manifold theorem). *Assume that the spectral properties of §6.1.1 and the gap condition (6.6) hold. Then, $\mathcal{M}$ is an N-dimensional Lipschitz invariant manifold, called a center-unstable manifold, for $\Phi(t)$, with*

$$\mathcal{M} = \{(\xi, h(\xi)) : \xi \in H_c\},$$

where $h : H_c \to H_s$ is a Lipschitz continuous mapping and $h(0) = 0$.

When the spectral part, σ_c, contains only eigenvalues of zero real part, this $\mathcal{M}$ is called a center manifold (a sort of slow manifold).

Proof. By Lemma 6.2, we solve the integral equation (6.9) in C_η^-. For any $\xi \in H_c$, define a nonlinear mapping $\mathcal{N}$ on C_η^- by

$$\mathcal{N}(u, \xi)(t) \triangleq e^{A_c t}\xi + \int_0^t e^{A_c(t-\tau)}\Pi_c F(u(\tau))\,d\tau + \int_{-\infty}^t e^{A_s(t-\tau)}\Pi_s F(u(\tau))\,d\tau$$

for $u \in C_\eta^-$. To solve the integral Equation (6.9), we only need to construct a fixed point of $\mathcal{N}$ in C_η^-. To this end, we first show $\mathcal{N}$ maps C_η^- into C_η^-. For $u \in C_\eta^-$,

$$\sup_{t\leq 0} e^{-\eta t} \left\| e^{A_c t}\xi + \int_0^t e^{A_c(t-\tau)}\Pi_c F(u(\tau))\,d\tau \right\|$$

$$\leq \sup_{t\leq 0} e^{(\alpha-\eta)t}\|\xi\| + \sup_{t\leq 0} L_F \int_0^t e^{(\alpha-\eta)(t-\tau)}\,d\tau |u|_{C_\eta^-}$$

$$\leq \|\xi\| + L_F \frac{1}{\alpha-\eta}|u|_{C_\eta^-},$$

and

$$\sup_{t\leq 0} e^{-\eta t} \left\| \int_{-\infty}^t e^{A_s(t-\tau)}\Pi_s F(u(\tau))\,d\tau \right\|$$

$$\leq L_F \int_{-\infty}^t e^{(-\beta-\eta)(t-\tau)}\,d\tau |u|_{C_\eta^-} \leq L_F \frac{1}{\beta+\eta}|u|_{C_\eta^-}.$$

So, $\mathcal{N}(u) \in C_\eta^-$. Then, let us show that $\mathcal{N}$ is a contraction mapping. For $\xi \in H_c$, and $u, \bar{u} \in C_\eta^-$,

$$|\mathcal{N}(u, \xi) - \mathcal{N}(\bar{u}, \xi)|_{C_\eta^-}$$

$$\leq \sup_{t\leq 0} \left\{ L_F\left(\int_0^t e^{(\alpha-\eta)(t-s)}ds + \int_\infty^t e^{(-\beta-\eta)(t-s)}ds \right) \right\} |u - \bar{u}|_{C_\eta^-}$$

$$\leq L_F \left(\frac{1}{\eta+\beta} + \frac{1}{\alpha-\eta} \right) |u - \bar{u}|_{C_\eta^-}.$$

Due to the gap condition (6.6), $\mathcal{N}$ is a contraction mapping on C_η^-. Thus, by the Banach fixed-point theorem, $\mathcal{N}$ has a unique fixed point $u^*(t, \xi) \in C_\eta^-$ that solves the integral equation (6.9). Define $h(\xi) \triangleq \Pi_s u^*(0, \xi)$, that is,

$$h(\xi) = \int_{-\infty}^{0} e^{-A_s \tau} \Pi_s F(u^*(\tau)) \, d\tau. \tag{6.14}$$

Notice that $\mathcal{N}(0, 0) = 0$ and, hence, $h(0) = 0$. Furthermore, by the same discussion for the contraction of $\mathcal{N}$, we get

$$|u^*(\cdot, \xi) - u^*(\cdot, \zeta)|_{C_\eta^-}$$

$$\leq \|\xi - \zeta\| + L_F \left(\frac{1}{\alpha - \eta} + \frac{1}{\beta + \eta} \right) |u^*(\cdot, \xi) - u^*(\cdot, \zeta)|_{C_\eta^-}.$$

Thus,

$$|u^*(\cdot, \xi) - u^*(\cdot, \zeta)|_{C_\eta^-} \leq \frac{1}{1 - L_F \left(\frac{1}{\alpha - \eta} + \frac{1}{\beta + \eta} \right)} \|\xi - \zeta\|. \tag{6.15}$$

This says that $u^*(t, \xi)$ is Lipschitz continuous in ξ. Noticing (6.14) and using the fact that F is Lipschitz continuous, we see that $h : H_c \to H_s$ is Lipschitz continuous. This completes the proof.   □

If the invariant manifold exponentially attracts all other orbits, then $\mathcal{M}$ becomes an *inertial manifold* [287, Ch. VIII]. The original nonlinear dynamical system (6.1) can then be reduced to this inertial manifold as the following lower-dimensional system:

$$\dot{u}_c(t) = A_c u_c(t) + F_c(u_c + h(u_c)),$$

where F_c is the projection of nonlinearity F to H_c.

6.2   Random Center-Unstable Manifolds for Stochastic Systems

Random invariant manifolds of SPDEs differ from their deterministic counterparts due to the influence of stochastic effects.

6.2.1   Linear Stochastic Systems

We first consider a linear stochastic system with linear multiplicative noise:

$$du(t) = Au(t) \, dt + u(t) \circ dW(t), \quad u(0) \in H, \tag{6.16}$$

where $W(t)$ is a scalar Wiener process, $\circ$ denotes Stratonovich stochastic differentiation, and the linear operator A has the spectral properties specified in §6.1.1. Then, by the properties of the Wiener process, for any solution $u = u_c + u_s$ in $H_c \oplus H_s$ to (6.16), we still have

$$\|(u_c, u_s) - (u_c, 0)\| \to 0, \tag{6.17}$$

with exponential rate $-\beta$ as $t \to \infty$. The subspace $H_c = \{(u_c, 0)\}$ is called the center-unstable invariant space for (6.16). We see that the stochastic term, the linear multiplicative noise $u \circ dW(t)$, does not change the invariant space of the deterministic system (6.5), since the strength of the noise is changing linearly with u. However, it is a different story for a linear stochastic system with additive noise

$$du(t) = Au(t)\,dt + dW(t), \quad u(0) \in H, \tag{6.18}$$

where A is the same as above. Assume that $W(t)$ is an H-valued, two-sided Wiener process with covariance operator Q, satisfying $\text{Tr}(Q) < \infty$. We consider the canonical probability space $(\Omega_0, \mathcal{F}_0, \mathbb{P})$, with Wiener shift $\{\theta_t\}$ (see (4.77)). Notice that $(0, 0)$ is not the stationary solution due to the additive noise. In this case, (6.18) has a unique stationary solution $\eta(\omega) = (\eta_c(\omega), \eta_s(\omega)) \in H_c \oplus H_s$, provided that either the lower bound of $\sigma(A_c)$ is strictly positive (i.e., $\alpha > 0$) or H_c is the null space (Problem 6.1). Then, for any solution $u = (u_c, u_s) \in H_c \oplus H_s$, we have

$$\|(u_c(t), u_s(t)) - (u_c(t), \eta_s(\theta_t \omega))\| \to 0, \quad t \to \infty,$$

with exponential rate $-\beta$. The set $\mathcal{M}(\omega) \triangleq \{(\xi, \eta_s(\omega)) : \xi \subset H_c\}$ is the random center-unstable invariant manifold of the stationary solution $\eta(\omega)$. Notice that this random invariant manifold does not coincide with H_c but is flat and is simply a random translation $(\eta_s(\omega))$ from H_c due to the additive noise.

6.2.2   Nonlinear Stochastic Systems

For the linear stochastic systems (6.16) and (6.18) in the previous subsection, the stochastic force on one mode does not affect any other mode. Hence, the dynamics are simple.

Consider a nonlinear stochastic system described by the following SPDE:

$$du(t) = [Au(t) + F(u(t))]\,dt + \sigma(u(t)) \circ dW(t), \quad u(0) = u_0 \in H, \tag{6.19}$$

where the linear operator A has the spectral properties in §6.1.1, the nonlinearity $F :$ $H \to H$ is a Lipschitz continuous mapping with Lipschitz constant L_F, and $F(0) = 0$. The Wiener process is specified later.

Due to nonlinear interactions between different modes or scales, the stochastic influence is now coupled between different scales. It is difficult to define a random dynamical system from (6.19) directly, but for some special form of stochastic force, for example, linear multiplicative noise or additive noise, this SPDE can be transformed to an evolutionary equation with random coefficients, which can be treated for almost all ω. This makes the Lyapunov–Perron method applicable to such an SPDE. Next we show how to apply this method to construct a random invariant manifold for a class of nonlinear SPDEs.

We consider the following SPDE with linear multiplicative noise:

$$du(t) = [Au(t) + F(u(t))]\, dt + u(t) \circ dW(t), \quad u(0) = u_0 \in H, \tag{6.20}$$

where $W(t)$ is a standard, two-sided scalar Wiener process on the canonical probability space $(\Omega_0, \mathcal{F}_0, \mathbb{P})$, defined in §4.10.

Random Dynamical Systems

Recall the Wiener shift (see (4.77)), $\{\theta_t\}$, defined on $(\Omega_0, \mathcal{F}_0, \mathbb{P})$ in §4.10. For $\omega \in \Omega_0$, we identify

$$W(\cdot, \omega) = \omega(\cdot)$$

and recall that

$$W(\cdot, \theta_t\omega) = W(\cdot + t, \omega) - W(t, \omega).$$

To study the random dynamics of (6.20), as in [109], we introduce the following stationary process $z(\omega)$, which solves a Langevin equation:

$$dz + z\,dt = dW, z(0) = 0. \tag{6.21}$$

Next we examine some growth properties of this stationary solution. Properties of the stationary solution of this linear Langevin equation have been studied by many authors, see, e.g., [62,109]. Here we follow the discussion in [109].

First, recall the law of iterated logarithm [24, Ch. 9.4]: There exists an $\Omega_1 \subset \Omega_0$ with $\mathbb{P}(\Omega_1) = 1$ and $\theta_t\Omega_1 = \Omega_1$ such that

$$\limsup_{t \to \pm\infty} \frac{|W(t, \omega)|}{\sqrt{2|t| \log\log|t|}} = 1, \quad \text{for all } \omega \in \Omega_1.$$

For $\omega \in \Omega_1$ and $z_0 \in \mathbb{R}$, consider a solution $z(t, \theta_{-t}\omega, z_0)$ with initial value $z_0 \in \mathbb{R}$. Then, by integration by parts and the definition of θ_t, we have

$$
\begin{aligned}
z(t, \theta_{-t}\omega, z_0) &= e^{-t}z_0 + \int_0^t e^{-t+s}dW(s, \theta_{-t}\omega) \\
&= e^{-t}z_0 + W(t, \theta_{-t}\omega) - \int_0^t e^{-t+s}[W(s-t, \omega) - W(-t, \omega)]\,ds \\
&= e^{-t}z_0 - W(-t, \omega) + \int_0^t e^{-t+s}W(s-t, \omega)\,ds + W(-t, \omega) \\
&= e^{-t}z_0 - \int_0^t e^{-t+s}W(s-t, \omega)\,ds.
\end{aligned}
$$

Taking the transformation $\tau = -t + s$, we conclude that

$$\int_0^t e^{-t+s}W(s-t, \omega)\,ds = \int_{-t}^0 e^{\tau}W(\tau, \omega)\,d\tau.$$

Then, passing the limit $t \to \infty$, for all $\omega \in \Omega_1$ we get

$$\lim_{t \to \infty} z(t, \theta_{-t}\omega, z_0) = -\int_{-\infty}^{0} e^\tau W(\tau, \omega)\, d\tau. \tag{6.22}$$

The Riemann integral on the right-hand side of Equation (6.22) is well defined for all $\omega \in \Omega_1$.

Define

$$z(\omega) \triangleq -\int_{-\infty}^{0} e^\tau W(\tau, \omega)\, d\tau. \tag{6.23}$$

We now show that z is a stationary solution (or stationary orbit). In fact, integrating by parts yields

$$\begin{aligned}
z(t, \omega, z(\omega)) &= e^{-t}z(\omega) + \int_{0}^{t} e^{-t+\tau}\, dW(\tau, \omega) \\
&= -\int_{-\infty}^{t} e^{-t+\tau} W(\tau, \omega)\, d\tau + W(t, \omega), \\
&= \int_{-\infty}^{0} e^t W(\tau + t, \omega)\, d\tau + W(t, \omega) \\
&= z(\theta_t\omega).
\end{aligned}$$

In the final step, we have used the change of time variable $(-t + \tau \mapsto \tau)$. Then the solution to (6.21) with initial value $z(\omega)$ is the unique stationary solution in the following form:

$$z(\theta_t\omega) = -\int_{-\infty}^{0} e^\tau W(\tau, \theta_t\omega)\, d\tau = -\int_{-\infty}^{0} e^\tau W(t+\tau, \omega)\, d\tau + W(t, \omega). \tag{6.24}$$

Furthermore, as in [109], by the law of iterated logarithm, for $\omega \in \Omega_1$ and every $t_0 \in \mathbb{R}$, the function

$$\tau \to e^\tau \sup_{[t_0-1, t_0+1]} |W(\tau + t_0, \omega)|$$

is an integrable upper bound for $e^\tau W(\tau + t, \omega)$ for $t \in [t_0-1, t_0+1]$, and $\tau \in (-\infty, 0]$. Hence, by the Lebesgue dominated convergence theorem, we have the continuity of $z(\theta_t\omega)$ at t_0. Also, by the law of iterated logarithm again, for $1/2 < \delta < 1$ and $\omega \in \Omega_1$ there exists a constant $C_{\delta,\omega} > 0$, such that

$$|W(\tau + t, \omega)| \le C_{\delta,\omega} + |t + \tau|^\delta \le C_{\delta,\omega} + |t|^\delta + |\tau|^\delta, \quad \tau \le 0.$$

Therefore,

$$\begin{aligned}
&\lim_{t \to \pm\infty} \left| \frac{1}{t} \int_{-\infty}^{0} e^\tau W(\tau + t, \omega)\, d\tau \right| \\
&\le \lim_{t \to \pm\infty} \frac{1}{|t|} \int_{-\infty}^{0} e^\tau (C_{\delta,\omega} + |t|^\delta + |\tau|^\delta)\, d\tau = 0.
\end{aligned}$$

Hence,

$$\lim_{t \to \pm\infty} \frac{|z(\theta_t \omega)|}{t} = 0, \quad \omega \in \Omega_1.$$

Since $\mathbb{E}z(\theta_t \omega) = 0$, by an ergodic theorem [179, Ch. 10], [24, Theorem 8.3.10], there is a θ_t invariant set $\Omega_2 \subset \Omega_0$ with $\mathbb{P}(\Omega_2) = 1$, such that

$$\lim_{t \to \pm\infty} \frac{1}{t} \int_0^t z(\theta_\tau \omega) \, d\tau = 0 \quad \text{for all} \quad \omega \in \Omega_2.$$

Now define a new sample space $\Omega = \Omega_1 \cap \Omega_2$, which is θ_t-invariant and $\mathbb{P}(\Omega) = 1$. Also, take a new σ-field

$$\mathcal{F} \triangleq \{\Omega \cap U : U \in \mathcal{F}_0\}.$$

For the rest of this section, we consider the probability space $(\Omega, \mathcal{F}, \mathbb{P})$.

We introduce the following transform:

$$u(t) = v(t)e^{z(\theta_t \omega)}. \tag{6.25}$$

Then SPDE (6.20) becomes a random partial differential equation (or random evolutionary equation)

$$\dot{v} = Av + G(\theta_t \omega, v) + z(\theta_t \omega)v, \ v(0) = x \in H, \tag{6.26}$$

where $G \triangleq e^{-z(\omega)} F(e^{z(\omega)}v) : H \to H$ is also a Lipschitz continuous mapping with the same Lipschitz constant L_F. In contrast to the original SPDE, no noise term appears in (6.26). For a fixed $\omega \in \Omega$, noticing the Lipschitz property of G, and by the deterministic method [276, Ch. 4.6] the random evolutionary system (6.26) has a unique solution in the following mild sense:

$$v(t, \omega, x) = e^{At + \int_0^t z(\theta_\tau \omega) \, d\tau} x + \int_0^t e^{A(t-s) + \int_s^t z(\theta_\tau \omega) \, d\tau} G(\theta_s \omega, v(s, \omega, x)) \, ds.$$

Hence, the solution mapping

$$(t, \omega, x) \to v(t, \omega, x)$$

is $(\mathcal{B}(\mathbb{R}) \otimes \mathcal{F} \otimes \mathcal{H}, \mathcal{F})$-measurable and generates a continuous random dynamical system $\varphi(t, \omega) : H \to H$, defined by $\varphi(t, \omega)x = v(t, \omega, x)$ for every $x \in H$ and $\omega \in \Omega$. Now, denote by $T(\omega)$ the transform (6.25), that is,

$$T(\omega) : H \to H, \quad T(\omega)x = xe^{z(\omega)}, \quad \text{for } x \in H, \omega \in \Omega.$$

Its inverse mapping is

$$T^{-1}(\omega) : H \to H, \quad T^{-1}(\omega)x = xe^{-z(\omega)}, \quad \text{for } x \in H, \omega \in \Omega.$$

Now, we have the following result [109].

Lemma 6.4. *Define $\hat{\varphi}(t, \omega) : H \to H$ by*

$$\hat{\varphi}(t, \omega)x \triangleq T^{-1}(\theta_t\omega)v(t, \omega, T(\omega)x).$$

Then $\hat{\varphi}(t, \omega)$ defines a continuous random dynamical system, and for every $x \in H$, $(t, \omega) \to u(t, \omega, x) = \hat{\varphi}(t, \omega)x$ solves Equation (6.20) uniquely.

Proof. By the definition of T, it is easy to check that $\hat{\varphi}(t, \omega)$ defines a continuous random dynamical system on H. Applying Itô formula to $T^{-1}(\theta_t\omega)v(t, \omega, T(\omega)x)$ shows that $u(t, \omega, x)$ is a solution of (6.20). Then, by the inverse transform T^{-1}, we obtain the uniqueness of u. This completes the proof. $\qquad\square$

Remark 6.5. An SPDE with additive noise

$$du(t) = [Au(t) + F(u(t))]\,dt + dW(t), \tag{6.27}$$

where $W(t)$ is an H-valued, two-sided Wiener process with covariance operator Q, can also be converted to a random evolutionary equation with random coefficients. Introduce a stationary process $z(\theta_t\omega)$ that solves the following linear Langevin SPDE in H

$$dz(t) = Az(t)\,dt + dW(t), \quad z(0) = 0.$$

By a transformation

$$v(t) \triangleq u(t) - z(\theta_t\omega),$$

SPDE (6.27) becomes the following random evolutionary equation

$$\dot{v} = Av + G(\theta_t\omega, v), \tag{6.28}$$

where $G(\theta_t\omega, v) \triangleq F(v(t)+z(\theta_t\omega)) : H \to H$ is also a Lipschitz continuous mapping with the same Lipschitz constant L_F. Thus, the following discussion about invariant manifolds for random evolutionary equations is applicable to system (6.28) as well.

Existence of Random Center-Unstable Manifold

Now we apply the Lyapunov–Perron method, introduced in §6.1.2, to construct a random center-unstable manifold for the random dynamical system defined by SPDE (6.20). This method has been adapted systematically to construct random invariant manifolds for SPDEs under various conditions, e.g., [109,110]. There are also some recent works on invariant manifolds for infinite dimensional random dynamical systems e.g., [60,213], which could deal with random partial differential equations directly.

Let d_H be the metric induced by the norm $\|\cdot\|$. Then (H, d_H) is a complete separable metric space. Let Π_c, Π_s be the projections to H_c and H_s, respectively. Projecting system (6.26) to H_c and then to H_s, we obtain

$$\dot{v}_c = A_c v_c + z(\theta_t\omega)v_c + g_c(\theta_t\omega, v_c + v_s), \tag{6.29}$$

$$\dot{v}_s = A_s v_s + z(\theta_t\omega)v_s + g_s(\theta_t\omega, v_c + v_s), \tag{6.30}$$

where

$$g_c(\theta_t\omega, v_c + v_s) = \Pi_c G(\theta_t\omega, v_c + v_s)$$

and

$$g_s(\theta_t\omega, v_c + v_s) = \Pi_s G(\theta_t\omega, v_c + v_s).$$

Now, define the following Banach space for η satisfying $-\beta < \eta < 0$:

$$C_{\eta,z}^- \triangleq \left\{ v : (-\infty, 0] \to H : v \text{ is continuous and} \right.$$

$$\left. \sup_{t\in(-\infty,0]} e^{-\eta t - \int_0^t z(\theta_\tau\omega)\, d\tau} \|v(t)\| < \infty \right\}, \tag{6.31}$$

with norm

$$|v|_{C_{\eta,z}^-} = \sup_{t\in(-\infty,0]} e^{-\eta t - \int_0^t z(\theta_s\omega)\, ds} \|v(t)\|.$$

Indeed, it can be verified that $C_{\eta,z}^-$ is a Banach space [109]. We construct a random center-unstable invariant manifold for $\varphi(t, \omega)$ by seeking a solution to (6.26) in $C_{\eta,z}^-$. First, we define a random set

$$\widetilde{\mathcal{M}}(\omega) \triangleq \{ x \in H : v(\cdot, \omega, x) \in C_{\eta,z}^- \}.$$

By uniqueness of the solution to (6.26), for $x \in \widetilde{\mathcal{M}}(\omega)$ and $s > 0$,

$$v(t, \theta_s\omega, v(s, \omega, x)) = v(t + s, \omega, x)$$

solves

$$\dot{v} = Av + z(\theta_t\theta_s\omega)v + G(\theta_t\theta_s\omega, v), \quad v(0) = v(s, \omega, x).$$

Therefore, by the definition of $\widetilde{\mathcal{M}}(\omega)$, $v(s, \omega, x)$ is in $\widetilde{\mathcal{M}}(\theta_s\omega)$. This yields the invariance of $\widetilde{\mathcal{M}}(\omega)$.

Next we show that $\widetilde{\mathcal{M}}(\omega)$ can be represented as the graph of a random Lipschitz continuous mapping, and thus, it is a random invariant manifold. In fact,

$$\widetilde{\mathcal{M}}(\omega) = \{ (\xi, h^s(\xi, \omega)) : \xi \in H_c \} \tag{6.32}$$

for some Lipschitz continuous random mapping $h^s : H_c \to H_s$. To this end, we need to prove the following fact: $v(\cdot, \omega, x)$, in $C_{\eta,z}^-$, is the solution of (6.26) if and only if $v(\cdot, \omega, x)$ satisfies

$$v(t, \omega, x) = e^{A_c t + \int_0^t z(\theta_\tau\omega)\, d\tau}\xi + \int_0^t e^{A_c(t-\tau) + \int_\tau^t z(\theta_\varsigma\omega)\, d\varsigma} g_c(\theta_\tau\omega, v_c + v_s)\, d\tau$$

$$+ \int_{-\infty}^t e^{A_s(t-\tau) + \int_\tau^t z(\theta_\varsigma\omega)\, d\varsigma} g_s(\theta_\tau\omega, v_c + v_s)\, d\tau, \tag{6.33}$$

with $\xi = \Pi_c x$. Indeed, if $v(t, \omega, x)$ is the solution to (6.26), by the variation of constants formula (6.10) we have

$$v_c(t, x, \omega) = e^{A_c t + \int_0^t z(\theta_\tau \omega) \, d\tau} \Pi_c x$$
$$+ \int_0^t e^{A_c(t-\tau) + \int_\tau^t z(\theta_\varsigma \omega) \, d\varsigma} g_c(\theta_\tau \omega, v(\tau, \omega, x)) \, d\tau,$$

$$v_s(t, x, \omega) = e^{A_s(t-t_0) + \int_{t_0}^t z(\theta_\tau \omega) \, d\tau} \Pi_s v(t_0, \omega, x)$$
$$+ \int_{t_0}^t e^{A_s(t-\tau) + \int_\tau^t z(\theta_\varsigma \omega) \, d\varsigma} g_s(\theta_\tau \omega, v(\tau, \omega, x)) \, d\tau,$$

for $t > t_0$ and $t_0 \le 0$. Since $v(\cdot, \omega, x) \in C_{\eta,z}^-$ and $\beta + \eta > 0$, we have that

$$\left\| e^{A_s(t-t_0) + \int_{t_0}^t z(\theta_\tau \omega) \, d\tau} \Pi_s v(t_0, \omega, x) \right\|$$
$$\le e^{-\beta t} e^{(\beta+\eta) t_0} e^{\int_0^t z(\theta_\tau \omega) \, d\tau} |v|_{C_{\eta,z}^-}$$
$$\to 0, \quad \text{as} \quad t_0 \to -\infty.$$

Taking the limit $t_0 \to -\infty$, we get

$$v_s(t, \omega, x) = \int_{-\infty}^t e^{A_s(t-\tau) + \int_\tau^t z(\theta_\varsigma \omega) \, d\varsigma} g_s(\theta_\tau \omega, v_c + v_s) \, d\tau,$$

which proves (6.33). Conversely, if $v(t, \omega, x)$ is of the form (6.33), direct computation yields that (6.33) is a solution of (6.26).

Now, by the definition of $\widetilde{\mathcal{M}}(\omega)$ and the above important fact we have proved, to represent the random invariant set as a graph of a random mapping, we have to solve the integral equation (6.33). To this end, for a fixed $\omega \in \Omega$, define a nonlinear mapping $\mathcal{N}_z$ by

$$\mathcal{N}_z(v, \xi)(t, \omega) \triangleq e^{A_c t + \int_0^t z(\theta_\tau \omega) \, d\tau} \xi$$
$$+ \int_0^t e^{A_c(t-\tau) + \int_\tau^t z(\theta_\varsigma \omega) \, d\varsigma} g_c(\theta_\tau \omega, v_c + v_s) \, d\tau$$
$$+ \int_{-\infty}^t e^{A_s(t-\tau) + \int_\tau^t z(\theta_\varsigma \omega) \, d\varsigma} g_s(\theta_\tau \omega, v_c + v_s) \, d\tau, \qquad (6.34)$$

where $\xi \in H_c$ and $v = (v_c, v_s) \in C_{\eta,z}^-$. We will show that $\mathcal{N}_z$ maps $C_{\eta,z}^-$ to itself. Note that, for $v \in C_{\eta,z}^-$,

$$\sup_{t \le 0} e^{-\eta t - \int_0^t z(\theta_\tau \omega) \, d\tau} \left\| e^{A_c t + \int_0^t z(\theta_\tau \omega) \, d\tau} \xi \right.$$
$$+ \int_0^t e^{A_c(t-\tau) + \int_\tau^t z(\theta_\varsigma \omega) \, d\varsigma} g_c(\theta_\tau \omega, v_c + v_s) \, d\tau \Big\|$$
$$\le \sup_{t \le 0} e^{(\alpha-\eta) t} \|\xi\| + \sup_{t \le 0} L_F \int_0^t e^{(\alpha-\eta)(t-\tau)} \, d\tau |v|_{C_{\eta,z}^-}$$
$$\le \|\xi\| + L_F \frac{1}{\alpha - \eta} |v|_{C_{\eta,z}^-},$$

and

$$\sup_{t \leq 0} e^{-\eta t - \int_0^t z(\theta_\tau \omega)\, d\tau} \left\| \int_{-\infty}^t e^{A_s(t-\tau) + \int_\tau^t z(\theta_\varsigma \omega)\, d\varsigma} g_s(\theta_\tau \omega, v_c + v_s)\, d\tau \right\|$$

$$\leq L_F \int_{-\infty}^t e^{(-\beta-\eta)(t-\tau)}\, d\tau\, |v|_{C_{\eta,z}^-} \leq L_F \frac{1}{\beta+\eta} |v|_{C_{\eta,z}^-}.$$

Thus, $\mathcal{N}_z(v) \in C_{\eta,z}^-$. For any given $\xi \in H_c$ and each $v, \bar{v} \in C_{\eta,z}^-$, we estimate

$$|\mathcal{N}_z(v,\xi) - \mathcal{N}_z(\bar{v},\xi)|_{C_{\eta,z}^-}$$

$$\leq \sup_{t \leq 0} \left\{ L_F \left(\int_0^t e^{(\alpha-\eta)(t-s)}\, ds + \int_\infty^t e^{(-\beta-\eta)(t-s)}\, ds \right) \right\} |v - \bar{v}|_{C_{\eta,z}^-}$$

$$\leq L_F \left(\frac{1}{\eta+\beta} + \frac{1}{\alpha-\eta} \right) |v - \bar{v}|_{C_{\eta,z}^-}. \tag{6.35}$$

As in the deterministic case in the previous section, if the gap condition (6.6) holds true, the mapping $\mathcal{N}_z$ is a contraction mapping on $C_{\eta,z}^-$. Hence, by the Banach fixed-point theorem,

$$v = \mathcal{N}_z(v) \tag{6.36}$$

has a unique solution $v^*(\cdot, \omega, x) \in C_{\eta,z}^-$. Since the solution is uniquely determined by $\xi \in H_c$, we write the solution as $v^*(\cdot, \omega, \xi)$. Furthermore, as in the calculation that $\mathcal{N}_z$ maps $C_{\eta,z}^-$ to itself, we have that, for any $\xi, \zeta \in H_c$,

$$|v^*(\cdot, \omega, \xi) - v^*(\cdot, \omega, \zeta)|_{C_{\eta,z}^-}$$

$$\leq \|\xi - \zeta\| + L_F \left(\frac{1}{\alpha-\eta} + \frac{1}{\beta+\eta} \right) |v^*(\cdot, \omega, \xi) - v^*(\cdot, \omega, \zeta)|_{C_{\eta,z}^-}.$$

Thus,

$$|v^*(\cdot, \omega, \xi) - v^*(\cdot, \omega, \zeta)|_{C_{\eta,z}^-} \leq \frac{1}{1 - L_F \left(\frac{1}{\alpha-\eta} + \frac{1}{\beta+\eta} \right)} \|\xi - \zeta\|, \tag{6.37}$$

and $v^*(\cdot, \omega, \xi)$ is measurable with respect to ω and ξ. Define $h^s(\xi, \omega) \triangleq \Pi_s v^*(0, \omega; \xi)$, that is,

$$h^s(\xi, \omega) = \int_{-\infty}^0 e^{-A_s \tau + \int_\tau^0 z(\theta_\varsigma \omega)\, d\varsigma} g_s(\theta_\tau \omega, v^*(\tau, \omega; \xi))\, d\tau. \tag{6.38}$$

Hence, $h^s(0, \omega) = 0$, and by (6.37),

$$\|h^s(\xi, \omega) - h^s(\zeta, \omega)\| \leq \frac{1}{1 - L_F \left(\frac{1}{\alpha-\eta} + \frac{1}{\beta+\eta} \right)} \|\xi - \zeta\|.$$

Thus $h^s(\xi, \omega)$ is Lipschitz continuous in ξ. Recall that $x \in \widetilde{\mathcal{M}}(\omega)$ if and only if there exists a $v(\cdot, \omega, x) \in C_{\eta,z}^-$ which is the solution to (6.33). Thus, we conclude that $x \in \widetilde{\mathcal{M}}(\omega)$ if and only if there exists a $\xi \in H_c$, such that $x = \xi + h^s(\xi, \omega)$. Hence, we have (6.32), the representation of $\widetilde{\mathcal{M}}(\omega)$, which is a Lipschitz random invariant manifold.

We draw the following result regarding the existence of a random invariant manifold for the random dynamical system $\varphi(t, \omega)$ generated by (6.26).

Lemma 6.6 (Random center-unstable manifold for a random PDE).   *Assume that A has the spectral properties in §6.1.1 and that the gap condition (6.6) holds. Then there exists an N-dimensional Lipschitz random invariant manifold $\widetilde{\mathcal{M}}(\omega)$ for $\varphi(t, \omega)$, and this manifold is represented as $\widetilde{\mathcal{M}}(\omega) \triangleq \{\xi + h^s(\xi, \omega) : \xi \in H_c\}$.*

By the transform (6.25), we have the following conclusion as well.

Lemma 6.7 (Random center-unstable manifold for a SPDE).   *Assume that A has the spectral properties in §6.1.1 and that the gap condition (6.6) holds. Then $\mathcal{M}(\omega) \triangleq \{\xi + e^{z(\omega)}h^s(e^{-z(\omega)}\xi, \omega) : \xi \in H_c\}$ is an N-dimensional Lipschitz random center-unstable manifold for SPDE (6.20).*

For simplicity in the following discussions, we introduce a notation $\bar{h}^s(\xi, \omega) \triangleq e^{z(\omega)}h^s(e^{-z(\omega)}\xi, \omega)$, for $\xi \subset H_c$.

Remark 6.8.   When the central spectral part, σ_c, contains only eigenvalues of zero real part, $\mathcal{M}(\omega)$ is called a random center manifold.

Remark 6.9.   The Hadamard graph transform method may also be utilized to construct a random invariant manifold for SPDEs [109]. In this case, a random fixed-point theorem is applied [274]. By the way, for SPDEs with a general nonlinear multiplicative noise, the construction of a random invariant manifold for SPDEs is still an open problem.

We should mention that the existence of random invariant manifolds depends heavily on the Lipschitz property of the nonlinearity. However, for local Lipschitz continuous nonlinearity and when the system is dissipative, a cutoff procedure can be introduced to construct a local random invariant manifold, which describes the local random dynamics of the system with high probability. This problem is treated in §6.4.

6.3   Random Center-Unstable Manifold Reduction

The existence of a random center-unstable manifold offers a possibility to describe a nonlinear dynamical system by restricting it to the random invariant manifold. In fact, if the random invariant manifold (e.g., a center-unstable invariant manifold) is attracting, then a lower-dimensional reduced system can be defined on the attracting random invariant manifold [299]. The reduced system describes long time dynamics of the original system in a lower dimensional space. If dim $H_c < \infty$, the attracting random center-unstable manifold is in fact a *stochastic* version of an inertial manifold. Moreover, the reduced model includes the random effects. This is treated in this section.

We now present a random invariant manifold reduction for Equation (6.20), which is an SPDE with linear multiplicative noise.

By the spectral properties of A in §6.1.1, dim $H_c = N$. For $u \in H$, we split $u = u_c + u_s$, with $u_c = \Pi_c u \in H_c$ and $u_s = \Pi_s u \in H_s$. Define a nonlinear mapping on H_c as follows:

$$F_c : H_c \to H_c$$

$$u_c \mapsto F_c(u_c) = \sum_{i=1}^{n} \langle F(u_c + 0), e_i \rangle e_i,$$

where 0 is the zero element in the vector space H_s. Now we prove the following result [299].

Theorem 6.10 (Random center-unstable manifold reduction). *Assume that the linear operator A has the spectral properties in §6.1.1 and that the Lipschitz nonlinearity F satisfies the gap condition (6.6). Also, let the Lipschitz constant L_F be small enough so that*

$$\alpha - \beta + 2L_F + \delta L_F + \delta^{-1} L_F < 0$$

for some $\delta > 0$. Then:

(i) *The random center-unstable manifold $\mathcal{M}(\omega)$ of SPDE (6.20) is an N-dimensional random inertial manifold.*

(ii) *There exists a positive random variable $D(\omega)$ and a positive constant $k = \beta - L_F - \delta^{-1} L_F$, such that: For any solution $u(t, \theta_{-t}\omega)$ of (6.20), there is an orbit $U(t, \theta_{-t}\omega)$ on the invariant manifold $\mathcal{M}(\omega)$, with the following approximation property*

$$\|u(t, \theta_{-t}\omega) - U(t, \theta_{-t}\omega)\| \leq D(\omega)\|u(0) - U(0)\| e^{-kt}, \quad t > 0, \quad \text{a.s.} \quad (6.39)$$

(iii) *Furthermore, $U(t, \omega)$ is represented as $u_c(t, \omega) + \bar{h}^s(u_c(t, \omega), \theta_t\omega)$, where u_c solves the reduced system on the random inertial manifold $\mathcal{M}(\omega)$,*

$$du_c(t) = \left[A_c u_c(t) + F_c(u_c(t) + \bar{h}^s(u_c(t), \theta_t\omega)) \right] dt + u_c(t) \circ dW(t),$$

$$(6.40)$$

and $\bar{h}^s(\cdot, \omega) : H_c \to H_s$, with $\bar{h}(0, \omega) = 0$, is the graph for this manifold.

Note that $\bar{h}^s$ depends on ω, which is driven by θ_t, so (6.40) in fact is a *nonautonomous* stochastic differential equation on H_c.

Remark 6.11. When the central spectrum, σ_c, contains only eigenvalues of zero real part, this theorem provides a random center manifold reduction. For more information, see [52,310].

Remark 6.12. The exact expression for a random invariant manifold is usually not available. However, an asymptotic expansion near a stationary orbit may be possible and useful in some cases—for example, for slow-fast SPDEs in the next section. For more general cases, the computer algebra method developed by Roberts [267] is an effective tool for deriving an asymptotic expansion of random invariant manifolds.

To prove this theorem, we need some preliminary results. First, we prove the attracting property of the random invariant manifold. To this end, we introduce some notions regarding random invariant manifolds for a random dynamical system $\varphi(t, \omega)$.

Definition 6.13 (Almost sure asymptotic completeness). An invariant manifold $\mathcal{M}(\omega)$ for a random dynamical system $\varphi(t, \omega)$ is called almost surely asymptotically complete if, for every $x \in H$, there exists $y \in \mathcal{M}(\omega)$, such that

$$\|\varphi(t, \omega)x - \varphi(t, \omega)y\| \le D(\omega)\|x - y\|e^{-kt}, \ t \ge 0,$$

for almost all $\omega \in \Omega$, where k is a positive constant and D is a positive random variable.

Now we introduce the *almost sure cone invariance* concept. For a positive random variable δ, define the following random set:

$$\mathcal{C}_\delta := \big\{(v, \omega) \in H \times \Omega : \|\Pi_s v\| \le \delta(\omega)\|\Pi_c v\|\big\}.$$

The fiber $\mathcal{C}_{\delta(\omega)}(\omega) = \{v : (v, \omega) \in \mathcal{C}_\delta\}$ is called a *random cone*.

Definition 6.14 (Almost sure cone invariance). A random dynamical system $\varphi(t, \omega)$ is said to have the cone invariance property for the random cone $\mathcal{C}_{\delta(\omega)}(\omega)$ if there exists a random variable $\bar{\delta} \le \delta$, almost surely, such that for all $x, y \in H$,

$$x - y \in \mathcal{C}_{\delta(\omega)}(\omega)$$

implies that

$$\varphi(t, \omega)x - \varphi(t, \omega)y \in \mathcal{C}_{\bar{\delta}(\theta_t \omega)}(\theta_t \omega), \quad \text{for almost all} \ \ \omega \in \Omega.$$

Remark 6.15. Both asymptotic completeness and cone invariance are important tools for studying inertial manifolds of deterministic, infinite-dimensional systems; see [194,268], and [269, Ch. 15]. Here we modified both concepts for random dynamical systems.

Remark 6.16. Almost sure asymptotic completeness describes the attracting property of $\mathcal{M}(\omega)$ for a random dynamical system $\varphi(t, \omega)$. When this property holds, the infinite-dimensional system, $\varphi(t, \omega)$, can be reduced to a finite dimensional system on $\mathcal{M}(\omega)$, and the asymptotic behavior of $\varphi(t, \omega)$ can be determined by that of the reduced system on $\mathcal{M}(\omega)$.

Now we prove asymptotic completeness of the random dynamical system $\varphi(t, \omega)$ defined by random evolutionary Equation (6.26). First, we prove the following cone invariance property.

Lemma 6.17. *For a sufficiently small Lipschitz constant L_F, the random dynamical system $\varphi(t, \omega)$, defined by the random evolutionary Equation (6.26), possesses the cone invariance property for a cone with a deterministic positive constant δ. Moreover, if there exists $t_0 > 0$, such that for $x, y \in H$ and*

$$\varphi(t_0, \omega)x - \varphi(t_0, \omega)y \notin \mathcal{C}_\delta(\theta_{t_0}\omega),$$

then

$$\|\varphi(t, \omega)x - \varphi(t, \omega)y\| \le D(\omega)e^{-kt}\|x - y\|, \ 0 \le t \le t_0,$$

where $D(\omega)$ is a positive tempered random variable and $k = \beta - L_F - \delta^{-1}L_F > 0$.

Proof.   The proof is via direct estimations. Let $v, \bar{v}$ be two solutions of (6.26) and $p \triangleq v_c - \bar{v}_c, q \triangleq v_s - \bar{v}_s$. Then,

$$\dot{p} = A_c p + z(\theta_t\omega)p + g_c(\theta_t\omega, v_c + v_s) - g_c(\theta_t\omega, \bar{v}_c + \bar{v}_s), \tag{6.41}$$

$$\dot{q} = A_s q + z(\theta_t\omega)q + g_s(\theta_t\omega, v_c + v_s) - g_s(\theta_t\omega, \bar{v}_c + \bar{v}_s). \tag{6.42}$$

Using the properties of A and F, we have

$$\frac{1}{2}\frac{d}{dt}\|p\|^2 \ge \alpha\|p\|^2 + z(\theta_t\omega)\|p\|^2 - L_F\|p\|^2 - L_F\|p\| \cdot \|q\|, \tag{6.43}$$

and

$$\frac{1}{2}\frac{d}{dt}\|q\|^2 \le -\beta\|q\|^2 + z(\theta_t\omega)\|q\|^2 + L_F\|q\|^2 + L_F\|p\| \cdot \|q\|. \tag{6.44}$$

By (6.44), and multiplying (6.43) by δ^2, we have

$$\frac{1}{2}\frac{d}{dt}(\|q\|^2 - \delta^2\|p\|^2)$$
$$\le -\beta\|q\|^2 + z(\theta_t\omega)\|q\|^2 + L_F\|q\|^2 + L_F\|p\| \cdot \|q\|$$
$$- \alpha\delta^2\|p\|^2 - z(\theta_t\omega)\delta^2\|p\|^2 + \delta^2 L_F\|p\|^2 + \delta^2 L_F\|p\| \cdot \|q\|.$$

Note that, if $(p, q) \in \partial\mathcal{C}_\delta(\omega)$ (the boundary of the cone $\mathcal{C}_\delta(\omega)$), then $\|q\| = \delta\|p\|$, and

$$\frac{1}{2}\frac{d}{dt}(\|q\|^2 - \delta^2\|p\|^2) \le (-\alpha - \beta + 2L_F + \delta L_F + \delta^{-1}L_F)\|q\|^2.$$

If L_F is small enough so that

$$-\alpha - \beta + 2L_F + \delta L_F + \delta^{-1}L_F < 0, \tag{6.45}$$

then $\|q\|^2 - \delta^2\|p\|^2$ is decreasing on $\partial\mathcal{C}_\delta(\omega)$. Thus, it is obvious that whenever $x - y \in \mathcal{C}_\delta(\omega)$, $\varphi(t, \omega)x - \varphi(t, \omega)y$ cannot leave $\mathcal{C}_\delta(\theta_t\omega)$.

To prove the second claim, suppose that there exists a $t_0 > 0$ such that $\varphi(t_0, \omega)x - \varphi(t_0, \omega)y \notin \mathcal{C}_\delta(\theta_{t_0}\omega)$. Then cone invariance property yields that

$$\varphi(t, \omega)x - \varphi(t, \omega)y \notin \mathcal{C}_\delta(\theta_t\omega), \ 0 \le t \le t_0,$$

that is,

$$\|q(t)\| > \delta\|p(t)\|, \ 0 \le t \le t_0.$$

By (6.44), we have

$$\frac{1}{2}\frac{d}{dt}\|q\|^2 \le -(\beta - L_F - \delta^{-1}L_F - z(\theta_t\omega))\|q\|^2, \ 0 \le t \le t_0.$$

Hence,

$$\|p(t)\|^2 < \frac{1}{\delta^2}\|q(t)\|^2 \le \frac{1}{\delta^2}e^{-2kt+\int_0^t z(\theta_s\omega)\,ds}, \ 0 \le t \le t_0.$$

Thus, by the definition of $z(\theta_t\omega)$, there exists a tempered random variable $D(\omega)$ such that

$$\|\varphi(t,\omega)x - \varphi(t,\omega)y\| \le D(\omega)e^{-kt}\|x-y\|, \ 0 \le t \le t_0.$$

The proof is complete. $\square$

The above result implies that if, for any solution v which does not lie on the random invariant manifold, there is a solution $\bar{v}$ on the random invariant manifold such that $v - \bar{v}$ does not stay in a random cone, then we have the almost sure asymptotic completeness of the random invariant manifold.

Before proving Theorem 6.10, we state a backward solvability result for system (6.26) restricted to the invariant manifold $\widetilde{\mathcal{M}}(\omega)$.

For any given final time $T_f > 0$, consider the following system for $t \in [0, T_f]$:

$$\dot{v}_c = A_c v_c + z(\theta_t\omega)v_c + g_c(\theta_t\omega, v_c + v_s), \ v_c(T_f) = \xi \in H_c, \tag{6.46}$$

$$\dot{v}_s = A_s v_s + z(\theta_t\omega)v_s + g_s(\theta_t\omega, v_c + v_s), \ v_s(0) = h^s(v_c(0)), \tag{6.47}$$

where h^s is defined in (6.38). Rewrite the above system in the following equivalent integral form for $t \in [0, T_f]$:

$$v_c(t) = e^{A_c(t-T_f)+\int_{T_f}^t z(\theta_\tau\omega)\,d\tau}\xi$$
$$+ \int_{T_f}^t e^{A_c(t-\tau)+\int_\tau^t z(\theta_s\omega)\,ds}g_c(\theta_\tau\omega, v(\tau))\,d\tau, \tag{6.48}$$

$$v_s(t) = e^{A_s t+\int_0^t z(\theta_\tau\omega)\,d\tau}h^s(v_c(0))$$
$$+ \int_0^t e^{A_s(t-\tau)+\int_\tau^t z(\theta_s\omega)\,ds}g_s(\theta_\tau\omega, v(\tau))\,d\tau. \tag{6.49}$$

Lemma 6.18. *Assume that the linear operator A has the spectral properties in §6.1.1 and that the gap condition (6.6) is satisfied. Then, for any $T_f > 0$, the stochastic system (6.48) and (6.49) has a unique solution $(v_c(\cdot), v_s(\cdot)) \in C(0, T_f; H_c \times H_s)$. Moreover, for every $t \ge 0$ and for almost all $\omega \in \Omega$, $(v_c(t, \theta_{-t}\omega), v_s(t, \theta_{-t}\omega)) \in \widetilde{\mathcal{M}}(\omega)$.*

Proof. The existence and uniqueness on a small time interval can be obtained by a contraction mapping argument [109, Lemma 3.3]. Then the solution can be extended to any time interval [109, Theorem 3.8]. We omit the details. $\square$

Proof of Theorem 6.10. It remains only to prove the almost sure asymptotic completeness of $\widetilde{\mathcal{M}}(\omega)$. Fix an $\omega \in \Omega$ and consider a solution

$$v(t, \theta_{-t}\omega) = (v_c(t, \theta_{-t}\omega), v_s(t, \theta_{-t}\omega))$$

of (6.26). For any $\tau > 0$, by Lemma 6.18 we can find a solution of (6.26) $\bar{v}(t, \theta_{-t}\omega)$, lying on $\widetilde{\mathcal{M}}(\omega)$, such that

$$\bar{v}_c(\tau, \theta_{-\tau}\omega) = v_c(\tau, \theta_{-\tau}\omega).$$

Then $\bar{v}(t, \theta_{-t}\omega)$ depends on $\tau > 0$. Write

$$\bar{v}_c(0; \tau, \omega) \triangleq \bar{v}_c(0, \omega)$$

and

$$\bar{v}_s(0; \tau, \omega) \triangleq \bar{v}_s(0, \omega).$$

By the construction of $\widetilde{\mathcal{M}}(\omega)$, we conclude that

$$
\begin{aligned}
\|\bar{v}_s(0; \tau, \omega)\| &\leq \int_0^\infty e^{-\beta r - \int_0^{-r} z(\theta_\varsigma\omega)\,d\varsigma} \|g_s(\theta_r\omega, v^*(-r))\| dr \\
&\leq L_F \int_0^\infty e^{-(\beta+\eta)r} e^{\eta r} e^{-\int_0^{-r} z(\theta_\varsigma\omega)\,d\varsigma} \|v^*(-r)\| dr \\
&\triangleq N_{L_F}(\omega) \\
&\leq L_F |v^*|_{C_\eta^-} \int_0^\infty e^{-(\beta+\eta)r}\,dr. \qquad (\beta + \eta > 0 \text{ by the choice of } \eta)
\end{aligned}
$$

Here we have introduced a random variable $N_{L_F}(\omega)$, which is a finite tempered random variable, and $N_{L_F}(\omega) \sim \mathcal{O}(L_F)$ almost surely. Since $\bar{v}_c(\tau, \theta_{-\tau}\omega) = v_c(\tau, \theta_{-\tau}\omega)$, by the cone invariance

$$v(t, \theta_{-t}\omega) - \bar{v}(t, \theta_{-t}\omega) \notin \mathcal{C}_\delta(\omega), \quad 0 \leq t \leq \tau.$$

Define $S(\omega) \triangleq \{\bar{v}_c(0; \tau, \omega) : \tau > 0\}$ and notice that

$$
\begin{aligned}
\|\bar{v}_c(0; \tau, \omega) - v_c(0, \omega)\| &< \frac{1}{\delta}\|\bar{v}_s(0; \tau, \omega) - v_s(0, \omega)\| \\
&\leq \frac{1}{\delta}\big(N_{L_F}(\omega) + \|v_s(0, \omega)\|\big).
\end{aligned}
$$

Then S is a random bounded set in a finite dimensional space, that is, for almost all $\omega \in \Omega$, $S(\omega)$ is a bounded set in $\mathbb{R}^N$, where $N = \dim H_c$. But the bound may not be uniform in $\omega \in \Omega$. However, for almost all $\omega \in \Omega$, we can pick up a sequence $\tau_m \to \infty$ such that

$$\lim_{m \to \infty} \bar{v}_c(0; \tau_m, \omega) = V_c(\omega).$$

Moreover, $V_c(\omega)$ is measurable with respect to ω. Define

$$V(t, \theta_{-t}\omega) \triangleq (V_c(t, \theta_{-t}\omega), V_s(t, \theta_{-t}\omega))$$

to be a solution of (6.26), with $V(0, \omega) = (V_c(\omega), h^s(V_c(\omega), \omega))$. Then $V(t, \theta_{-t}\omega) \in \widetilde{\mathcal{M}}(\omega)$, and it is clear, by a contradiction argument, that

$$v(t, \theta_{-t}\omega) \| - V(t, \theta_{-t}\omega) \notin C_\delta(\omega), \ 0 \leq t < \infty,$$

which implies the almost sure asymptotic completeness of $\widetilde{\mathcal{M}}(\omega)$.

Now, by the representation of $\widetilde{\mathcal{M}}(\omega)$ in Lemma 6.6 and the transform (6.25), restricting SPDE (6.20) to $\mathcal{M}(\omega)$, we have the reduced system (6.40). This completes the proof of Theorem 6.10. $\qquad\qquad\qquad\qquad\qquad\qquad\qquad\qquad\qquad\qquad\qquad\qquad\qquad$ □

Remark 6.19. Although we have considered SPDEs with linear multiplicative noise, the above discussion is also applicable to the additive noise case. We leave this as an exercise. Moreover, the assumption that the lower bound of σ_c is nonnegative is not necessary; the discussion is applicable for any α with $\alpha > -\beta$, as seen in [109]. But in this chapter, we are interested in random center or center-unstable manifolds.

6.4 Local Random Invariant Manifold for SPDEs

One key assumption in the approach to construct a random invariant manifold is the global Lipschitz property of the nonlinearity, as in the previous section. However, many physical models have local Lipschitz nonlinearities, so the results in §6.3 do not apply directly to these models. The existence of a random invariant manifold for SPDEs with local Lipschitz nonlinearity may still be proven in many cases. One method is to cut the nonlinearity outside an absorbing set for a dissipative system and thus obtain a local random invariant manifold. Recent work by Blömker and Wang [47] applied a cutoff technique to study the local random invariant manifold for a class of SPDEs with quadratic nonlinearity. The shape of the random invariant manifold is determined with a high probability. Here we give a brief introduction to this cut-off technique for some SPDEs with a quadratic nonlinearity.

Consider the following SPDE in a separable Hilbert space $(H, \| \cdot \|, \langle \cdot, \cdot \rangle)$,

$$\partial_t u = Au + vu + B(u, u) + \sigma u \circ \dot{W}, \ u(0) = u_0 \in H, \qquad\qquad (6.50)$$

where v and σ are positive constants and W is a standard scalar Brownian motion. The linear operator A has the spectral properties of §6.1.1, and furthermore, we assume $\lambda_i = 0$, for $i = 1, 2, \cdots, N$, and let $\lambda^* = \lambda_{N+1} < 0$. We also assume that the nonlinearity $B(u, u)$ satisfies the following assumption:

(A) For some $\alpha \in (0, 1)$, let $B : H \times H \to H^{-\alpha}$ be a bounded bilinear and symmetric operator, i.e., $B(u, \bar{u}) = B(\bar{u}, u)$, and let there be a constant $C_B > 0$ such that $\|B(u, \bar{u})\|_{-\alpha} \leq C_B \|u\| \|\bar{u}\|$. Also, denote by $\langle \cdot, \cdot \rangle$ the dual paring between $H^{-\alpha}$ and H^α, and further assume that for $B(u) \triangleq B(u, u)$,

$$\langle B(u), u \rangle = 0 \quad \text{for all } u \in H^\alpha.$$

As usual, denote by Π_c and Π_s the orthogonal projections from H to H_c and H_s respectively. We introduce the interpolation spaces H^α, $\alpha > 0$, as the domain of $(-A)^{\alpha/2}$ endowed with scalar product $\langle u, v \rangle_\alpha = \langle u, (1 - A)^\alpha v \rangle$ and corresponding norm $\| \cdot \|_\alpha$. We identify $H^{-\alpha}$ with the dual of H^α with respect to the scalar product in H.

Here the symmetry of B is not necessary, but it simplifies expansions of $B(u + v)$. Moreover, for simplicity, we assume that B has quadratic form.

We work in the probability space $(\Omega, \mathcal{F}, \mathbb{P})$ constructed in §6.2 and use the transform (6.25), that is,

$$v(t) = u(t)e^{-z(\theta_t \omega)}.$$

Then we have a random system

$$\partial_t v = Av + zv + vv + e^z B(v, v), \tag{6.51}$$

which is an evolutionary equation with random stationary coefficients. For almost all $\omega \in \Omega$, by a similar discussion as for the well-posedness of deterministic evolutionary equations [276, Ch. 4.6], for any $t_0 < T$ and $v_0 \in H$, there is a unique solution $v(t, \omega; t_0, v_0) \in C(t_0, T; H)$ of Equation (6.51), with $v(t_0) = v_0$ and solution mapping $v_0 \mapsto v(t, \omega; t_0, v_0)$, is continuous for all $t \geq t_0$. Then, by the above transform, (6.50) generates a continuous RDS $\varphi(t, \omega)$ on H.

Due to the local Lipschitz property of B, the result in §6.2 and §6.3 can not be applied directly to Equation (6.51). So, we define an R-Ball in H by $B_R(0) \triangleq \{u \in H : \|u\| < R\}$ and define the following local random invariant manifold (LRIM).

Definition 6.20. A random dynamical system $\varphi(t, \omega)$, defined on H, is said to have a local random invariant manifold with radius R if there is a random set $\mathcal{M}^R(\omega)$, which is defined by the graph of a random Lipschitz continuous function $h(\omega, \cdot)$: $\overline{B_R(0)} \cap H_c \to H_s$, such that for all bounded sets B in $B_R(0)$,

$$\varphi(t, \omega)[\mathcal{M}^R(\omega) \cap B] \subset \mathcal{M}^R(\theta_t \omega)$$

for all $t \in (0, \tau_0(\omega))$, where

$$\tau_0(\omega) = \tau_0(\omega, B) = \inf\{t \geq 0 : \varphi(t, \omega)[\mathcal{M}^R(\omega) \cap B] \not\subset B_R(0)\}. \tag{6.52}$$

In order to construct a LRIM for system (6.50), we introduce the following cut-off technique.

Definition 6.21 (Cut-off). Let $\chi : H \to \mathbb{R}^1$ be a bounded smooth function such that $\chi(u) = 1$ if $\|u\| \leq 1$ and $\chi(u) = 0$ if $\|u\| \geq 2$. For every $R > 0$, define $\chi_R(u) = \chi(u/R)$ for all $u \in H$. Given a radius $R > 0$, we further define a new nonlinear mapping

$$B^{(R)}(u) \triangleq \chi_R(u) B(u, u).$$

Now, by assumption (**A**), for a given α with property $0 < \alpha < 1$, the nonlinear mapping $B^{(R)}$ is globally Lipschitz continuous from H to $H^{-\alpha}$ with Lipschitz constant

$$\text{Lip}_{H,H^{-\alpha}}(B^{(R)}) = L_R \triangleq 2RC_B. \tag{6.53}$$

Next, we consider the following cut-off system

$$\partial_t u = Au + vu + B^{(R)}(u) + \sigma u \circ \dot{W}, \qquad u(0) = u_0. \tag{6.54}$$

By the transform (6.25), we have

$$v_t = Av + zv + vv + e^{-z} B^{(R)}(e^z v), \qquad v(0) = u_0 e^{-z(0)}. \tag{6.55}$$

By projections Π_c and Π_s, Equation (6.55) is split into

$$\partial_t v_c = vv_c + zv_c + \Pi_c e^{-z} B^{(R)}(e^z v), \qquad v_c(0) = \Pi_c u_0 e^{-z(0)},$$
$$\partial_t v_s = -A_s v_s + vv_s + zv_s + \Pi_s e^{-z} B^{(R)}(e^z v), \qquad v_s(0) = \Pi_s u_0 e^{-z(0)}.$$

Denote by $\varphi^R(t, \omega)$ the continuous random dynamical system defined by Equation (6.54) on H. Then, by the Lyapunov–Perron method of §6.2, we obtain a (global) random invariant manifold $\mathcal{M}_{\text{cut}}^R(\omega)$ for $\varphi^R(t, \omega)$, which is the graph of a Lipschitz continuous random mapping $h^{\text{cut}}(\omega, \cdot) : H_c \to H_s$ with

$$h^{\text{cut}}(\omega, \xi) = e^{z(\omega)} \int_{-\infty}^0 e^{(A_s - v)\tau + \int_\tau^0 z(r)\, dr} e^{-z(\tau)}$$
$$\times \Pi_s B^{(R)}(v^*(\tau, e^{-z(\omega)}\xi)e^{z(\tau)})\, d\tau,$$

provided that R is sufficiently small, such that for some $0 \prec \eta + v \prec \lambda \prec -\lambda_*$,

$$L_R\left[\frac{C_\alpha}{\eta + v} + M_{\alpha,\lambda}\frac{\Gamma(1-\alpha)}{(\lambda - \eta - v)^{1-\alpha}}\right] < 1. \tag{6.56}$$

Here, $\Gamma(\cdot)$ denotes the usual Gamma function and $M_{\alpha,\lambda}$ is a positive constant such that for $u \in H_s$,

$$\|e^{A_s t} u\| \le M_{\alpha,\lambda} e^{-t\lambda} t^{-\alpha} \|u\|_{-\alpha},$$

and $C_\alpha > 0$ is a constant such that $\|\Pi_c v\| \le C_\alpha \|\Pi_c v\|_{-\alpha}$.

Remark 6.22. The above inequality holds true for λ sufficiently close to $-\lambda_*$ and $\lambda < -\lambda_*$; see [287, p. 538].

Now define a Lipschitz continuous random mapping by

$$h(\omega, \cdot) : H_c \cap B_R(0) \to H_s$$
$$\xi \mapsto h(\omega, \xi) = h^{\text{cut}}(\omega, \xi).$$

Then,

$$\mathcal{M}^R(\omega) = \text{graph}(h(\omega, \cdot)) = \mathcal{M}_{\text{cut}}^R(\omega) \cap B_R(0)$$

defines an LRIM of the random dynamical system $\varphi(t, \omega)$ in H.

Remark 6.23. The above local random invariant manifold depends heavily on the exit time $\tau_0(\omega)$ of the orbit from the R-Ball $B_R(0)$. So, an estimate on $\tau_0(\omega)$ is important for further understanding the random dynamics of SPDE (6.50).

We are ready to show that this local random invariant manifold $\mathcal{M}^R(\omega)$ is locally exponentially attracting. First, we have the following cone invariance property.

Lemma 6.24. *Fix $\delta > 0$, and define the cone*

$$\mathcal{K}_\delta \triangleq \{u \in H \ : \ \|u_s\| < \delta\|u_c\|\}.$$

Suppose that R is sufficiently small such that

$$\lambda_* \geq 2\left(1 + \tfrac{1}{\delta}\right)^2 L_R^2 + 4(1+\delta)L_R, \tag{6.57}$$

and

$$\lambda_* > 4\nu + 2L_R^2\left(1 + \tfrac{1}{\delta}\right)^2. \tag{6.58}$$

Let $v, \bar{v}$ be two solutions of (6.55) with initial value $v_0 = u_0 e^{z(\omega)}$ and $\bar{v}_0 = \bar{u}_0 e^{z(\omega)}$. If $v(t_0) - \bar{v}(t_0) \in \mathcal{K}_\delta$, then $v_c(t) - \bar{v}_c(t) \in \mathcal{K}_\delta$ for all $t \geq t_0$. Moreover, if $v - \bar{v}$ is outside of $\mathcal{K}_\delta$ at some time t_0, then

$$\|v_s(t, \omega) - \bar{v}_s(t, \omega)\|^2$$
$$\leq \|u_0 - \bar{u}_0\|^2 \exp\left\{-\tfrac{1}{2}\lambda_* t + z(\omega) + 2\int_0^t z(\theta_\tau \omega)\,d\tau\right\}, \tag{6.59}$$

for all $t \in [0, t_0]$.

In the proof, we will see that 4ν is not optimal in (6.58), but for simplicity, we keep the 4ν.

Proof. Define

$$p \triangleq v_c - \bar{v}_c \quad \text{and} \quad q \triangleq v_s - \bar{v}_s.$$

Then,

$$\partial_t p = \nu p + zp + e^{-z}\Pi_c B^{(R)}(ve^z) - e^{-z}\Pi_c B^{(R)}(\bar{v}e^z),$$
$$\partial_t q = -A_s q + \nu q + zq + e^{-z}\Pi_s B^{(R)}(ve^z) - e^{-z}\Pi_s B^{(R)}(\bar{v}e^z).$$

By the properties of linear operator A and the Lipschitz property of $B^{(R)}$, we obtain, for some positive constant c_1, depending on λ_* and α, that

$$\frac{1}{2}\frac{d}{dt}\|p\|^2 \geq \nu\|p\|^2 + z\|p\|^2 - L_R\|p\|^2 - L_R\|p\|\|q\|, \tag{6.60}$$

and

$$\frac{1}{2}\frac{d}{dt}\|q\|^2 \leq -\|q\|_1^2 + \nu\|q\|^2 + z\|q\|^2 + L_R\|q\|\|q\|_\alpha + L_R\|p\|\|q\|_\alpha. \tag{6.61}$$

Thus,

$$\frac{1}{2}\frac{d}{dt}(\|q\|^2 - \delta^2\|p\|^2)$$
$$\leq -\|q\|_1^2 + \nu\|q\|^2 + z\|q\|^2 + L_R\|q\|\|q\|_\alpha + L_R\|p\|\|q\|_\alpha$$
$$- \nu\delta^2\|p\|^2 - z\delta^2\|p\|^2 + \delta^2 L_R\|p\|^2 + \delta^2 L_R\|p\|\|q\|.$$

Now, if $v - \bar{v} \in \mathcal{K}_\delta$ (that is $\delta\|p\| = \|q\|$) for some t, then

$$\frac{1}{2}\frac{d}{dt}(\|q\|^2 - \delta^2\|p\|^2)$$

$$\leq -\|q\|_1^2 + L_R\|q\|\|q\|_\alpha + L_R\|p\|\|q\|_\alpha + \delta^2 L_R\|p\|^2 + \delta^2 L_R\|p\|\|q\|$$

$$\leq -\|q\|_1^2 + L_R(\|p\| + \|q\|)\|q\|_1 + \delta^2 L_R\|p\|^2 + \delta^2 L_R\|p\|\|q\|$$

$$\leq -\|q\|_1^2 + L_R\left(1 + \tfrac{1}{\delta}\right)\|q\|\|q\|_1 + (1+\delta)L_R\|q\|^2$$

$$\leq -\tfrac{1}{2}\|q\|_1^2 + \left(\tfrac{1}{2}L_R^2\left(1 + \tfrac{1}{\delta}\right)^2 + (1+\delta)L_R\right)\|q\|^2,$$

where we have used Young's inequality in the last step.

By (6.57), via the Poincáre inequality, we obtain

$$\frac{d}{dt}(\|q\|^2 - \delta^2\|p\|^2) \leq -\tfrac{1}{2}\lambda_*\|q\|^2,$$

which yields the desired cone invariance property.

For the second claim, consider now that if $p+q$ is outside the cone at time t_0 (that is $\|q(t_0)\| > \delta\|p(t_0)\|$), then, by the first result, we have $\|q(t)\| > \delta\|p(t)\|$ for $t \in [0, t_0]$. By (6.61), we estimate

$$\frac{1}{2}\frac{d}{dt}\|q\|^2 \leq -\|q\|_1^2 + (v+z)\|q\|^2 + L_R\|q\|\|q\|_1 + L_R\|p\|\|q\|_1$$

$$\leq -\|q\|_1^2 + (v+z)\|q\|^2 + L_R\left(1 + \tfrac{1}{\delta}\right)\|q\|\|q\|_1$$

$$\leq -\tfrac{1}{2}\|q\|_1^2 + \left(v + z + \tfrac{1}{2}L_R^2\left(1 + \tfrac{1}{\delta}\right)^2\right)\|q\|^2.$$

By (6.58),

$$\frac{d}{dt}\|q(t, \omega)\|^2 \leq \left(-\tfrac{1}{2}\lambda_* + 2z(\theta_t\omega)\right)\|q(t, \omega)\|^2.$$

A comparison principle implies that, for almost all ω,

$$\|q(t, \omega)\|^2 \leq \|q(0, \omega)\|^2 \exp\left\{-\tfrac{1}{2}\lambda_* t + 2\int_0^t z(\theta_\tau\omega)\,d\tau\right\}.$$

Finally, $\|q(0, \omega)\| = \|v_s(0, \omega) - \bar{v}_s(0, \omega)\| \leq \|u_0 - \bar{u}_0\|e^{z(\omega)}$, which yields the second claim. This completes the proof. $\qquad\square$

To prove the attracting property of $\mathcal{M}_{cut}^R$, we also need the following backward solvability result.

Lemma 6.25. *For any given $T > 0$, the following terminal value problem,*

$$\dot{v}_c = v v_c + z v_c + \Pi_c e^{-z} B^{(R)}((v_c + v_s)e^z), \quad v_c(T) = \xi \in H_c,$$

$$\dot{v}_s = (-A_s + v + z)v_s + \Pi_s e^{-z} B^{(R)}((v_c + v_s)e^z), \quad v_s(0) = h^s(v_c(0)),$$

has a unique solution $(v_c(t, \omega), v_s(t, \omega))$ in $C(0, T; H_c \times H_s)$, which lies on the manifold $\mathcal{M}_{cut}^R(\theta_t\omega)$, a.s.

By the discussion in §6.3, we have the following attracting property of the random invariant manifold $\mathcal{M}_{cut}^{R}$:

Theorem 6.26 (Local random center manifold reduction). *Assume that the linear operator A has the spectral properties in §6.1.1 and that (6.56)–(6.58) hold. For every solution $u(t, \omega)$ of the cut-off system (6.54), there exists an orbit $U(t, \omega)$ on $\mathcal{M}_{cut}^{R}(\theta_t \omega)$, where $\Pi_c U(t, \omega)$ solves the following equation:*

$$\partial_t u_c = \nu u_c + \Pi_c B^{(R)}(u_c + e^{z(\theta_t \omega)} h(\theta_t \omega, e^{-z(\theta_t \omega)} u_c)) + \sigma u_c \circ \dot{W}(t),$$

and satisfies

$$\|u(t, \omega) - U(t, \omega)\| \leq D(t, \omega) \|u(0, \omega) - U(0, \omega)\| e^{-\lambda_* t},$$

with $D(t, \omega)$ being a tempered increasing process defined by

$$D(t, \omega) = e^{z(\theta_t \omega) + \int_0^t z(\theta_\tau \omega)\, d\tau}. \tag{6.62}$$

This is a local random center manifold reduction theorem.

For the original system (6.50), we have the following local approximation result.

Corollary 6.27. *There exists a local random invariant manifold $\mathcal{M}^R(\omega)$ for system (6.50), in a small ball $B(0, R)$, with R and ν sufficiently small. Moreover, for $\|u_0\| < R$,*

$$\mathrm{dist}(\varphi(t, \omega)u_0, \mathcal{M}(\omega)) \leq 2RD(t, \omega)e^{-\lambda_* t},$$

for all $t \in [0, \tau_0)$, where $\tau_0 = \inf\{t > 0 : \varphi(t, \omega)u_0 \notin B_r(0)\}$.

6.5   Random Slow Manifold Reduction for Slow-Fast SPDEs

In this section, we consider a random invariant manifold reduction, which is also called a random slow manifold reduction, for SPDEs with slow and fast time scales. The separation between slow time and fast time scales is measured by a small parameter $\epsilon > 0$. Some complex systems, as discussed in Chapter 5, have separated time scales and are described by the following slow-fast system in a separable Hilbert space H,

$$du^\epsilon(t) = [A_1 u^\epsilon(t) + f(u^\epsilon(t), v^\epsilon(t))]\, dt, \tag{6.63}$$

$$dv^\epsilon(t) = \frac{1}{\epsilon}[A_2 v^\epsilon(t) + g(u^\epsilon(t), v^\epsilon(t))]\, dt + \frac{1}{\sqrt{\epsilon}} dW(t), \tag{6.64}$$

with initial conditions $u^\epsilon(0) = u_0 \in H_1$, $v^\epsilon(0) = v_0 \in H_2$. Here the Hilbert space $H = H_1 \bigoplus H_2$ is the product of two separable Hilbert spaces $(H_1, |\cdot|_1)$ and $(H_2, |\cdot|_2)$, with norm $\|\cdot\| = |\cdot|_1 + |\cdot|_2$.

For a fixed $\epsilon > 0$, if the linear operator $A = (A_1, A_2/\epsilon)^T : D(A) \subset H \to H$ has the spectral properties in §6.1.1, and the nonlinearity $(f, g/\epsilon)^T : H \to H$ is Lipschitz continuous, with Lipschitz constant satisfying the gap condition (6.6), then the Lyapunov–Perron method is applicable here for constructing a random invariant manifold $\mathcal{M}^\epsilon$ for system (6.63) and (6.64). Furthermore, if H_1 is finite dimensional, then we have a random invariant manifold reduction on $\mathcal{M}^\epsilon$ by the method in §6.3. Now, for the slow-fast system (6.63) and (6.64), an interesting and important issue is to describe the random invariant manifold $\mathcal{M}^\epsilon$ and the reduced system on $\mathcal{M}^\epsilon$ for small ϵ. We call the approximation of the random invariant manifold $\epsilon \to 0$ and reduction of the system to the approximated random invariant manifold, as will be seen in Lemma 6.32 and Theorem 6.33, random slow manifold reduction.

The slow manifold is a special invariant manifold that describes slow dynamics of a system having slow and fast time scales [222]. Having a slow manifold, all fast variables are eliminated by "slaving" to the slow ones, and thus a reduced system on the slow manifold captures the effective dynamics of the original slow-fast system. Another interesting issue is the relation between this slow manifold reduction and averaging principles (see (6.85) below).

We first look at a linear example.

Example 6.28. Consider the following linear system with $H_1 = H_2 = \mathbb{R}^1$

$$\dot{u}^\epsilon(t) = au^\epsilon(t) + v^\epsilon(t), \tag{6.65}$$

$$dv^\epsilon(t) = -\frac{1}{\epsilon}v^\epsilon(t)\,dt + \frac{1}{\sqrt{\epsilon}}dW(t), \tag{6.66}$$

where a is a constant and $W(t)$ is a scalar Wiener process with covariance σ. Also, work in the canonical probability space $(\Omega_0, \mathcal{F}_0, \mathbb{P})$ defined by the sample paths of W, as in §4.10. By the averaging method from Chapter 5, we have an averaged equation for (6.65) and (6.66) on any finite time interval $[0, T]$, for $\epsilon \to 0$,

$$\dot{\bar{u}}(t) = a\bar{u}(t).$$

Notice that for any fixed $\epsilon > 0$, Equation (6.66) has a unique stationary solution $\eta^\epsilon(\theta_t\omega)$, which is exponentially stable, that is,

$$|v^\epsilon(t) - \eta^\epsilon(t)|_2 \to 0, \quad t \to \infty,$$

with exponential rate. However, direct verification yields that, for any fixed $\epsilon > 0$,

$$\mathcal{M}^\epsilon(\omega) \triangleq \{(\xi, \eta^\epsilon(\omega)) : \xi \in \mathbb{R}^1\}$$

is, in fact, a random invariant manifold for the system (6.65) and (6.66). Moreover,

$$\|(u^\epsilon(t), v^\epsilon(t)) - (u^\epsilon(t), \eta^\epsilon(\theta_t\omega))\| \to 0, \quad \text{exponentially as } t \to \infty.$$

That is, the random invariant manifold $\mathcal{M}^\epsilon(\omega)$ is exponentially attracting all orbits. Thus, we have the following reduced system

$$\dot{\bar{u}}^\epsilon(t) = a\bar{u}^\epsilon(t) + \eta^\epsilon(\theta_t\omega), \tag{6.67}$$

which is an equation with random force $\eta^\epsilon(\theta_t\omega)$.

On a slower time scale $t \mapsto t/\epsilon$, (6.65) and (6.66) are transformed to

$$\dot{\tilde{u}}^\epsilon(t) = \epsilon[a\tilde{u}^\epsilon(t) + \tilde{v}^\epsilon(t)], \tag{6.68}$$

$$d\tilde{v}^\epsilon(t) = -\tilde{v}^\epsilon(t)\, dt + d\tilde{W}(t), \tag{6.69}$$

where $\tilde{W}(t)$ is a scaled version of $W(t)$, which is also a scalar Wiener process with covariance σ. In this time scale, similarly, we also have a random invariant manifold, for fixed $\epsilon > 0$,

$$\tilde{M}(\omega) = \{(\xi, \eta(\omega)) : \xi \in \mathbb{R}^1\},$$

where $\eta(\theta_t\omega)$ is the unique stationary solution of (6.69). Again, we have the following reduced system on the random invariant manifold:

$$\dot{\tilde{u}}^\epsilon \triangleq \epsilon[a\hat{u}^\epsilon(t) + \eta(\theta_t\hat{\omega})]. \tag{6.70}$$

Here, $\hat{\omega} \in \Omega$, and this is due to the time-scale transformation (also see (6.77) below). We see that $\tilde{M}$ is in fact independent of ϵ, whereas $\mathcal{M}^\epsilon$ is highly oscillating due to $\eta^\epsilon(\omega)$ for ϵ small. Moreover, in this slower time scale, the reduced system (6.70) trivially becomes

$$\dot{\hat{u}} = 0, \text{ as } \epsilon \to 0.$$

By Problem 6.4 or the deviation discussion in §5.3, the reduced system (6.67) becomes a system driven by a white noise

$$\dot{\bar{u}}^\epsilon = a\bar{u}^\epsilon + \sqrt{\epsilon}\dot{\bar{W}} + \mathcal{O}(\epsilon),$$

where $\bar{W}$ is a scalar Wiener process with covariance σ. So, the reduced system (6.67) provides more information regarding the behavior of slow dynamics on a longer time scale.

Remark 6.29. The random invariant manifold $\mathcal{M}^\epsilon(\omega)$ has the same distribution as that of $\tilde{M}(\omega)$ for any fixed $\epsilon > 0$. For more discussion about these two kinds of random invariant manifolds, see [275].

In the previous example of a linear finite dimensional system, the random invariant manifold is easily constructed. For nonlinear systems, this is generally impossible. However, an approximation to the random invariant manifold is possible and effective in many cases. We next present a slow manifold reduction method for slow-fast system (6.63) and (6.64) for small $\epsilon > 0$.

Assume that $W(t)$ is an H_2-valued, two-sided Wiener process with covariance operator Q, defined on the canonical probability space $(\Omega, \mathcal{F}, \mathbb{P})$ associated with $W(t)$, with the measure-preserving Wiener shift $\{\theta_t\}_{t \in \mathbb{R}}$ on Ω. Furthermore, we make the following assumptions:

($\mathbf{A}_1$) *There exist constants $\beta > 0$ and $\alpha > -\beta$ such that, for every $u \in H_1$ and $v \in H_2$,*

$$|e^{A_1 t}u|_1 \leq e^{\alpha t}|u|_1, \quad t \leq 0 \quad \text{and} \quad |e^{A_2 t}v|_2 \leq e^{-\beta t}|v|_2, \quad t \geq 0.$$

(A$_2$) $f : H_1 \times H_2 \to H_1$ and $g : H_1 \times H_2 \to H_2$ are both Lipschitz continuous with Lipschitz constants L_f and L_g, respectively. Moreover, $f(0, 0) = 0$, $g(0, 0) = 0$, and for every $x, y \in \mathbb{R}$,

$$|f(x, y)| \leq L_f(|x| + |y| + 1), \quad |g(x, y)| \leq L_g(|x| + |y| + 1).$$

(A$_3$) The following spectral gap condition holds for every ϵ with $0 < \epsilon \leq 1$

$$L_f \frac{1}{\alpha - \lambda} + L_g \frac{1}{\epsilon\lambda + \beta} < 1,$$

for some $-\beta < \epsilon\lambda < \alpha$.

(A$_4$) The covariance operator Q is of trace class, that is, $\text{Tr}(Q) < \infty$.

Now, for any fixed $\epsilon > 0$, we rewrite equations (6.63) and (6.64) in the form (6.27). Let $H = H_1 \oplus H_2$, $U = (u, v)^T$, $W = (0, 1/\sqrt{\epsilon}W)^T$, and $F(U) = (f(u, v)$, $g(u, v)/\epsilon)^T$. Define a linear operator on H by

$$\mathcal{A}U \triangleq \text{diag}\left(A_1 u, \frac{1}{\epsilon} A_2 v\right).$$

Then (6.63) and (6.64) can be rewritten in the following form:

$$dU = [\mathcal{A}U + F(U)] \, dt + dW(t).$$

Introduce a stationary process $\eta^\epsilon(t) = \eta^\epsilon(\theta_t \omega)$ that solves

$$d\eta^\epsilon(t, \omega) = \frac{1}{\epsilon} A_2 \eta^\epsilon(t, \omega) \, dt + \frac{1}{\sqrt{\epsilon}} dW(t, \omega),$$

and set

$$(X^\epsilon, Y^\epsilon)^T = U - (0, \eta^\epsilon)^T.$$

Therefore, we have the following random evolutionary equations:

$$\dot{X}^\epsilon(t, \omega) = A_1 X^\epsilon(t, \omega) + f(X^\epsilon(t, \omega), Y^\epsilon(t, \omega) + \eta^\epsilon(\theta_t \omega)), \tag{6.71}$$

$$\dot{Y}^\epsilon(t, \omega) = \frac{1}{\epsilon} A_2 Y^\epsilon(t, \omega) + \frac{1}{\epsilon} g(X^\epsilon(t, \omega), Y^\epsilon(t, \omega) + \eta^\epsilon(\theta_t \omega)). \tag{6.72}$$

For λ satisfying $-\beta/\epsilon < \lambda < 0$, introduce the following Banach space similar to (6.7):

$$C_\lambda^- \triangleq \left\{ v : (-\infty, 0] \to H_1 \times H_2 : v \text{ is continuous and} \right.$$

$$\left. \sup_{t \in (-\infty, 0]} e^{-\lambda t} \|v(t)\| < \infty \right\}$$

with norm

$$|v|_{C_\lambda^-} = \sup_{t \in (-\infty, 0]} e^{-\lambda t} \|v(t)\|.$$

Then, as in §6.3, for the system (6.63) and (6.64), we have the following result.

Theorem 6.30 (Random slow manifold reduction).   *Assume that* $(\mathbf{A}_1)$–$(\mathbf{A}_4)$ *hold. Then the system (6.63) and (6.64) defines a continuous random dynamical system* $\varphi^\epsilon(t, \omega)$ *on* $H_1 \oplus H_2$. *Moreover,* $\varphi^\epsilon(t, \omega)$ *has a random invariant manifold*

$$\mathcal{M}^\epsilon(\omega) \triangleq \{(u, h^\epsilon(u, \omega) + \eta^\epsilon(\omega)) : u \in H_1\},$$

where $h^\epsilon(\cdot, \omega) : H_1 \to H_2$ *is a Lipschitz continuous random mapping defined by*

$$h^\epsilon(u, \omega) \triangleq \frac{1}{\epsilon} \int_{-\infty}^0 e^{-A_2 s / \epsilon} g(X^{\epsilon*}(s, \omega), Y^{\epsilon*}(s, \omega) + \eta^\epsilon(\theta_s \omega))\, ds, \qquad (6.73)$$

and

$$X^{\epsilon*}(t, \omega) = e^{A_1 t} u + \int_0^t e^{(t-s)A} f(X^{\epsilon*}(s), Y^{\epsilon*}(s) + \eta^\epsilon(\theta_s \omega))\, ds,$$

$$Y^{\epsilon*}(t, \omega) = \frac{1}{\epsilon} \int_{-\infty}^t e^{A_2(t-s)/\epsilon} g(X^{\epsilon*}(s), Y^{\epsilon*}(s) + \eta^\epsilon(\theta_s \omega))\, ds,$$

are the unique solutions in C_λ^- *of (6.71) and (6.72). Furthermore, if* H_1 *is finite dimensional and*

$$\epsilon \alpha + \epsilon L_f + \epsilon \delta L_f - \beta + L_g + \delta^{-1} L_g < 0,$$

then the random invariant manifold $\mathcal{M}^\epsilon(\omega)$ *is almost surely asymptotically complete. That is, for every solution* $(u^\epsilon(t), v^\epsilon(t))$ *to (6.63) and (6.64), there exists a solution* $(\bar{u}^\epsilon(t), h^\epsilon(\bar{u}^\epsilon(t), \omega) + \eta^\epsilon(\omega))$, *lying on* $\mathcal{M}^\epsilon(\omega)$, *which is governed by the following differential equation:*

$$\dot{\bar{u}}^\epsilon(t, \omega) = A_1 \bar{u}^\epsilon(t, \omega) + f(\bar{u}^\epsilon(t, \omega), h^\epsilon(\bar{u}^\epsilon(t, \omega), \theta_t \omega) + \eta^\epsilon(\theta_t \omega)), \qquad (6.74)$$

such that, for almost all ω *and* $t \geq 0$,

$$\|(u^\epsilon(t, \omega), v^\epsilon(t, \omega)) - (\bar{u}^\epsilon(t, \omega), h^\epsilon(u^\epsilon(t, \omega), \omega))\|$$
$$\leq D \|(u_0, v_0) - (\bar{u}^\epsilon(0), h^\epsilon(\bar{u}^\epsilon(0), \omega))\| e^{-\frac{\gamma}{\epsilon} t},$$

with $\gamma = \beta - L_g - \delta^{-1} L_g > 0$ *and a deterministic constant* $D > 0$.

If space H_1 is an infinite dimensional space, a bounded set may not be compact, so we cannot follow the approach in §6.3 to reduce (6.63) and (6.64) to the random invariant manifold $\mathcal{M}^\epsilon$ to obtain (6.74). Very recent work by Fu and Duan [138] reduced (6.63) and (6.64) onto the random invariant manifold $\mathcal{M}^\epsilon$, for ϵ sufficiently small, but without assuming that H_1 is finite dimensional.

Furthermore, if we are concerned with the case of small ϵ, we will still derive a slow manifold reduced system by approximating $\mathcal{M}^\epsilon$. A similar result is also obtained by Fu and Duan [138] for a slower time scale.

To approximate the random manifold $\mathcal{M}^\epsilon(\omega)$ for small ϵ, we consider the following random evolutionary equations:

$$\dot{\bar{X}}^\epsilon(t, \omega) = \epsilon A_1 \bar{X}^\epsilon(t, \omega) + \epsilon f(\bar{X}^\epsilon(t, \omega), \bar{Y}^\epsilon(t, \omega) + \eta(\theta_t \omega)), \qquad (6.75)$$

$$\dot{\bar{Y}}^\epsilon(t, \omega) = A_2 \bar{Y}^\epsilon(t, \omega) + g(\bar{X}^\epsilon(t, \omega), \bar{Y}^\epsilon(t, \omega) + \eta(\theta_t \omega)), \qquad (6.76)$$

where $\eta(t) = \eta(\theta_t \omega)$ is the unique stationary solution of the following linear equation:

$$d\eta(t, \omega) = A_2 \eta(t, \omega) + dW(t, \omega).$$

Notice that equations (6.75) and (6.76) are equations (6.71) and (6.72) on time scale ϵt. However, we show that equations (6.71) and (6.72) do not coincide with (6.75) and (6.76) for the same $\omega \in \Omega$, because the time scale transform $t \to \epsilon t$ also transforms $\theta_t \omega$ to $\theta_{\epsilon t} \omega$. In fact,

$$\eta^\epsilon(\omega) = \frac{1}{\sqrt{\epsilon}} \int_{-\infty}^{0} e^{-A_2 s/\epsilon} dW(s, \omega)$$

$$= \int_{-\infty}^{0} e^{-A_2 s} d \frac{1}{\sqrt{\epsilon}} W(\epsilon s, \omega)$$

$$= \int_{-\infty}^{0} e^{-A_2 s} dW^\epsilon(s, \omega).$$

By the scaling property of Wiener process, $W^\epsilon(t, \omega) = \frac{1}{\sqrt{\epsilon}} W(\epsilon s, \omega)$ is still a Wiener process with the same distribution as $W(t, \omega)$, but $W(t, \omega) \neq W^\epsilon(t, \omega)$ for almost all $\omega \in \Omega$. For this, we introduce a new measure-preserving transform $\omega \mapsto \psi^\epsilon \omega$, on probability space $(\Omega, \mathcal{F}, \mathbb{P})$, given by

$$\psi^\epsilon \omega(t) = \frac{1}{\sqrt{\epsilon}} \omega(\epsilon t), \quad \omega \in \Omega.$$

Then

$$W^\epsilon(t, \omega) = \frac{1}{\sqrt{\epsilon}} W(\epsilon t, \omega) = \frac{1}{\sqrt{\epsilon}} \omega(\epsilon t) = \psi^\epsilon \omega(t) = W(t, \psi^\epsilon \omega). \tag{6.77}$$

Therefore,

$$\eta^\epsilon(\omega) = \int_{-\infty}^{0} e^{-A_2 s} dW(s, \psi^\epsilon \omega) = \eta(\psi^\epsilon \omega).$$

Moreover, by the definitions of θ_t and ψ^ϵ, we have, for almost all $\omega \in \Omega$,

$$\eta^\epsilon(\theta_t \omega) = \eta(\theta_{t/\epsilon} \psi^\epsilon \omega).$$

Now, by (6.73), for almost all $\omega \in \Omega$,

$$h^\epsilon(u, \omega) = \int_{-\infty}^{0} e^{-A_2 s} g(X^{\epsilon*}(\epsilon s, \omega), Y^{\epsilon*}(\epsilon s, \omega) + \eta^\epsilon(\theta_{\epsilon s} \omega))$$

$$= \int_{-\infty}^{0} e^{-A_2 s} g(X^{\epsilon*}(\epsilon s, \omega), Y^{\epsilon*}(\epsilon s, \omega) + \eta(\theta_s \psi^\epsilon \omega))$$

$$= \int_{-\infty}^{0} e^{-A_2 s} g(\bar{X}^{\epsilon*}(s, \psi^\epsilon \omega), \bar{Y}^{\epsilon*}(s, \psi^\epsilon \omega) + \eta(\theta_s \psi^\epsilon \omega)),$$

where $(\bar{X}^{\epsilon*}(t), \bar{Y}^{\epsilon*}(t))$ is the unique solution of (6.75) and (6.76) in $C_{\epsilon\lambda}^-$. This shows that $\mathcal{M}^\epsilon(\omega)$ is a random invariant manifold for random dynamical system $\bar{\varphi}^\epsilon(t, \psi^\epsilon \omega)$, defined by (6.75) and (6.76).

To approximate $h^\epsilon(u, \omega)$, we introduce

$$h^0(u, \omega) = \int_{-\infty}^0 e^{-A_2 s} g(u, \bar{Y}^*(s) + \eta(\theta_s \omega)) \, ds, \tag{6.78}$$

where

$$\bar{Y}^*(t, \omega) = \int_{-\infty}^t e^{A_2(t-s)} g(u, \bar{Y}^*(s) + \eta(\theta_s \omega)) \, ds, t \leq 0 \tag{6.79}$$

is the unique solution, in C_λ^-, of

$$\dot{\bar{Y}}(t, \omega) = A_2 \bar{Y}(t, \omega) + g(u, \bar{Y}(t, \omega) + \eta(\theta_t \omega)). \tag{6.80}$$

In fact, by $(\mathbf{A}_2)$, $\beta + L_g < 0$, so for every fixed $u \in H_1$, the solution of (6.80), with initial value $h^0(u, \omega)$, is the unique stationary solution of (6.80).

Next we show that for small ϵ, $h^\epsilon(u, \omega)$ is approximated by $h^0(u, \psi^\epsilon \omega)$. For this, we need some energy estimates on the solution (u^ϵ, v^ϵ). In fact, we have the following result.

Lemma 6.31. *For every $(u_0, v_0) \in H$, the unique solution (u^ϵ, v^ϵ) of (6.63) and (6.64) has the following property: For every $T > 0$, there exists a constant $C_T > 0$ such that*

$$\sup_{0 \leq t \leq T} \mathbb{E}\left[|u^\epsilon(t)|_1^2 + |v^\epsilon(t)|_2^2\right] \leq C_T \left(|u_0|_1^2 + |v_0|_2^2\right).$$

Proof. Denote by $\langle \cdot, \cdot \rangle_1$ and $\langle \cdot, \cdot \rangle_2$ the scalar products in space H_1 and H_2, respectively. By the Lipschitz property and the facts that $f(0, 0) = 0$ and $g(0, 0) = 0$, we have

$$\begin{aligned}
\frac{1}{2}\frac{d}{dt}|u^\epsilon|_1^2 &= \langle A_1 u^\epsilon, u^\epsilon \rangle_1 + \langle f(u^\epsilon, v^\epsilon) - f(0, v^\epsilon u^\epsilon) \rangle_1 \\
&\quad + \langle f(0, v^\epsilon) - f(0, 0), u^\epsilon \rangle_1 \\
&\leq \alpha |u^\epsilon|_1^2 + L_f \|u^\epsilon\|_1^2 + L_f |u^\epsilon|_1 |v^\epsilon|_2,
\end{aligned} \tag{6.81}$$

and

$$\begin{aligned}
\frac{1}{2}\frac{d}{dt}|v^\epsilon|_2^2 &= \frac{1}{\epsilon}\langle A_2 v^\epsilon, v^\epsilon \rangle_2 + \frac{1}{\epsilon}\langle g(u^\epsilon, v^\epsilon) - g(u^\epsilon, 0), v^\epsilon \rangle_2 \\
&\quad + \frac{1}{\epsilon}\langle g(u^\epsilon, 0) - g(0, 0), v^\epsilon \rangle_2 + \frac{1}{2\epsilon}\mathrm{Tr}(Q) + \frac{1}{\sqrt{\epsilon}}\langle \dot{W}, v^\epsilon \rangle_2 \\
&\leq -\frac{1}{\epsilon}\beta |v^\epsilon|_2^2 + \frac{1}{\epsilon}L_g |v^\epsilon|_2^2 + \frac{1}{\epsilon}L_g |u^\epsilon|_1 |v^\epsilon|_2 + \frac{1}{2\epsilon}\mathrm{Tr}(Q) + \frac{1}{\epsilon}\langle \dot{W}, v^\epsilon \rangle_2.
\end{aligned} \tag{6.82}$$

Adding ϵ times (6.82) to (6.81) and taking expectation imply

$$\frac{d}{dt}\mathbb{E}\left[|u^\epsilon|_1^2 + \epsilon|v^\epsilon|_2^2\right] \leq (\alpha + L_f)\mathbb{E}|u^\epsilon|_1^2 - (\beta - L_g)\mathbb{E}|v^\epsilon|_2^2$$
$$+ (L_f + L_g)\mathbb{E}[|u^\epsilon|_1|v^\epsilon|_2] + \mathrm{Tr}(Q).$$

For every $\varepsilon > 0$, by Hölder inequality there is $C_\varepsilon > 0$, such that

$$(L_f + L_g)\mathbb{E}[|u^\epsilon|_1|v^\epsilon|_2] \leq C_\varepsilon|u^\epsilon|_1^2 + \varepsilon|v^\epsilon|_2^2.$$

Choose $\varepsilon > 0$ so that $-\beta + L_g + \varepsilon < 0$. Hence, for some positive constant C_1, we have

$$\frac{1}{2}\frac{d}{dt}\mathbb{E}\left[|u^\epsilon|_1^2 + \epsilon|v^\epsilon|_2^2\right] \leq (\alpha + L_f + C_\varepsilon)\mathbb{E}|u^\epsilon|_1^2 - (\beta - L_g - \varepsilon)|v^\epsilon|_2^2 + \mathrm{Tr}(Q)$$
$$\leq C_1\mathbb{E}[|u^\epsilon|_1^2 + \epsilon|v^\epsilon|_2^2] + \mathrm{Tr}(Q).$$

Thus, Gronwall inequality yields

$$\mathbb{E}\left[|u^\epsilon(t)|_1^2 + \epsilon|v^\epsilon(t)|_2^2\right] \leq C_T\left(|u_0|_1^2 + |v_0|_2^2\right), \quad 0 \leq t \leq T,$$

for some positive constant C_T. Now, going back to Equation (6.82), we still choose $\varepsilon > 0$ as above; then we have, for some $C'_\varepsilon > 0$,

$$\frac{1}{2}\frac{d}{dt}\mathbb{E}|v^\epsilon|_2^2 \leq -\frac{1}{\epsilon}(\beta - L_g + \varepsilon)|v^\epsilon|_2^2 + \frac{1}{\epsilon}C_\varepsilon|u^\epsilon|_1^2 + \frac{1}{\epsilon}\mathrm{Tr}(Q).$$

Again, by Gronwall inequality and the estimate on $|u^\epsilon|_1^2$, we conclude that

$$\mathbb{E}|v^\epsilon(t)|_2^2 \leq C_T\left(|u_0|_1^2 + |v_0|_2^2\right), \quad 0 \leq t \leq T,$$

for some constant $C_T > 0$. This completes the proof. $\qquad\square$

We now prove the following result regarding approximation of the random slow manifold.

Lemma 6.32.   *Assume that* $(\mathbf{A}_1)$–$(\mathbf{A}_4)$ *hold. Then, for almost all* $\omega \in \Omega$,

$$|h^\epsilon(u, \omega) - h^0(u, \psi^\epsilon\omega)|_2 = \mathcal{O}(\epsilon), \quad \text{as} \quad \epsilon \to 0,$$

which is uniform for u on any bounded set of the Hilbert space H_1.

Proof.   For a Lipschitz continuous mapping $h : H_1 \to H_2$ and $u \in H_1$, consider the solution $(\bar{X}^\epsilon(t, \omega, (u, h(u))), \bar{Y}^\epsilon(t, \omega, (u, h(u))))$ of the Equations (6.75) and (6.76), with initial value $(u, h(u) + \eta(\omega))$. By Lemma 6.31, for almost all ω,

$$\bar{X}^\epsilon(t, \omega, (u, h(u))) = e^{\epsilon A_1 t}u + \epsilon\int_0^t e^{\epsilon A_1(t-s)}f(\bar{X}^\epsilon(s, \omega, (u, h(u))),$$
$$\bar{Y}^\epsilon(s, \omega, (u, h(u))))\,ds \to u, \quad \epsilon \to 0,$$

uniformly in u from bounded set and $t \in [0, T]$.

Denote by $\bar{Y}(t, \omega, h(u))$ the solution of (6.80) with initial value $h(u)$. Note that

$$
|\bar{Y}^\epsilon(t, \omega, (u, h(u))) - \bar{Y}(t, \omega, h(u))|_2
$$

$$
\leq \int_0^t \left| e^{A_2(t-s)} \left[g(\bar{X}^\epsilon(s, \omega, (u, h(u))), \bar{Y}^\epsilon(s, \omega, (u, h(u)))) \right. \right.
$$

$$
\left. \left. - g(u, \bar{Y}(s, \omega, h(u))) \right] \right|_2 ds
$$

$$
\leq L_g \int_0^t e^{-\beta(t-s)} |\bar{X}^\epsilon(s, \omega, (u, h(u))) - u|_1 ds
$$

$$
+ L_g \int_0^t e^{-\beta(t-s)} |\bar{Y}^\epsilon(s, \omega, (u, h(u))) - \bar{Y}(s, \omega, h(u))|_2 ds.
$$

Using Gronwall inequality, we obtain that, uniformly for u in any bounded set,

$$
\bar{Y}^\epsilon(t, \omega, (u, h(u))) \to \bar{Y}(t, \omega, h(u)), \quad a.s. \quad \epsilon \to 0.
$$

Now, by the invariance of $(u, h^\epsilon(u, \omega))$ and $(u, h^0(u, \omega))$,

$$
|h^\epsilon(u, \omega) - h^0(u, \psi^\epsilon \omega)|_2
$$

$$
= |\bar{Y}^\epsilon(t, \theta_{-t}\psi^\epsilon \omega, (u, h^\epsilon(u, \omega))) - \bar{Y}(t, \theta_{-t}\psi^\epsilon \omega, h^0(u, \psi^\epsilon \omega))|_2
$$

$$
\leq |\bar{Y}^\epsilon(t, \theta_{-t}\psi^\epsilon \omega, (u, h^\epsilon(u, \omega))) - \bar{Y}^\epsilon(t, \theta_{-t}\psi^\epsilon \omega, h^0(u, \psi^\epsilon \omega))|_2
$$

$$
+ |\bar{Y}^\epsilon(t, \theta_{-t}\psi^\epsilon \omega, h^0(u, \psi^\epsilon \omega)) - \bar{Y}(t, \theta_{-t}\psi^\epsilon \omega, h^0(u, \psi^\epsilon \omega))|_2.
$$

Let $V^\epsilon(t, \omega) \triangleq \bar{Y}^\epsilon(t, \theta_{-t}\psi^\epsilon \omega, (u, h^\epsilon(u, \omega))) - \bar{Y}^\epsilon(t, \theta_{-t}\psi^\epsilon \omega, h^0(u, \psi^\epsilon \omega))$. From (6.76), and by choosing $\kappa > 0$ such that $-\beta + L_g + \kappa < 0$, we deduce that, for some $C_\kappa > 0$,

$$
\frac{1}{2}\frac{d}{dt}|V^\epsilon(t, \omega)|_2^2 \leq (-\beta + L_g + \kappa)|V^\epsilon(t, \omega)|_2^2
$$

$$
+ C_\kappa |\bar{X}^\epsilon(t, \theta_{-t}\omega, (u, h^\epsilon(u, \omega)))
$$

$$
- \bar{X}^\epsilon(t, \theta_{-t}\omega, (u, h^0(u, \psi^\epsilon \omega)))|_1^2.
$$

Next, choose $T > 0$ such that $e^{(-\beta + L_g + \kappa)T} < 1/4$, and using Cauchy–Schwarz inequality, we get

$$
|V^\epsilon(T, \omega)|_2 \leq \frac{1}{2}|h^\epsilon(u, \omega) - h^0(u, \psi^\epsilon \omega)|_2
$$

$$
+ 2\left[\int_0^T e^{(-\beta + L_g + \kappa)(T-s)} |\bar{X}^\epsilon(s, \theta_{-s}\omega, (u, h^\epsilon(u, \omega))) \right.
$$

$$
\left. - \bar{X}^\epsilon(s, \theta_{-s}\omega, (u, h^0(u, \psi^\epsilon \omega)))|_1^2 ds \right]^{1/2}.
$$

Hence,

$$
|h^\epsilon(u, \omega) - h^0(u, \psi^\epsilon \omega)|_2
$$

$$
\leq |\bar{Y}^\epsilon(T, \theta_{-T}\psi^\epsilon \omega, (u, h^0(u, \psi^\epsilon \omega))) - \bar{Y}(T, \theta_{-T}\psi^\epsilon \omega, h^0(u, \psi^\epsilon \omega))|_2
$$

$$+ 2\left[\int_0^T e^{(-\beta+L_g+\kappa)(T-s)} |\bar{X}^\epsilon(s, \theta_{-s}\omega, (u, h^\epsilon(u, \omega)))\right.$$

$$\left. - \bar{X}^\epsilon(s, \theta_{-s}\omega, (u, h^0(u, \psi^\epsilon\omega)))|_1^2 \, ds\right]^{1/2} \to 0, \quad a.s. \quad \epsilon \to 0.$$

This completes the proof.   □

For the reduced system (6.74) and by Lemma 6.32, we have

$$\dot{\bar{u}}^\epsilon(t, \omega) = A_1\bar{u}^\epsilon(t, \omega) + f(\bar{u}^\epsilon(t, \omega), h^\epsilon(\bar{u}^\epsilon(t, \omega), \theta_t\omega) + \eta^\epsilon(\theta_t\omega))$$

$$= A_1\bar{u}^\epsilon(t, \omega) + f(\bar{u}^\epsilon(t, \omega), h^0(\bar{u}^\epsilon(t, \omega), \psi^\epsilon\theta_t\omega)) + \mathcal{O}(\epsilon) + \eta(\psi^\epsilon\theta_t\omega))$$

$$= A_1\bar{u}^\epsilon(t, \omega) + f(\bar{u}^\epsilon(t, \omega), h^0(\bar{u}^\epsilon(t, \omega), \theta_{\frac{t}{\epsilon}}\psi^\epsilon\omega)$$

$$+ \eta(\theta_{\frac{t}{\epsilon}}\psi^\epsilon\omega)) + \mathcal{O}(\epsilon).$$

Thus, we have the following approximate random slow manifold reduction theorem.

Theorem 6.33 (Approximate random slow manifold reduction). *Assume that* $(\mathbf{A}_1)$–$(\mathbf{A}_4)$ *hold. Then random slow manifold reduction for the slow-fast* SPDEs *(6.63) and (6.64), up to the order of* $\mathcal{O}(\epsilon)$*, is*

$$d\hat{u}^\epsilon(t) = \left[A_1\hat{u}^\epsilon(t) + f\left(\hat{u}^\epsilon(t), h^0\left(\hat{u}^\epsilon(t), \theta_{\frac{t}{\epsilon}}\psi^\epsilon\omega\right) + \eta\left(\theta_{\frac{t}{\epsilon}}\psi^\epsilon\omega\right)\right)\right] dt \quad (6.83)$$

with initial value $\hat{u}^\epsilon(0) = u_0$.

The above result shows that the slow manifold reduction has an error of order ϵ instead of $\sqrt{\epsilon}$, as in the averaging approach. However, for small $\epsilon > 0$, the slow manifold reduced model (6.83) is in fact a differential equation with oscillating random coefficients that are strong, mixing with exponential decay rate. Then, by the averaging results for random differential equations with oscillating coefficients in §5.6, we have the following averaged equation:

$$d\bar{u} = [A_1\bar{u} + \bar{f}(\bar{u})] dt, \quad (6.84)$$

where $\bar{f}(\bar{u}) \triangleq \mathbb{E}f(\bar{u}, h^0(\bar{u}, \omega) + \eta(\omega))$. This again yields an averaging result for slow-fast SPDEs. Furthermore, by the deviation consideration in Chapter 5, a reduced model with an error of ϵ can be derived in the sense of distribution. We omit the details here.

So, if the random slow invariant manifold $\mathcal{M}^\epsilon(\omega)$ is almost surely asymptotically complete, we can reduce the system onto $\mathcal{M}^\epsilon(\omega)$, which is an approximation up to an error of order $\mathcal{O}(\epsilon)$. Furthermore, if ϵ is small, we then can approximate the slow manifold by an averaging model, which is up to an error of $\mathcal{O}(\sqrt{\epsilon})$. Then we formally deduce the following relation between slow manifold reduction and averaging reduction for the slow-fast system (6.63) and (6.64):

$$(u^\epsilon(t), v^\epsilon(t)) = (\bar{u}^\epsilon(t), h^\epsilon(\bar{u}^\epsilon(t), \theta_t\omega) + \eta^\epsilon(\theta_t\omega)) + \mathcal{O}(\exp\{-\lambda t\})$$

for some $\lambda > 0$, a.s. $\omega \in \Omega$ and any $t > 0$

(almost sure asymptotic completeness)(Theorem 6.30)

$$= \left(\hat{u}^\epsilon(t), h^0\left(\hat{u}^\epsilon(t), \theta_{\frac{t}{\epsilon}} \psi^\epsilon \omega \right) + \eta\left(\theta_{\frac{t}{\epsilon}} \psi^\epsilon \omega \right) \right) + \mathcal{O}(\epsilon)$$

on any finite time interval $[0, T]$(Theorem 6.33)

$$= (\bar{u}(t), \bar{h}^0(\bar{u}(t))) + \sqrt{\epsilon} \text{ fluctuation} + \mathcal{O}(\epsilon)$$

in the sense of distribution on finite time interval

$[0, T]$(averaging and deviation result),   (6.85)

where $\bar{u}$ solves (6.84), and $\bar{h}(\bar{u}) \triangleq \mathbb{E}h^0(\bar{u}, \omega)$.

These relations imply that random slow manifold reduction ($\mathcal{O}(\epsilon)$) is more effective than averaging reduction ($\mathcal{O}(\sqrt{\epsilon})$). Moreover, random slow manifold reduction gives an approximation at any time, whereas the averaging approximation is effective with any fixed end time T. The inclusion of a deviation estimate improves the effectiveness of the averaging approximation in the sense of distribution with an error up to $\mathcal{O}(\epsilon)$. One advantage of the averaging and deviation is that the reduced equation is a stochastic differential system, which is easier to treat than a differential equation with random coefficients (random slow manifold reduced equation).

Now we present one example to illustrate slow manifold reduction for slow-fast SPDEs.

Example 6.34. Consider the following slow-fast SPDEs:

$$\dot{u}^\epsilon = au^\epsilon + f(u^\epsilon) + v^\epsilon, \quad u^\epsilon(0) = u_0 \in L^2(D),$$

$$dv^\epsilon = \frac{1}{\epsilon}\left[v^\epsilon_{xx} + u^\epsilon \right] dt + \frac{1}{\sqrt{\epsilon}}dW(t), \quad v^\epsilon(0) = v_0 \in L^2(D),$$

for $x \in D = (-l, l), l > 0$, with zero Dirichlet boundary conditions, with $a \geq 0$, and $W(t)$ an L^2-valued Wiener process. Denote by $-\lambda_1$ the first eigenvalue of ∂_{xx} on $(-l, l)$ with zero Dirichlet boundary conditions. The nonlinear term $f : \mathbb{R}^1 \to \mathbb{R}^1$ is Lipschitz continuous with Lipschitz constant L_f such that, for every ϵ with $0 < \epsilon \leq 1$,

$$L_f \frac{1}{a - \lambda} + \frac{1}{\epsilon\lambda + \lambda_1} < 1$$

for $-\lambda_1 < \epsilon\lambda < a$. Assumption ($\mathbf{A}_3$) is thus satisfied.

The above slow-fast stochastic system has an almost sure asymptotic complete stochastic slow manifold, which can be represented as the graph of a mapping $h^\epsilon(\cdot, \omega) : L^2(D) \to L^2(D)$. Thus, we have the following random invariant manifold reduction:

$$\dot{u}^\epsilon(t) = au^\epsilon(t) + f(u^\epsilon(t)) + h^\epsilon(u^\epsilon(t), \theta_t\omega).$$

For small $\epsilon > 0$, by Lemma 6.32, $h^\epsilon(u, \omega)$ has the following expansion in u for almost all $\omega \in \Omega$:

$$h^\epsilon(u, \omega) = A^{-1}u + \eta^\epsilon(\omega) + \mathcal{O}(\epsilon)$$

$$= h^0(u, \psi^\epsilon \omega) + \mathcal{O}(\epsilon),$$

where $h^0(u, \omega) \triangleq A^{-1}u + \eta(\omega)$, $A \triangleq -\Delta$ with zero Dirichlet boundary conditions on D, $\eta^\epsilon(t, \omega) = \eta^\epsilon(\theta_t\omega)$ and $\eta(t, \omega) = \eta(\theta_t\omega)$ are the stationary solutions of

$$d\eta^\epsilon(t, \omega) = \frac{1}{\epsilon}\eta^\epsilon_{xx}(t, \omega)\, dt + \frac{1}{\sqrt{\epsilon}}dW(t, \omega)$$

and

$$d\eta(t, \omega) = \eta_{xx}(t, \omega)\, dt + dW^{\epsilon}(t, \omega),$$

respectively. Here $W^{\epsilon}(t, \omega) = \frac{1}{\sqrt{\epsilon}} W(\epsilon t, \omega)$, which is an L^2-valued Wiener process with the same distribution as that of $W(t, \omega)$.

Then we have the following random slow manifold reduction up to an error of $\mathcal{O}(\epsilon)$:

$$\dot{\bar{u}}^{\epsilon}(t) = a\bar{u}^{\epsilon}(t) + f(\bar{u}^{\epsilon}(t)) + A^{-1}\bar{u}^{\epsilon}(t) + \eta^{\epsilon}(\theta_t \omega)$$
$$= a\bar{u}^{\epsilon}(t) + f(\bar{u}^{\epsilon}(t)) + A^{-1}\bar{u}^{\epsilon}(t) + \eta(\theta_{\frac{t}{\epsilon}} \psi^{\epsilon} \omega).$$

Furthermore, by applying the averaging method, we have an averaged system up to an error of $\sqrt{\epsilon}$:

$$\dot{\bar{u}}(t) = a\bar{u}(t) + f(\bar{u}(t)) + A^{-1}\bar{u}(t).$$

By Problem 6.4, in the sense of distribution, the limit of $\frac{1}{\sqrt{\epsilon}} \int_0^t \eta^{\epsilon}(s)\, ds, \epsilon \to 0$, behaves like a Wiener process $\bar{W}(t)$. Therefore, we have a reduced system, up to an error of $\mathcal{O}(\epsilon)$, in the sense of distribution

$$\dot{\bar{u}}(t) = a\bar{u}(t) + f(\bar{u}(t)) + A^{-1}\bar{u}(t) + \sqrt{\epsilon}\dot{\bar{W}}(t).$$

Notice that here the fluctuation is independent of $\bar{u}$, so we can write down the above equation. For a general case, where the fast part and slow part are nonlinearly coupled, the above equation is not obvious [219].

6.6 A Different Reduction Method for SPDEs: Amplitude Equation

We apply stochastic slow manifold reduction to study the amplitude of solutions for a class of SPDEs near a change of stability that has been studied by asymptotic expansions [43]. Consider the following stochastic reaction-diffusion equation on $[0, \pi]$:

$$dw = [w_{xx} + w + \epsilon\gamma w + f(w)]\, dt + \sqrt{\epsilon}dW(t), \tag{6.86}$$
$$w^{\epsilon}(0, t) = 0, w^{\epsilon}(\pi, t) = 0, \tag{6.87}$$

where $\gamma > -1$ is a bifurcation parameter. We consider a special degenerate Wiener process W, that is, $W(x, t) = \beta_2(t) \sin 2x$, where $\beta_2(t)$ is a standard scalar Brownian motion. The amplitude equation for SPDEs with more general degenerate noise is discussed in [46]. Additionally, we assume that $f(0) = 0$, $f(w)w < 0$, and f is Lipschitz continuous, with Lipschitz constant L_f satisfying

$$\gamma - L_f > -1.$$

Furthermore, we assume that for any $x \in \mathbb{R}^1$, $\frac{1}{\alpha^3} f(\alpha x)$ is continuous in α, and the limit $h(x) \triangleq \lim_{\alpha \to 0} \frac{1}{\alpha^3} f(\alpha x)$ exists and is Lipschitz continuous with Lipschitz constant L_f.

Remark 6.35.   An example of the nonlinear function f is $f(x) = 6r(\sin x - x)$ on $(-1/2, 1/2)$, but it vanishes outside of $(-1, 1)$ for a small $r > 0$. Then $h(x) = -rx^3$ on $(-1/2, 1/2)$ and vanishes outside of $(-1, 1)$. So, for this special f, $f(w) \sim -rw^3$ when we consider a small solution. Thus, the properties of the small solutions of Equation (6.86) are similar to those of Equation (5.86) with cubic nonlinearity. Next we show this by random slow manifold reduction.

We are interested in the small solution on a long time scale ϵ^{-1}. Decomposing the solution as $w(t) \triangleq \sqrt{\epsilon} u^\epsilon(t') + \sqrt{\epsilon} v^\epsilon(t')$ in slow time scale $t' = \epsilon t$ and by omitting the primes, we have

$$u_t^\epsilon = \gamma u^\epsilon + \mathcal{P}_1 f^\epsilon(u^\epsilon, v^\epsilon), \tag{6.88}$$

$$dv^\epsilon = \frac{1}{\epsilon} \left[v_{xx}^\epsilon + v^\epsilon + \epsilon \gamma v^\epsilon + \epsilon \mathcal{Q}_1 f^\epsilon(u^\epsilon, v^\epsilon) \right] dt + \frac{1}{\sqrt{\epsilon}} dW^\epsilon(t), \tag{6.89}$$

where $f^\epsilon(u^\epsilon, v^\epsilon) \triangleq \frac{1}{\epsilon\sqrt{\epsilon}} f(\sqrt{\epsilon}u^\epsilon + \sqrt{\epsilon}v^\epsilon)$, $W^\epsilon(t) = \sqrt{\epsilon} W(\epsilon^{-1}t)$, $\mathcal{P}_1$ is the projection from $H = L^2(0, \pi)$ to space $\{a \sin x : a \in \mathbb{R}\}$, and $\mathcal{Q}_1 = Id_H - \mathcal{P}_1$. Equations (6.88) and (6.89) are of the form (6.63) and (6.64), with $\alpha = \gamma$, $\beta = 1 - \epsilon\gamma$, and $L_g = \epsilon L_f$. Then, for small ϵ and $\gamma - L_f > -1$, Assumption ($\mathbf{A}_3$) holds for $-1 + \epsilon\gamma < \lambda < \gamma$. By the discussion in the previous section, we have the following random invariant manifold reduction for (6.88) and (6.89):

$$\bar{u}_t^\epsilon(t) = \gamma \bar{u}^\epsilon(t) + \mathcal{P}_1 f^\epsilon(\bar{u}^\epsilon(t), h^\epsilon(\bar{u}^\epsilon(t), \theta_t\omega) + \eta^\epsilon(\theta_t\omega)), \tag{6.90}$$

where $h^\epsilon(\cdot, \omega) : \mathcal{P}_1 H \to \mathcal{Q}_1 H$ and

$$d\eta^\epsilon = \frac{1}{\epsilon} \left[\partial_{xx} + 1 \right] \eta^\epsilon dt + \frac{1}{\sqrt{\epsilon}} dW^\epsilon(t).$$

Now we approximate $h^\epsilon(u, \omega) + \eta^\epsilon(\omega)$ by noticing the special form of the Equations (6.89). First, by the assumption $f(w)w < 0$, for any $T > 0$, the same discussion for (5.94) yields a uniform estimate in ϵ for $(u^\epsilon(t), v^\epsilon(t))$ in $L^2(\Omega, C(0, T; H) \times L^2(0, T; H))$. Then, for small $\epsilon > 0$, $\epsilon \gamma v^\epsilon + \epsilon \mathcal{Q}_1 f^\epsilon(u^\epsilon, v^\epsilon)$ is of $\mathcal{O}(\epsilon)$ in $L^2(\Omega, H)$; and by the same discussion for (5.96), $h^\epsilon(u, \omega)$, on any bounded set of $\mathcal{P}_1 H$, is approximated by $\eta^\epsilon(\omega) \in \mathcal{Q}_1 H$ up to an error of $\mathcal{O}(\epsilon)$ in the sense of mean square, or by $\eta(\psi^\epsilon\omega)$, which solves

$$d\eta = \left[\partial_{xx} + 1 \right] \eta \, dt + dW(t).$$

Remark 6.36.   For the above approximation of a random invariant manifold, a formal explanation is that in slow time scale t', slow variable u^ϵ varies on the order of $\mathcal{O}(\epsilon)$, so up to an error of $\mathcal{O}(\epsilon)$, the slow manifold is close to v^ϵ with any fixed slow variable u^ϵ in (6.89). Furthermore, v^ϵ is approximated by η^ϵ due to *a priori* estimates of the solutions. Then we have the approximation of the random invariant manifold by $\eta^\epsilon(\omega)$.

Now we have the following random slow manifold reduction up to an error of $\mathcal{O}(\epsilon)$ for almost all $\omega \in \Omega$:

$$
\begin{aligned}
\hat{u}_t^\epsilon(t) &= \gamma \hat{u}^\epsilon(t) + \mathcal{P}_1 f^\epsilon(\hat{u}^\epsilon(t), \eta^\epsilon(\theta_t \omega)) \\
&= \gamma \hat{u}^\epsilon(t) + \mathcal{P}_1 f^\epsilon\left(\hat{u}^\epsilon(t), \eta\left(\theta_{\frac{t}{\epsilon}} \psi^\epsilon \omega\right)\right).
\end{aligned}
\tag{6.91}
$$

To compare with the averaged model, we consider the following example.

Example 6.37. Consider $f(x) = 6r(\sin x - x)$ on $(-1/2, 1/2)$ but vanishes outside of $(-1, 1)$ for small $r > 0$. Then, for small $\epsilon > 0$, by the expansion of $\sin x$,

$$
f^\epsilon(x) = -rx^3 + \mathcal{O}(\epsilon)x^5 + \cdots.
$$

Thus, Equation (6.91) can further be reduced to, up to an error of $\mathcal{O}(\epsilon)$ in $L^2(\Omega, C(0, T; H))$,

$$
\bar{u}_t^\epsilon = \gamma \bar{u}^\epsilon - r\mathcal{P}_1(\bar{u}^\epsilon + \eta^\epsilon)^3,
\tag{6.92}
$$

which is a differential equation with random oscillating coefficients. By an averaging result for random differential equation with oscillating coefficients in Chapter 5, we arrive at the averaged reduced equation

$$
\partial_t \bar{u} = \gamma \bar{u} - r\mathcal{P}_1 \bar{u}^3 - r\frac{3\sigma^2}{6}\mathcal{P}_1(\bar{u} \sin^2 2x),
\tag{6.93}
$$

which is exactly (5.97). This also reveals the relationship between slow manifold reduction and averaging reduction for slow-fast SPDEs.

Remark 6.38. In the above example, we use the approximation of $f^\epsilon(u)$ by $-u^3$ for small ϵ. In fact, this makes sense only on bounded set of $u \in H$ with small ϵ. So, the reduced system (6.92) is effective for $t < t_R(\omega)$ with stopping time t_R defined as

$$
t_R(\omega) \triangleq \inf\{t > 0 : \|\bar{u}^\epsilon(t)\| \geq R\pi\}
$$

for every R with $0 < R < \frac{1}{2}$.

6.7   Looking Forward

In this chapter, we have considered a random center-unstable manifold reduction for a system of SPDEs and a random slow manifold reduction for SPDEs with slow and fast time scales. The effective dynamics is described by a reduced system on a random inertial manifold, random center manifold, or random slow manifold.

A further issue is the investigation of the structure or geometric shape [70] of random slow manifolds to facilitate better understanding of the reduced systems on such manifolds.

It is also interesting to investigate slow manifold reduction for fully coupled slow-fast SPDEs (5.128) and (5.129) near a stationary orbit.

6.8   Problems

We assume that the Wiener process $W(t)$, $t \in \mathbb{R}$, takes values in Hilbert space $H = L^2(0, l)$, $l > 0$, and it has a trace class covariance operator Q that shares the same eigenfunctions with the Laplace operator ∂_{xx} under appropriate boundary conditions. Additionally, we take ϵ to be a small positive real parameter.

6.1. Stationary solution

Let $0 < \lambda_1 \le \lambda_2 \cdots \le \lambda_N \le \cdots$ be the eigenvalues of $-\partial_{xx}$ with zero Dirichlet boundary conditions on $(0, l)$. Consider the following linear SPDE:

$$du = [\partial_{xx}u + \lambda u]\,dt + dW(t), \quad u(0, t) = u(l, t) = 0,$$

where $\lambda_N > \lambda > \lambda_{N-1}$ for some $N \ge 1$ and $\lambda_0 = 0$. Determine the unique stationary solution of this linear SPDE.

6.2. Random invariant manifold for SPDEs with additive noise

Apply the Lyapunov–Perron method to derive a random invariant manifold for the following nonlinear stochastic heat equation with additive noise on $(0, l)$,

$$du = [\partial_{xx}u + \lambda u + f(u)]\,dt + dW(t), \quad u(0, t) = u(l, t) = 0,$$

with λ a positive constant and f a Lipschitz continuous nonlinear function with Lipschitz constant small enough. Then reduce the system to the random invariant manifold.

6.3. Random invariant manifold for slow-fast SPDEs

Consider the following system of slow-fast SPDEs on $(0, l)$,

$$\dot{u}^\epsilon = au^\epsilon + f(u^\epsilon, v^\epsilon),$$
$$\epsilon\,dv^\epsilon = [\partial_{xx}v^\epsilon + g(u^\epsilon)]\,dt + \sqrt{\epsilon}\,dW(t), \quad v^\epsilon(0, t) = v^\epsilon(l, t) = 0,$$

where f and g are Lipschitz continuous with Lipschitz constants small enough. What is the formulation for the random invariant manifold? Then, for sufficiently small $\epsilon > 0$, find an asymptotic expansion of the random invariant manifold up to the order of $\mathcal{O}(\epsilon)$. Finally, find the reduced system on the random slow manifold.

6.4. Ornstein–Uhlenbeck process

Let $\eta^\epsilon(t)$ be a stationary process solving the following Langevin SPDE on $(0, l)$

$$\epsilon d\eta^\epsilon(t) = \partial_{xx}\eta^\epsilon\,dt + \sqrt{\epsilon}\,dW(t),$$

with zero Dirichlet boundary conditions and $\epsilon \in (0, 1)$.

(a) Show that $\frac{1}{\sqrt{\epsilon}}\int_0^t \eta^\epsilon(s)\,ds$ converges to a Wiener process in distribution as $\epsilon \to 0$.

(b) Back to Problem 6.3: Assume that $f(u, v) = u + v$. Further reduce the SPDEs of Problem 6.3.

6.5. Random slow manifold

Consider the following SPDEs on $(0, \pi)$:

$$du^\epsilon = [\partial_{xx} u^\epsilon + \epsilon \gamma u^\epsilon + 6(\sin u^\epsilon - v^\epsilon + \partial_{xx} v^\epsilon)] \, dt,$$
$$dv^\epsilon = [\partial_{xx} v^\epsilon - v^\epsilon + u^\epsilon] \, dt + dW(t),$$

with zero Dirichlet boundary conditions, and W is an $L^2(0, \pi)$-valued Q-Wiener process. To be specific, assume that $W(x, t) = w_2(t) \sin 2x$, with $w_2(t)$ being a standard scalar Brownian motion. Present an effective reduction for the above SPDEs up to the order of $\mathcal{O}(\epsilon)$ using a random slow manifold.

6.6. Random manifold and Galerkin approximation

Consider the following SPDE on a bounded interval $(0, l)$:

$$du = [\partial_{xx} u + f(u)] \, dt + dW, \quad u(0, t) = u(l, t) = 0.$$

Let $\{e_k\}$ be the eigenfunctions of ∂_{xx} on $(0, l)$ with zero Dirichlet boundary conditions. Assume that $W(t) = \sum_{k=1}^{\infty} w_k(t) e_k$, with $\{w_k\}$ mutually independent scalar Brownian motions, and $f : L^2(0, l) \to L^2(0, l)$ being Lipschitz continuous with a small enough Lipschitz constant $L_f > 0$. Expand $u = \sum_{k=1}^{\infty} u_k e_k$, and denote the Nth order Galerkin approximation by $u_N^G = \sum_{k=1}^{N} u_k e_k$. The Galerkin equation is

$$du_N^G = [\partial_{xx} u_N^G + f_N(u_N^G)] \, dt + dW_N,$$

where $W_N(t) = \sum_{k=1}^{N} w_k(t) e_k$ and f_N is the projection of f to the linear subspace $H_N = \text{span}\{e_1, e_2, \dots, e_N\}$.

Try to compare the Galerkin approximated system with the random manifold reduced system for some appropriate N.

6.7. Random invariant manifold for a stochastic damped wave equation

Consider the following damped stochastic wave equation on $(0, l)$,

$$\nu d\partial_t u + \partial_t u \, dt = [\partial_{xx} u + f(u)] \, dt + dW, \quad u(0) = u_0, \quad \partial_t u(0) = u_1, \quad (6.94)$$

with zero Dirichlet boundary conditions, $\nu > 0$ is a small parameter, and $f : H \to H$ is Lipschitz continuous with a small Lipschitz constant. Show that the above equation has a random invariant manifold for ν small enough.

7 Stochastic Homogenization

Homogenization techniques for stochastic partial differential equations; slow-fast stochastic systems; random dynamical boundary conditions; homogenized systems; effective dynamics

Instead of averaging in time we consider homogenization in space in order to extract effective or homogenized dynamics for SPDEs with multiple spatial scales [297,298].

For partial differential equations (PDEs) with random coefficients (so-called random PDEs but not SPDEs), homogenization issues have been considered recently; see, e.g., [14,25,26,27,37,59,73,122,133,175,169,205,208,216,228,250,251,279,311,318]. A basic assumption in some of these works is either ergodicity or recurrence in time (i.e., periodicity) for the random coefficients. Some recent progress appears in the context of random environment [49,285]. There are also recent works on deterministic PDEs on randomly perforated domains [54], [175, Ch. 8], [309,319,320]. A homogenization result for SPDEs with rapidly oscillating coefficients was recently obtained in [98,261].

In this chapter we consider homogenization for SPDEs, i.e., partial differential equations with noises. These are *different* from random partial differential equations, which are partial differential equations with random coefficients but not containing noises, which are modeled as generalized time derivative of Wiener processes. More precisely, we consider a microscopic heterogeneous system under random influence, which is described by SPDEs defined on a perforated domain (a domain with small holes or obstacles), and derive a homogenized macroscopic model for this microscopic heterogeneous stochastic system. This homogenized or effective model is a new stochastic partial differential equation defined on a unified domain without small holes.

Sometimes randomness enters a system at the physical boundary of small-scale obstacles as well as at the interior of the physical medium. This system is then modeled by an SPDE defined on a domain perforated with small holes (obstacles or heterogeneities), together with some boundary conditions—for example, random dynamical boundary conditions on the boundaries of these small holes. In this case, the random fluctuation on the boundary of small holes (microscopic fluctuation) may affect the system evolution at the macroscopic level. In fact, in our homogenization procedure, the random dynamical boundary conditions are homogenized out, whereas the impact of random forces on the small holes' boundaries is quantified as an extra stochastic term in the homogenized stochastic partial differential equation. Moreover, the validity of the homogenized model is justified by showing that the solutions of the microscopic model converge to those of the effective macroscopic model in distribution as the size of small holes diminishes to zero.

Effective Dynamics of Stochastic Partial Differential Equations. http://dx.doi.org/10.1016/B978-0-12-800882-9.00007-X

This chapter is organized as follows: Some results about deterministic homogenization are reviewed in §7.1. Stochastic homogenized dynamical systems or effective dynamical systems are derived for linear systems and nonlinear systems in §7.2 and in §7.3, respectively. Finally, some further research topics are discussed in §7.4.

7.1 Deterministic Homogenization

Homogenization is a procedure for extracting macroscopic behaviors of a system that has microscopic heterogeneities. Such microscopic heterogeneities appear in, for example, composite materials and fluid flow through porous media. Homogenization aims to obtain the macroscopic description of the system by taking the microscopic properties into account. Roughly speaking, we consider a process u^ϵ modeled by a partial differential equation, with $\epsilon > 0$ measuring the separation of large and small spatial scales. The goal is to determine the limit

$$u = \lim_{\epsilon \to 0} u^\epsilon$$

in some sense or in a specific space and find a new partial differential equation that u satisfies. This new partial differential equation is called the *homogenized* or *effective model*. To obtain a homogenized system, the first step is usually to examine the compactness of $\{u^\epsilon\}$ in a specific Hilbert or Banach space so that we can extract a convergent subsequence u^{ϵ_k} which converges to a process u as $\epsilon \to 0$. The second step is to determine the equation satisfied by u, which is the homogenized equation. The second step usually involves various homogenization procedures.

Homogenization for deterministic systems has been investigated extensively—for example, for heat transfer in composite materials [53,241,242,280,286], for wave propagation in composite materials [53,85,87], for fluid flow [161,216,230], for variational inequalities [233], and for mesoscopic diffusion processes [209]. For systematic discussions about homogenization for deterministic systems, see [25,31,32,84,175,224,272].

In the rest of this section, we briefly recall some basic homogenization results for deterministic partial differential equations on perforated domains.

7.1.1 Perforated Domains

We are interested in some partial differential equations defined on a perforated domain. The geometry of the domain is described in the following: Let the physical medium D be an open bounded domain in $\mathbb{R}^n$ with smooth boundary ∂D, and let $\epsilon > 0$ be a small parameter. Let $Y = [0, l_1) \times [0, l_2) \times \cdots \times [0, l_n)$ be a representative elementary cell in $\mathbb{R}^n$, and let S be an open subset of Y with smooth boundary ∂S such that $\overline{S} \subset Y$. The elementary cell Y, and the small cavity or hole S inside it, are used to model small-scale obstacles or heterogeneities in a physical medium D. Write $l = (l_1, l_2, \ldots, l_n)$. Define $\epsilon S := \{\epsilon y : y \in S\}$. Denote by $S_{\epsilon,k}$ the translated image of ϵS by $kl, k \in \mathbb{Z}^n$, with $kl := (k_1 l_1, k_2 l_2, \ldots, k_n l_n)$. Moreover, let S_ϵ be the set of all the holes contained in D, and let $D_\epsilon := D \backslash S_\epsilon$. Then D_ϵ is a periodically perforated domain with holes of the same size as period ϵ. We remark that the holes are assumed to have no intersection

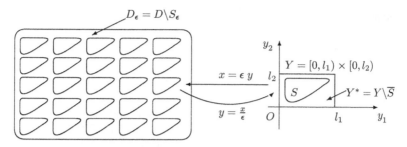

Figure 7.1 A periodically perforated domain D_ϵ in $\mathbb{R}^2$.

with the boundary ∂D, which implies that $\partial D_\epsilon = \partial D \cup \partial S_\epsilon$; see Fig. 7.1 for the case $n = 2$. This assumption is only needed to avoid technical complications, and the results remain valid without this assumption [5].

In the sequel, we use the notation

$$Y^* \triangleq Y \backslash \overline{S}, \quad \vartheta := \frac{|Y^*|}{|Y|},$$

with $|Y|$ and $|Y^*|$ the Lebesgue measure of Y and Y^*, respectively. Denote by χ, the indicator function that takes value 1 on Y^* and value 0 on $Y \backslash Y^*$. In particular, let χ_A be the indicator function of $A \subset \mathbb{R}^n$. Due to the periodic arrangement of the perforated holes, χ_{D_ϵ} is Y periodic. Furthermore,

$$\chi_{D_\epsilon} \rightharpoonup \frac{1}{|D|} \int_D \chi_{D_\epsilon} \, dx = \frac{|Y^*|}{|Y|} = \vartheta, \quad \text{in } L^2(D) \tag{7.1}$$

and

$$\chi_{D_\epsilon} \rightharpoonup \vartheta, \quad \text{in } L^\infty(D). \tag{7.2}$$

Also, denote by $\tilde{u}$, the zero extension to the entire D for a function u defined on D_ϵ:

$$\tilde{u} \triangleq \chi_{D_\epsilon} u = \begin{cases} u & \text{on } D_\epsilon, \\ 0 & \text{on } S_\epsilon. \end{cases}$$

7.1.2 Homogenization on a Periodically Perforated Domain

We consider the following partial differential equation on the perforated domain D_ϵ:

$$\dot{u}^\epsilon(x, t) = \text{div} \left(a \left(\frac{x}{\epsilon} \right) \nabla u^\epsilon(x, t) \right), \quad \text{on } D_\epsilon \times (0, T), \tag{7.3}$$

$$u^\epsilon(x, t) = 0, \quad \text{on } \partial D \times (0, T), \tag{7.4}$$

$$\frac{\partial u^\epsilon(x, t)}{\partial \nu_a} = 0, \quad \text{on } \partial S_\epsilon \times (0, T), \tag{7.5}$$

where $a : D \to \mathbb{R}^1$ is bounded, continuous, and Y-periodic with a positive lower bound (i.e., $a(y) \geq \alpha > 0$ for some constant α); ν is the exterior unit normal vector on the boundary ∂D_ϵ; and

$$\frac{\partial u^\epsilon(x, t)}{\partial \nu_a} \triangleq \nu \cdot a\left(\frac{x}{\epsilon}\right) \nabla u^\epsilon.$$

For simplicity, we assume a zero initial value for u^ϵ, that is, $u^\epsilon(0, x) = 0$ on D_ϵ, and define $D_T \triangleq D \times (0, T)$.

We examine the limit u of u^ϵ and find the effective equation satisfied by u, which is a homogenized model for the heterogeneous system (7.3)–(7.5). To this end, we consider two homogenization methods: An oscillating test function method and a two-scale convergence method.

Oscillating Test Function Method

Noticing that the domain is changing with ϵ, we introduce a linear extension operator. Denote by $\mathbf{L}(\mathcal{X}, \mathcal{Y})$, the space of bounded linear operators from Banach space $\mathcal{X}$ to Banach space $\mathcal{Y}$. Denote by $\oplus_n$, the direct sum of the Hilbert spaces with usual direct sum norm.

Lemma 7.1. *There exists a bounded linear operator*

$$\hat{Q} \in \mathbf{L}(H^k(Y^*), H^k(Y)), \quad k = 0, 1,$$

such that

$$\|\nabla \hat{Q}v\|_{\oplus_n L^2(Y)} \leq C\|\nabla v\|_{\oplus_n L^2(Y^*)}, \quad v \in H^1(Y^*)$$

for some constant $C > 0$.

For the proof of Lemma 7.1, see [85, Remark 1.1].

We define an extension operator P_ϵ in terms of the above bounded linear operator $\hat{Q}$ in the following lemma [85, Theorem 1.3]:

Lemma 7.2. *There exists an extension operator*

$$P_\epsilon \in \mathbf{L}\big(L^2(0, T; H^k(D_\epsilon)), L^2(0, T; H^k(D))\big), \quad k = 0, 1,$$

such that, for any $v \in H^k(D_\epsilon)$:

1. $P_\epsilon v = v$ on $D_\epsilon \times (0, T)$;
2. $|P_\epsilon v|_{L^2(0,T;H)} \leq C_T |v|_{L^2(0,T;H_\epsilon)}$;
3. $|a(x/\epsilon)\nabla(P_\epsilon v)|_{L^2(0,T;\oplus_n L^2(D))} \leq C_T |a(x/\epsilon)\nabla v|_{L^2(0,T;\oplus_n L^2(D_\epsilon))}$,

where C_T is a constant, independent of ϵ.

Proof. For $\varphi \in H^k(D_\epsilon)$,

$$\varphi_\epsilon(y) = \frac{1}{\epsilon}\varphi(\epsilon\, y)$$

belongs to $H^k(Y_l^*)$, with Y_l^* the translation of Y^* for some $l \in \mathbb{R}^n$. Define

$$\hat{Q}_\epsilon \varphi(x) := \epsilon(\hat{Q}\varphi_\epsilon)\left(\frac{x}{\epsilon}\right). \tag{7.6}$$

Now, for $\varphi \in L^2(0, T; H^k(D_\epsilon))$, we define

$$(P_\epsilon \varphi)(x, t) = [\hat{Q}_\epsilon \varphi(\cdot, t)]\left(\frac{x}{\epsilon}\right) = \epsilon[\hat{Q}\varphi_\epsilon(\cdot, t)]\left(\frac{x}{\epsilon}\right).$$

Then the operator $P_\epsilon \in \mathbf{L}\big(L^2(0, T; H^k(D_\epsilon)), L^2(0, T; H^k(D))\big)$ for $k = 0, 1$ and satisfies the conditions (1)–(3) listed in the lemma. This completes the proof. $\qquad\square$

To apply the linear operator P_ϵ to u^ϵ, we first need the following estimate.

Lemma 7.3.   *For every $T > 0$,*

$$|u^\epsilon|_{L^2(0,T;H^1(D_\epsilon))} + |\dot{u}^\epsilon|_{L^2(0,T;H^{-1}(D_\epsilon))} \leq C_T \tag{7.7}$$

for some constant $C_T > 0$.

The proof is straightforward and omitted here.

Since the operator P_ϵ is defined on $L^2(0, T; H^k(D_\epsilon))$, $k = 0, 1$, we set

$$P_\epsilon \dot{u}^\epsilon(x, t) \triangleq \operatorname{div}\left(a\left(\frac{x}{\epsilon}\right)\nabla(P_\epsilon u^\epsilon(x, t))\right), \quad \text{on} \quad D_T.$$

By the properties of P_ϵ and the estimate on u^ϵ,

$$P_\epsilon \dot{u}^\epsilon = (P_\epsilon u^\epsilon)\dot{}, \quad \text{on} \quad D_\epsilon \times (0, T),$$

and

$$|P_\epsilon \dot{u}^\epsilon|_{L^2(0,T;H^{-1}(D))} \leq |\dot{u}^\epsilon|_{L^2(0,T;H^{-1}(D_\epsilon))}.$$

By Lemma 3.7, $\{P_\epsilon u^\epsilon\}_{0<\epsilon<1}$ is compact in $L^2(0, T; L^2(D))$. So, we can extract a subsequence of u^ϵ, still denoted by u^ϵ, such that

$$P_\epsilon u^\epsilon \rightharpoonup u \quad \text{weakly* in } L^\infty(0, T; L^2(D)), \tag{7.8}$$
$$P_\epsilon u^\epsilon \rightharpoonup u \quad \text{weakly in } L^2(0, T; H^1), \tag{7.9}$$
$$P_\epsilon u^\epsilon \to u \quad \text{strongly in } L^2(0, T; L^2(D)), \tag{7.10}$$
$$P_\epsilon \dot{u}^\epsilon \rightharpoonup \dot{u} \quad \text{weakly in } L^2(0, T; H^{-1}), \tag{7.11}$$

for some $u \in L^2(0, T; L^2(D))$. Define $\xi^\epsilon \triangleq a(x/\epsilon)\nabla u^\epsilon$, which satisfies

$$\operatorname{div}\xi^\epsilon = \dot{u}^\epsilon, \quad \text{on} \quad D_\epsilon \times (0, T),$$
$$\xi^\epsilon \cdot \nu = 0, \quad \text{on} \quad \partial S_\epsilon \times (0, T).$$

Notice that by the estimate on u^ϵ, $\{\tilde{\xi}^\epsilon\}$, is bounded in $L^2(0, T; \oplus_n L^2(D))$ and thus it has a subsequence, still denoted by $\tilde{\xi}^\epsilon$, such that

$$\tilde{\xi}^\epsilon \rightharpoonup \xi \quad \text{weakly in} \quad L^2(0, T; \oplus_n L^2(D))$$

for some $\xi \in L^2(0, T; \oplus_n L^2(D))$. Hence, for any $\varphi \in C_0^\infty(0, T)$ and $v \in H_0^1(D)$,

$$\int_0^T \int_D \tilde{\xi}_\epsilon \cdot \nabla v \, dx\varphi \, dt = \int_0^T \int_D P_\epsilon u^\epsilon \chi_{D_\epsilon} v \, dx\dot{\varphi} \, dt. \tag{7.12}$$

Passing the limit $\epsilon \to 0$ and noticing (7.1), we have

$$\int_0^T \int_D \xi \cdot \nabla v \, dx \varphi \, dt = \int_0^T \int_D \vartheta u v \, dx \dot\varphi \, dt. \tag{7.13}$$

Therefore,

$$\operatorname{div} \xi(x, t) = \vartheta \dot u, \quad \text{on} \quad D_T. \tag{7.14}$$

Now we need to identify the limit ξ; this is the key step in homogenization. For any $\lambda \in \mathbb{R}^n$, let w_λ be the solution of the following elliptic system:

$$-\sum_{j=1}^n \frac{\partial}{\partial y_j} \left(\sum_{i=1}^n a(y) \frac{\partial w_\lambda}{\partial y_i} \right) = 0, \quad \text{on} \quad Y^* \tag{7.15}$$

$$w_\lambda - \lambda \cdot y \quad \text{is } Y - \text{periodic} \tag{7.16}$$

$$\frac{\partial w_\lambda}{\partial \nu_a} = 0, \quad \text{on} \quad \partial S \tag{7.17}$$

and define

$$w_\lambda^\epsilon = \epsilon (\hat Q w_\lambda)\left(\frac{x}{\epsilon}\right), \tag{7.18}$$

where $\hat Q$ is in Lemma 7.1. Then we have [84, p.139], as $\epsilon \to 0$

$$w_\lambda^\epsilon \rightharpoonup \lambda \cdot x \quad \text{weakly in } H^1(D), \tag{7.19}$$

$$\nabla w_\lambda^\epsilon \rightharpoonup \lambda \quad \text{weakly in } \oplus_n L^2(D). \tag{7.20}$$

Now define

$$\eta_{\lambda, j}(y) \triangleq a(y) \frac{\partial w_\lambda(y)}{\partial y_j}, \quad y \in Y^*, \quad j = 1, 2, \ldots n$$

and $\eta_\lambda^\epsilon(x) := (\eta_{\lambda_j}(x/\epsilon)) = a(x/\epsilon) \nabla w_\lambda^\epsilon$. Note that

$$-\operatorname{div} \tilde\eta_\lambda^\epsilon = 0 \quad \text{on} \quad D. \tag{7.21}$$

Due to (7.19) and (7.20) and periodicity,

$$\tilde\eta_\lambda^\epsilon \rightharpoonup \frac{1}{|Y|} \int_{Y^*} \eta_\lambda \, dy \quad \text{weakly in } L^2(D). \tag{7.22}$$

In fact, the above limit is given by the classical homogenized matrix, that is,

$$\frac{1}{|Y|} \int_{Y^*} \eta_\lambda \, dy \triangleq \bar A^* \lambda = (\bar a_{ji}) \lambda$$

with

$$\bar a_{ij} \triangleq \frac{1}{|Y|} \int_{Y^*} a(y) \frac{\partial w_{e_i}}{\partial y_j} dy. \tag{7.23}$$

Using the test function $\varphi v w_\lambda^\epsilon$ with $\varphi \in C_0^\infty(0, T)$, $v \in C_0^\infty(D)$ in (7.12), and multiplying both sides of (7.21) by $\varphi v P_\epsilon u^\epsilon$, we have

$$\int_0^T \int_D \tilde{\xi}^\epsilon \cdot \nabla v \varphi w_\lambda^\epsilon \, dx \, dt + \int_0^T \int_{D_\epsilon} \xi^\epsilon \cdot \nabla w_\lambda^\epsilon v \varphi \, dx \, dt$$

$$- \int_0^T \int_D \tilde{\eta}_\lambda^\epsilon \cdot \nabla v \varphi P_\epsilon u^\epsilon \, dx \, dt - \int_0^T \int_D \tilde{\eta}_\lambda^\epsilon \cdot \nabla (P_\epsilon u^\epsilon) v \varphi \, dx \, dt$$

$$= \int_0^T \int_D P_\epsilon u^\epsilon \chi_{D_\epsilon} \dot{\varphi} v w_\lambda^\epsilon \, dx \, dt.$$

Moreover, by the definitions of ξ^ϵ, η_λ^ϵ and weak convergence, we obtain

$$\int_0^T \int_D \xi \cdot \nabla(v\lambda \cdot x) \varphi \, dx \, dt - \int_0^T \int_D \xi \cdot \lambda v \varphi \, dx \, dt$$

$$- \int_0^T \int_D \bar{A}^T \lambda \cdot \nabla v \varphi u \, dx \, dt$$

$$= \int_0^T \int_D \vartheta u v \dot{\varphi} \lambda \cdot x \, dx \, dt.$$

Again, using the test function $v\lambda \cdot x\varphi$ in (7.13), we conclude that

$$\int_0^T \int_D \xi \cdot \lambda v \varphi \, dx \, dt = \int_0^T \int_D \bar{A}^T \lambda \cdot \nabla u \varphi v \, dx \, dt,$$

which yields

$$\xi \cdot \lambda = \bar{A}^T \lambda \cdot \nabla u = \bar{A} \nabla u \cdot \lambda.$$

Therefore, $\xi = \bar{A} \nabla u$, since λ is arbitrary. Finally, we have the following homogenized equation for (7.3)–(7.5):

$$\vartheta \dot{u} = \text{div}(\bar{A} \nabla u) \quad \text{on} \quad D_T, \tag{7.24}$$
$$u(x, t) = 0 \quad \text{on} \quad \partial D \times (0, T), \tag{7.25}$$
$$u(0, x) = 0 \quad \text{on} \quad D \tag{7.26}$$

with

$$P_\epsilon u^\epsilon \to u \quad \text{in} \quad L^2(0, T; L^2(D)), \quad \epsilon \to 0.$$

Remark 7.4. The homogenized matrix $\bar{A}$ is symmetric and positive definite [84, Ch. 6.3]. In fact, $\bar{a}_{ij}$ has the following expression [84, Proposition 6.8]:

$$\bar{a}_{ij} = \frac{1}{|Y|} \int_{Y^*} a(y)(\nabla_y w_{e_i}) \cdot (\nabla_y w_{e_j}) dy. \tag{7.27}$$

This is seen by the weak form of the cell problem (7.15)–(7.17). For any Y-periodic φ, we have

$$\int_{Y^*} a(y) \nabla_y w_{e_i} \nabla_y \varphi \, dy = 0.$$

Then, choosing $\varphi = w_{\mathbf{e}_j} - \mathbf{e}_j \cdot y$, we get

$$\int_{Y^*} a(y)\nabla_y w_{\mathbf{e}_i}(\nabla w_{\mathbf{e}_j} - \mathbf{e}_j)dy = 0,$$

that is,

$$\int_{Y^*} a\nabla_y w_{\mathbf{e}_i} \cdot \nabla_y w_{\mathbf{e}_j} \, dy = \int_{Y^*} a\nabla_y w_{\mathbf{e}_i} \cdot \mathbf{e}_j \, dy = \bar{a}_{ij}.$$

Let us now consider the two-scale convergence method for homogenization.

Two-Scale Convergence Method

The two-scale convergence method is efficient in homogenizing periodic heterogeneous systems. It was introduced by Allaire [4] and Nguetseng [243] and further developed by others. The main idea of the two-scale convergence method comes from the following two-scale asymptotic expansion:

$$u^\epsilon(x,t) = u_0(x,t) + \epsilon u_1(x,y,t) + \epsilon^2 u_2(x,y,t) + \cdots$$

with $y := x/\epsilon$. Each term $u_i(x,y,t)$ depends on x, the macroscopic variable and also periodically relies on y, the microscopic variable. To pass limit $\epsilon \to 0$ to $u^\epsilon(x,y,t)$, we have to consider weak convergence of two scales, that is, two-scale convergence. One difference with the oscillating test function method is that we do not need the complex oscillating test function w_λ^ϵ. To apply the two-scale convergence method, we first introduce some basic concepts.

In the following, we denote by $C_{\text{per}}^\infty(Y)$, the space of infinitely differentiable functions in $\mathbb{R}^n$ that are periodic in Y. We also denote by $L_{\text{per}}^2(Y)$ or $H_{\text{per}}^1(Y)$ the completion of $C_{\text{per}}^\infty(Y)$ in the usual norm, in $L^2(Y)$ or $H^1(Y)$, respectively. We introduce the space $H_{\text{per}}^1(Y)/\mathbb{R}$, which is the space of the equivalence classes of $u \in H_{\text{per}}^1(Y)$, under the following equivalence relation:

$$u \sim v \Leftrightarrow u - v = \text{constant}.$$

Definition 7.5.   A sequence of functions $u^\epsilon(x,t)$ in $L^2(D_T)$ is said to be two-scale convergent to a limit $u(x,y,t) \in L^2(D_T \times Y)$ if, for every function $\varphi(x,y,t) \in C_0^\infty(D_T, C_{\text{per}}^\infty(Y))$,

$$\lim_{\epsilon \to 0} \int_{D_T} u^\epsilon(x,t)\varphi\left(x, \frac{x}{\epsilon}, t\right) dx \, dt = \frac{1}{|Y|} \int_{D_T} \int_Y u(x,y,t)\varphi(x,y,t)dy \, dx \, dt.$$

This two-scale convergence is written as $u^\epsilon \xrightarrow{2-s} u$.

The following result ensures the existence of a two-scale limit [4], [84, Theorem 9.7].

Lemma 7.6.   *Let u^ϵ be a bounded sequence in $L^2(D_T)$. Then there exist a function $u \in L^2(D_T \times Y)$ and a subsequence u^{ϵ_k} with $\epsilon_k \to 0$ as $k \to \infty$, such that u^{ϵ_k} two-scale converges to u.*

Remark 7.7.   Taking φ independent of y in the definition of two-scale convergence, then $u^\epsilon \xrightarrow{2-s} u$ implies that u^ϵ weakly converges to its spatial average

$$u^\epsilon(x, t) \rightharpoonup \bar{u}(x, t) := \frac{1}{|Y|} \int_Y u(x, y, t)dy.$$

So, for a given bounded sequence in $L^2(D_T)$, the two-scale limit $u(x, y, t)$ contains more information than the weak limit $u(x, t)$ does: u carries information of the periodic oscillations of u^ϵ, whereas $\bar{u}$ is just the spatial average with respect to y. Another advantage of using two-scale convergence is that we do not need an extension operator as used in the homogenization method of the oscillating test functions.

We also need the following result in order to consider two-scale convergence of products of two convergent sequences [4], [84, Theorem 9.8].

Lemma 7.8.   *Let v^ϵ be a sequence in $L^2(D_T)$ that two-scale converges to a limit $v(x, y) \in L^2(D_T \times Y)$. Furthermore, assume that*

$$\lim_{\epsilon \to 0} \int_{D_T} |v^\epsilon(x, t)|^2 \, dx \, dt = \frac{1}{|Y|} \int_{D_T} \int_Y |v(x, y, t)|^2 \, dy \, dx \, dt. \tag{7.28}$$

Then, for every sequence $u^\epsilon \in L^2(D_T)$, which two-scale converges to a limit $u \in L^2(D_T \times Y)$, we have the weak convergence of the product $u^\epsilon v^\epsilon$:

$$u^\epsilon v^\epsilon \rightharpoonup \frac{1}{|Y|} \int_Y u(\cdot, \cdot, y)v(\cdot, \cdot, y)dy, \quad as \quad \epsilon \to 0 \quad in \quad L^2(D_T).$$

Remark 7.9.   Condition (7.28) always holds for a sequence of functions $\varphi(x, x/\epsilon, t)$, with $\varphi(x, y, t) \in L^2(D_T; C_{per}(Y))$. Such functions v^ϵ are called *admissible test functions*. With the additional condition (7.28), two-scale convergence of v^ϵ is also called *strong two-scale convergence*; see [4, Remark 1.9].

Now we consider $u^\epsilon(x, y, t)$ for the system (7.3)–(7.5). Since the domain is changing with ϵ, we consider $\tilde{u}^\epsilon$, the zero extension of u^ϵ. By Lemma 7.8, $\tilde{u}^\epsilon$, and $\widetilde{\nabla_x u^\epsilon}$ are bounded sequences in $L^2(0, T; H^1(D))$ and $L^2(0, T; \oplus_n L^2(D))$, respectively, for any $T > 0$. Thus, we have the following result concerning the two-scale limits of $\tilde{u}^\epsilon$ and $\widetilde{\nabla_x u^\epsilon}$ [4, Theorem 2.9].

Lemma 7.10.   *There exist $u(x, t) \in H_0^1(D_T)$, $u_1(x, y, t) \in L^2(D_T; H_{per}^1(Y))$ and a subsequence u^{ϵ_k} with $\epsilon_k \to 0$ as $k \to \infty$, such that*

$$\tilde{u}^{\epsilon_k}(x, t) \xrightarrow{2-s} \chi(y)u(x, t), \quad k \to \infty,$$

and

$$\widetilde{\nabla_x u^{\epsilon_k}} \xrightarrow{2-s} \chi(y)[\nabla_x u(x,t) + \nabla_y u_1(x,y,t)], \quad k \to \infty.$$

With this Lemma, we note that by Remark 7.7,

$$\tilde{u}^{\epsilon_j}(x,t) \rightharpoonup \frac{1}{|Y|} \int_Y \chi(y)u(x,t)dy = \vartheta u(x,t), \quad \text{weakly in} \quad L^2(D_T).$$

By Lemma 7.8 again, $\{\tilde{u}^{\epsilon}\}$ is compact in $L^2(D_T)$, and $\tilde{u}^{\epsilon}$ strongly converges to ϑu. To apply two-scale convergence, we choose a test function $\varphi^{\epsilon}(x,t) := \varphi(x,t) + \epsilon \Phi(x, x/\epsilon, t)$ with $\varphi(x,t) \in C_0^{\infty}(D_T)$ and $\Phi(x,y,t) \in C_0^{\infty}(D_T; C_{per}^{\infty}(Y))$. Thus,

$$\int_0^T \int_{D_\epsilon} \dot{u}^{\epsilon}(x,t)\varphi^{\epsilon}(x,t)dx\,dt = -\int_0^T \int_{D_\epsilon} u^{\epsilon}(x,t)\dot{\varphi}^{\epsilon}(x,t)dx\,dt$$

$$= -\int_0^T \int_D \tilde{u}^{\epsilon}(x,t)\dot{\varphi}(x,t)dx\,dt - \epsilon \int_0^T \int_D \tilde{u}^{\epsilon}(x,t)\dot{\Phi}\left(x,\frac{x}{\epsilon},t\right)dx\,dt$$

$$\to -\int_0^T \int_D \vartheta u(x,t)\dot{\varphi}(x,t)dx\,dt = \int_0^T \int_D \vartheta \dot{u}(x,t)\varphi(x,t)dx\,dt. \quad (7.29)$$

By the choice of φ^{ϵ},

$$\nabla_x \varphi(x,t) + \nabla_y \Phi(x,\frac{x}{\epsilon},t) \xrightarrow{2-s} \nabla_x \varphi(x,t) + \nabla_y \Phi(x,y,t), \quad \epsilon \to 0,$$

and

$$\lim_{\epsilon \to 0} \left\| \nabla_x \varphi(x,t) + \nabla_y \Phi\left(x,\frac{x}{\epsilon},t\right) \right\|_{[L^2(D_T)]^n}$$

$$= \frac{1}{|Y|} \int_{D_T \times Y} \left| \nabla_x \varphi(x,t) + \nabla_y \Phi(x,y,t) \right|^2 dy\,dx\,dt.$$

Hence by Lemma 7.8, and the assumption on $a(y)$, we conclude that

$$\int_0^T \int_{D_\epsilon} a\left(\frac{x}{\epsilon}\right) \nabla u^{\epsilon}(x,t) \cdot \nabla \varphi^{\epsilon}(x,t)dx\,dt$$

$$= \int_0^T \int_{D_\epsilon} a\left(\frac{x}{\epsilon}\right) \nabla u^{\epsilon}(x,t) \cdot \left(\nabla_x \varphi(x,t) + \nabla_y \Phi\left(x,\frac{x}{\epsilon},t\right) \right) dx\,dt$$

$$= \int_0^T \int_D a\left(\frac{x}{\epsilon}\right) \widetilde{\nabla u^{\epsilon}}(x,t) \cdot \left(\nabla_x \varphi(x,t) + \nabla_y \Phi\left(x,\frac{x}{\epsilon},t\right) \right) dx\,dt$$

$$\to \frac{1}{|Y|} \int_0^T \int_D \int_Y \chi(y)a(y)[\nabla_x u(x,t) + \nabla_y u_1(x,y,t)]$$

$$\times [\nabla_x \varphi(x,t) + \nabla_y \Phi(x,y,t)]dy\,dx\,dt$$

$$= \frac{1}{|Y|} \int_0^T \int_D \int_{Y^*} a(y)[\nabla_x u(x,t) + \nabla_y u_1(t,x,y)]$$

$$\times [\nabla_x \varphi(x,t) + \nabla_y \Phi(x,y,t)]dy\,dx\,dt. \quad (7.30)$$

Thus, combining (7.29) and (7.30) and using a density argument, we obtain

$$
\int_0^T \int_D \vartheta \dot{u}(x,t)\varphi(x,t)dx\,dt
$$
$$
= \frac{1}{|Y|} \int_0^T \int_D \int_{Y^*} a(y)\big[\nabla_x u(x,t) + \nabla_y u_1(t,x,y)\big]
$$
$$
\times \big[\nabla_x \varphi(x,t) + \nabla_y \Phi(x,y,t)\big]dy\,dx\,dt
$$

for any $\varphi \in H_0^1(D_T)$ and $\Phi \in L^2(D_T; H_{\mathrm{per}}^1(Y)/\mathbb{R})$. Integrating by parts, we have the following two-scaled homogenized system:

$$
\vartheta \dot{u} = \mathrm{div}\left[\frac{1}{|Y|} \int_{Y^*} a(y)[\nabla u(x,t) + \nabla_y u_1(x,y,t)]dy\right] \quad \text{on} \quad D_T, \tag{7.31}
$$
$$
u(x,t) = 0 \quad \text{on} \quad \partial D, \tag{7.32}
$$
$$
\mathrm{div}_y(a(y)[\nabla u(x,t) + \nabla_y u_1(x,y,t)]) = 0 \quad \text{on} \quad D \times Y^*, \tag{7.33}
$$
$$
(a(y)[\nabla u(x,y,t) + \nabla_y u_1(x,y,t)]) \cdot \nu = 0 \quad \text{on} \quad \partial Y^* - \partial Y, \tag{7.34}
$$
$$
y \to u_1(x,y,t) \quad Y\text{-periodic.} \tag{7.35}
$$

Notice that Equations (7.33)–(7.35) are a form of the cell problem (7.15)–(7.17). In fact, (7.33)–(7.35) are solved in $L^2(D_T; H_{\mathrm{per}}^1(Y^*)/\mathbb{R})$ by

$$
u_1(x,y,t) = \sum_{k=1}^n \frac{\partial u}{\partial x_k}(w_{\mathbf{e}_k}(y) - \mathbf{e}_k \cdot y),
$$

where $w_{\mathbf{e}_k}$ solves the cell problem (7.15)–(7.17) with $\lambda = \mathbf{e}_k$. Substituting the expression of $u_1(x,y,t)$ into Equation (7.31), we again arrive at the homogenized Equation (7.24).

We see that both the oscillating test function method and the two-scale convergence method provide the same homogenized model (7.24) for the original heterogeneous system (7.3)–(7.5).

Remark 7.11. We have just presented a brief introduction to two classical homogenization methods, the oscillating test function method and the two-scale convergence method, with many details omitted. For example, we have not mentioned the well-posedness of the system (7.3)–(7.5) and of the cell problem. Detailed discussions for deterministic homogenization are available in many references; see, e.g., [4,84]. For some more recent research on deterministic homogenization, see, e.g., [40,154,249].

7.2 Homogenized Macroscopic Dynamics for Stochastic Linear Microscopic Systems

We now investigate homogenization for stochastic systems with two spatial scales, described by SPDEs on a perforated domain. Our goal is to derive a homogenized, effective equation, which is a new SPDE (Theorems 7.13, 7.25, 7.29, 7.30, and 7.31) for the microscopic heterogenous system, by stochastic homogenization techniques. The solution of the microscopic or heterogeneous system is shown to converge to that of the macroscopic or homogenized system as the scale parameter $\epsilon \downarrow 0$ in distribution. This means that the distribution of solutions weakly converges, in some appropriate space, to the distribution of a stochastic process that solves the macroscopic effective equation.

7.2.1   Homogenization Under Neumann Boundary Conditions

To present the homogenization approach for a stochastic system, we first consider a simple system described by a linear SPDE on a perforated domain, with no-flux or zero Neumann boundary conditions on the boundary of small holes. This system may describe heat conduction under stochastic perturbation on a perforated domain.

For $T > 0$, we consider the following Itô-type nonautonomous SPDE defined on the perforated domain D_ϵ in $\mathbb{R}^n$:

$$du^\epsilon(x, t) = \left(\text{div}\left(A^\epsilon(x)\nabla u^\epsilon(x, t)\right) + f^\epsilon(x, t)\right)dt + g^\epsilon(t)dW(t)$$

$$\text{on}\quad D_\epsilon \times (0, T), \tag{7.36}$$

$$u^\epsilon = 0 \quad \text{on}\quad \partial D \times (0, T), \tag{7.37}$$

$$\frac{\partial u_\epsilon}{\partial \nu_{A^\epsilon}} = 0 \quad \text{on}\quad \partial S_\epsilon \times (0, T), \tag{7.38}$$

$$u^\epsilon(0) = u_0^\epsilon \quad \text{in}\quad D_\epsilon, \tag{7.39}$$

where the matrix A^ϵ is defined by

$$A^\epsilon \triangleq \left(a_{ij}\left(\frac{x}{\epsilon}\right)\right)_{ij}$$

and

$$\frac{\partial \cdot}{\partial \nu_{A^\epsilon}} \triangleq \sum_{ij} a_{ij}\left(\frac{x}{\epsilon}\right)\frac{\partial \cdot}{\partial x_j}\nu_i$$

with ν the exterior unit normal vector on the boundary ∂D_ϵ.

We make the following assumptions on the coefficients:

1. $a_{ij} \in L^\infty(\mathbb{R}^n), \quad i, j = 1, \ldots, n$;
2. $\sum_{i,j=1}^n a_{ij}\xi_i\xi_j \geq \alpha \sum_{i=1}^n \xi_i^2$ for $\xi \in \mathbb{R}^n$ and α a positive constant;
3. a_{ij} are Y-periodic.

Furthermore, we assume that

$$f^\epsilon \in L^2(D_\epsilon \times [0, T]), \tag{7.40}$$

and for $0 \leq t \leq T$, $g_\epsilon(t)$ is a linear operator from ℓ^2 to $L^2(D_\epsilon)$ defined by

$$g^\epsilon(t)k \triangleq \sum_{i=1}^\infty g_\epsilon^i(x, t)k_i, \quad k = (k_1, k_2, \ldots) \in \ell^2.$$

Here $g_i^{\epsilon}(x, t) \in L^2(D_{\epsilon} \times [0, T])$, $i = 1, 2, \cdots$, are measurable functions with

$$\sum_{i=1}^{\infty} \|g_i^{\epsilon}(x, t)\|_{L^2(D_{\epsilon})}^2 < C_T, \quad t \in [0, T], \tag{7.41}$$

for some constant $C_T > 0$ independent of ϵ. Moreover,

$$W(t) = (W_1(t), W_2(t), \ldots)$$

is a Wiener process in ℓ^2 with covariance operator $Q = Id_{\ell^2}$. Note that $\{W_i(t) : i = 1, 2, \cdots\}$ are mutually independent, real-valued standard Wiener processes on a probability space $(\Omega, \mathcal{F}, \mathbb{P})$ with a canonical filtration $(\mathcal{F}_t)_{t \geq 0}$. Then,

$$\|g^{\epsilon}(t)\|_{\mathcal{L}_2^Q}^2 = \sum_{i=1}^{\infty} \|g_i^{\epsilon}(x, t)\|_{L^2(D_{\epsilon})}^2 < C_T, \quad t \in [0, T]. \tag{7.42}$$

Here we use the notation $\mathcal{L}_2^Q \triangleq L_2(Q^{1/2}L^2(D), L^2(D))$, which consists of Hilbert–Schmidt operators from $Q^{1/2}L^2(D)$ to $L^2(D)$ as in [94, p. 418] or [162, p. 15], and the norm is defined in §3.6.

We also assume that

$$\tilde{f}^{\epsilon} \rightharpoonup f, \quad \text{weakly in } L^2(0, T; L^2(D)), \tag{7.43}$$

and

$$\tilde{g}_i^{\epsilon} \rightharpoonup g_i, \quad \text{weakly in } L^2(0, T; L^2(D)). \tag{7.44}$$

We recently noticed a homogenization result [168] proved for SPDEs with periodic coefficients, but not on perforated domains.

Effective Homogenization on Finite Time Interval

To homogenize the stochastic system (7.36)–(7.39), we first examine the compactness of the solution $\{u^{\epsilon}\}$ in some space. To this end, we need to understand more about the microscopic model.

Let $H = L^2(D)$ and $H_{\epsilon} = L^2(D_{\epsilon})$. Define the following space:

$$V_{\epsilon} \triangleq \{u \in H^1(D_{\epsilon}), \quad u|_{\partial D} = 0\}$$

equipped with the norm

$$\|v\|_{V_{\epsilon}} \triangleq \|\nabla_{A^{\epsilon}} v\|_{\oplus_n H_{\epsilon}} = \left\| \left(\sum_{j=1}^{n} a_{ij}\left(\frac{x}{\epsilon}\right) \frac{\partial v}{\partial x_j} \right)_{i=1}^{n} \right\|_{\oplus_n H_{\epsilon}}.$$

This norm is equivalent to the usual $H^1(D_{\epsilon})$-norm, with an embedding constant independent of ϵ, due to the assumptions on a_{ij}. Let

$$\mathcal{D}(\mathcal{A}^{\epsilon}) \triangleq \left\{ v \in V_{\epsilon} : \text{div}(A^{\epsilon}\nabla v) \in H_{\epsilon} \quad \text{and} \quad \frac{\partial v}{\partial \nu_{A^{\epsilon}}}\Big|_{\partial S_{\epsilon}} = 0 \right\}$$

with linear operator $\mathcal{A}^\epsilon v := \mathrm{div}(A^\epsilon \nabla v)$ for $v \in \mathcal{D}(\mathcal{A}^\epsilon)$. Thus, we rewrite the stochastic system (7.36)–(7.39) in the form of an abstract stochastic evolutionary equation,

$$du^\epsilon = (\mathcal{A}^\epsilon u^\epsilon + f^\epsilon)dt + g^\epsilon \, dW, \quad u^\epsilon(0) = u_0^\epsilon. \tag{7.45}$$

By the assumptions on a_{ij}, operator $\mathcal{A}^\epsilon$ generates a strongly continuous semigroup $S^\epsilon(t)$ on H_ϵ. So, we can further rewrite (7.45) in the mild form

$$u^\epsilon(t) = S^\epsilon(t)u_0^\epsilon + \int_0^t S^\epsilon(t-s)f^\epsilon(s)ds + \int_0^t S^\epsilon(t-s)g^\epsilon(s)dW(s). \tag{7.46}$$

The variational weak formulation becomes

$$\langle du^\epsilon(t), v \rangle_{H_\epsilon^{-1}, V_\epsilon}$$
$$= \left(-\int_{D_\epsilon} A^\epsilon(x)\nabla u^\epsilon(x,t)\nabla v(x)dx + \int_{D_\epsilon} f^\epsilon(x,t)v(x)dx \right)dt$$
$$+ \int_{D_\epsilon} g^\epsilon(x,t)v(x)dW(t), \quad \text{in} \quad \mathcal{D}'(0,T), \quad v \in V_\epsilon \tag{7.47}$$

with $u^\epsilon(0,x) = u_0^\epsilon(x)$.

We prove the following result for (7.45).

Lemma 7.12. *Assume that (7.40) and (7.42) hold. Let u_0^ϵ be an $(\mathcal{F}_0, \mathcal{B}(H_\epsilon))$-measurable random variable. Then system (7.45) has a unique mild solution u^ϵ, in $L^2(\Omega, C(0,T; H_\epsilon) \cap L^2(0,T; V_\epsilon))$, which is also a weak solution in the following sense:*

$$\langle u^\epsilon(t), v \rangle_{H_\epsilon} = \langle u_0^\epsilon, v \rangle_{H_\epsilon} + \int_0^t \langle \mathcal{A}^\epsilon u^\epsilon(s), v \rangle_{H_\epsilon} ds$$
$$+ \int_0^t \langle f^\epsilon, v \rangle_{H_\epsilon} ds + \int_0^t \langle g^\epsilon \, dW, v \rangle_{H_\epsilon} \tag{7.48}$$

for $t \in [0,T)$ and $v \in V_\epsilon$. Moreover, if u_0^ϵ is independent of $W(t)$ with $\mathbb{E}\|u_0^\epsilon\|_{H_\epsilon}^2 < \infty$, then

$$\mathbb{E}\|u^\epsilon(t)\|_{H_\epsilon}^2 + \mathbb{E}\int_0^t \|u^\epsilon(s)\|_{V_\epsilon}^2 \, ds \le \mathbb{E}\|u_0^\epsilon\|_{H_\epsilon}^2 + C_T, \quad for \quad t \in [0,T], \tag{7.49}$$

and

$$\mathbb{E}\int_0^t \|\dot{u}^\epsilon(s)\|_{H_\epsilon^{-1}}^2 \, ds \le C_T(\mathbb{E}\|u_0^\epsilon\|_{H_\epsilon}^2 + 1), \quad for \quad t \in [0,T]. \tag{7.50}$$

Furthermore, if

$$\|\nabla_{A^\epsilon} g^\epsilon(t)\|_{\mathcal{L}_2^Q}^2 = \sum_{i=1}^\infty |\nabla_{A^\epsilon} g_i^\epsilon(t)|_{\oplus_n H_\epsilon}^2 \le C_T, \quad for \quad t \in [0,T] \tag{7.51}$$

and $u_0^\epsilon \in V_\epsilon$ with $\mathbb{E}\|u_0^\epsilon\|_{V_\epsilon}^2 < \infty$, then

$$\mathbb{E}\|u^\epsilon(t)\|_{V_\epsilon}^2 + \mathbb{E}\int_0^t \|\mathcal{A}^\epsilon u^\epsilon(s)\|_{H_\epsilon}^2 \, ds \le \mathbb{E}\|u_0^\epsilon\|_{V_\epsilon}^2 + C_T, \quad for \quad t \in [0, T]. \quad (7.52)$$

Finally, system (7.45) is globally well-posed on $[0, \infty)$ when

$$f^\epsilon \in L^2(0, \infty; H_\epsilon) \quad and \quad g^\epsilon \in L^2(0, \infty; \mathcal{L}_2^Q). \quad (7.53)$$

Proof.   By Assumption (7.42), we have

$$\|g^\epsilon(t)\|_{\mathcal{L}_2^Q}^2 = \sum_{i=1}^\infty \|g_i^\epsilon(x, t)\|_{H_\epsilon}^2 < \infty.$$

Then the classical result of [94, Theorem 7.4] yields the local existence of u^ϵ. Applying the stochastic Fubini theorem, the local mild solution is also a weak solution.

Now we derive the following *a priori* estimates, which yield the existence of weak solution on $[0, T]$, provided that both (7.40) and (7.42) hold. Applying Itô's formula to $\|u^\epsilon\|_{H_\epsilon}^2$, we get

$$d\|u^\epsilon(t)\|_{H_\epsilon}^2 - 2\langle \mathcal{A}^\epsilon u^\epsilon, u^\epsilon \rangle_{H_\epsilon} dt$$
$$= 2\langle f^\epsilon, u^\epsilon \rangle_{H_\epsilon} dt + 2\langle g^\epsilon \, dW, u^\epsilon \rangle_{H_\epsilon} + \|g^\epsilon\|_{\mathcal{L}_2^Q}^2 \, dt. \quad (7.54)$$

By the assumption on a_{ij}, we see that

$$-\langle \mathcal{A}^\epsilon u^\epsilon, u^\epsilon \rangle_{H_\epsilon} \ge \lambda \|u^\epsilon\|_{H_\epsilon}^2$$

for some constant $\lambda > 0$, independent of ϵ. Then, integrating (7.54) with respect to t yields

$$\|u^\epsilon(t)\|_{H_\epsilon}^2 + \int_0^t \|u^\epsilon\|_{V_\epsilon}^2 \, ds$$
$$\le \|u_0^\epsilon\|_{H_\epsilon}^2 + \lambda^{-1}\|f^\epsilon\|_{L^2(0,T;H_\epsilon)}^2 + \int_0^t \langle g^\epsilon \, dW, u^\epsilon \rangle_{H_\epsilon} \, ds + \int_0^t \|g^\epsilon\|_{\mathcal{L}_2^Q}^2 \, ds.$$

Taking the expectation of both sides of the above inequality, we arrive at (7.49).

In a similar way, application of Itô's formula to $\|u^\epsilon\|_{V_\epsilon}^2 = \|\nabla_{A^\epsilon} u^\epsilon\|_{\oplus_n H_\epsilon}^2$ results in the relation

$$d\|u^\epsilon(t)\|_{V_\epsilon}^2 + 2\langle \mathcal{A}^\epsilon u^\epsilon, \mathcal{A}^\epsilon u^\epsilon \rangle_{H_\epsilon} dt$$
$$= -2\langle f^\epsilon, \mathcal{A}^\epsilon u^\epsilon \rangle_{H_\epsilon} dt - 2\langle g^\epsilon \, dW, \mathcal{A}^\epsilon u^\epsilon \rangle_{H_\epsilon} + \|\nabla_{A^\epsilon} g^\epsilon\|_{\mathcal{L}_2^Q}^2 \, dt. \quad (7.55)$$

Integrating both sides of (7.55) and using the Cauchy–Schwarz inequality, we obtain

$$\|u^\epsilon(t)\|_{V_\epsilon}^2 + \int_0^t \|\mathcal{A}^\epsilon u^\epsilon\|_{H_\epsilon}^2 \, ds$$
$$\le \|u^\epsilon(0)\|_{V_\epsilon}^2 + \|f^\epsilon\|_{L^2(0,T;H_\epsilon)}^2 - 2\int_0^t \langle g^\epsilon \, dW, \mathcal{A}^\epsilon u^\epsilon \rangle_{H_\epsilon} \, ds$$
$$+ \int_0^t \|\nabla_{A^\epsilon} g^\epsilon\|_{\mathcal{L}_2^Q}^2 \, ds.$$

Then, taking the expectation, we get (7.52). By (7.47) and the properties of stochastic integrals, we have (7.50).

Thus, by the above estimates, the solution can be extended to $[0, \infty)$, provided that (7.53) holds. The proof is complete.   □

As we mentioned in the beginning of this section, we need to consider the limit of the distribution of u^ϵ as $\epsilon \to 0$. However, the domain D_ϵ is changing with ϵ, that is, u^ϵ defines a family of probability measures on a family of functional spaces that depends on ϵ. So, we apply the extension operator defined by Lemma 7.2 and consider $P_\epsilon u^\epsilon$, which defines a family of probability measures, denoted by $\mathcal{L}(P_\epsilon u^\epsilon)$, on $L^2(0, T; H)$. By Lemma 7.12 and Lemma 3.7, $\{\mathcal{L}(P_\epsilon u^\epsilon)\}_\epsilon$ is tight in $L^2(0, T; H)$. That is, for any $\delta > 0$, there is a compact set $K_\delta \in L^2(0, T; H)$ such that

$$\mathbb{P}\{P_\epsilon u^\epsilon \in K_\delta\} > 1 - \delta.$$

Furthermore, by the Prokhorov theorem and the Skorohod embedding theorem [94, p. 32–33], for any sequence $\{\epsilon_j\}$ with $\epsilon_j \to 0$ as $j \to \infty$, there exists a subsequence $\{\epsilon_{j(k)}\}$, random variables $\{\hat{u}^{\epsilon_{j(k)}}\} \subset L^2(0, T; H_{\epsilon_{j(k)}})$, and $u \in L^2(0, T; H)$ defined on a new probability space $(\widehat{\Omega}, \widehat{\mathcal{F}}, \widehat{\mathbb{P}})$ such that

$$\mathcal{L}(P_{\epsilon_{j(k)}} \hat{u}^{\epsilon_{j(k)}}) = \mathcal{L}(P_{\epsilon_{j(k)}} u^{\epsilon_{j(k)}})$$

and

$$P_{\epsilon_{j(k)}} \hat{u}_{\epsilon_{j(k)}} \to u \quad \text{in} \quad L^2(0, T; H) \quad \text{as} \quad k \to \infty$$

for almost all $\omega \in \widehat{\Omega}$. Moreover, $P_{\epsilon_{j(k)}} \hat{u}^{\epsilon_{j(k)}}$ solves the system (7.36)–(7.39), with W replaced by Wiener process $\widehat{W}_k$, defined on probability space $(\widehat{\Omega}, \widehat{\mathcal{F}}, \widehat{\mathbb{P}})$, with the same distribution as that of W.

Recall a result on convergence in distribution (or convergence in law) from [113], Proposition 9.3.1: Given a sequence X^ϵ of random variables. If every subsequence of X^ϵ has a subsubsequence that converges in distribution to the same X as $\epsilon \to 0$, then X^ϵ itself converges in distribution to X as $\epsilon \to 0$. In this case, the law (i.e., probability distribution, which is a probability measure) of X^ϵ converges weakly to the law of X as $\epsilon \to 0$.

Next we determine the limit equation (homogenized effective equation) that u satisfies. Finally, we show that the effective equation is well-posed, so the limit u is unique. After all these are done, we see that $\mathcal{L}(u^\epsilon)$ weakly converges to $\mathcal{L}(u)$ as $\epsilon \downarrow 0$.

Having the tightness of $\mathcal{L}(P_\epsilon u^\epsilon)$, the next step is to pass the limit $\epsilon \to 0$ into (7.36)–(7.39). Define a new probability space $(\Omega_\delta, \mathcal{F}_\delta, \mathbb{P}_\delta)$ as

$$\Omega_\delta \triangleq \{\omega \in \Omega : u_\epsilon(\omega) \in K_\delta\},$$

$$\mathcal{F}_\delta \triangleq \{F \cap \Omega_\delta : F \in \mathcal{F}\},$$

and

$$\mathbb{P}_\delta(F) \triangleq \frac{\mathbb{P}(F \cap \Omega_\delta)}{\mathbb{P}(\Omega_\delta)}, \quad \text{for} \quad F \in \mathcal{F}_\delta.$$

Denote by $\mathbb{E}_\delta$ the expectation operator with respect to $\mathbb{P}_\delta$. Now we restrict the system to the probability space $(\Omega_\delta, \mathcal{F}_\delta, \mathbb{P}_\delta)$. In the following discussion we aim at obtaining $L^2(\Omega_\delta)$ convergence for any $\delta > 0$, which means the convergence in probability [38, p. 24], [113, p. 261].

Since K_δ is compact in $L^2(0, T; H)$, there exists a subsequence of u^ϵ in K_δ, still denoted by u^ϵ, such that for a fixed $\omega \in \Omega_\delta$,

$$P_\epsilon u^\epsilon \rightharpoonup u \quad \text{weakly* in} \quad L^\infty(0, T; H),$$
$$P_\epsilon u^\epsilon \rightharpoonup u \quad \text{weakly in} \quad L^2(0, T; H^1),$$
$$P_\epsilon u^\epsilon \to u \quad \text{strongly in} \quad L^2(0, T; H),$$
$$P_\epsilon \dot{u}_\epsilon \rightharpoonup \dot{u} \quad \text{weakly in} \quad L^2(0, T; H^{-1}).$$

Now we follow the oscillating test function method, except passing the limit $\epsilon \to 0$ in the stochastic sense.

We introduce

$$\xi^\epsilon \triangleq \left(\sum_{j=1}^n a_{ij}\left(\frac{x}{\epsilon}\right) \frac{\partial u^\epsilon}{\partial x_j} \right) \triangleq A^\epsilon \nabla u^\epsilon,$$

which satisfies

$$-\text{div}\xi^\epsilon = f^\epsilon + g^\epsilon \dot{W} - \dot{u}^\epsilon \quad \text{in} \quad D_\epsilon \times (0, T),$$
$$\xi^\epsilon \cdot \nu = 0 \quad \text{on} \quad \partial S_\epsilon \times (0, T).$$

By the estimates in Lemma 7.12, $\tilde{u}^\epsilon$ is bounded in $L^2(0, T; H_0^1)$. Thus, there is a subsequence of ξ^ϵ, still denoted by ξ^ϵ, such that

$$\tilde{\xi}^\epsilon \rightharpoonup \xi \quad \text{weakly in} \quad L^2(0, T; \oplus_n H).$$

Notice that for any $v \in H_0^1(D)$ and $\varphi \in C_0^\infty(0, T)$,

$$\int_0^T \int_D \tilde{\xi}^\epsilon \cdot \nabla v \varphi \, dx \, dt = \int_0^T \int_D \tilde{f}^\epsilon \epsilon v \varphi \, dx \, dt + \sum_{i=1}^\infty \int_0^T \int_D \tilde{g}_i^\epsilon v \, dx \varphi \, dW_i(t)$$
$$+ \int_0^T \int_D P_\epsilon u^\epsilon \chi_{D_\epsilon} \dot{\varphi} v \, dx \, dt. \tag{7.56}$$

By the estimate

$$\mathbb{E}\left| \sum_{i=1}^\infty \int_0^T \int_D \tilde{g}_i^\epsilon v \, dx \varphi \, dW_i(t) \right|^2 \leq \sum_{i=1}^\infty |\tilde{g}_i^\epsilon|^2_{L^2(0,T;H)} |v\varphi|^2_{L^2(0,T;H)}$$

and assumption (7.44), we have

$$\sum_{i=1}^\infty \int_0^T \int_D \tilde{g}_i^\epsilon v \, dx \varphi \, dW_i(t) \to \sum_{i=1}^\infty \int_0^T \int_D g_i v \, dx \varphi \, dW_i(t), \quad \text{in} \quad L^2(\Omega).$$

Now let $\epsilon \to 0$ in (7.56). By (7.2) and the fact $L^2(\Omega_\delta) \subset L^2(\Omega)$, we have in $L^2(\Omega_\delta)$,

$$\int_0^T \int_D \xi \cdot \nabla v \varphi \, dx \, dt = \int_0^T \int_D f v \varphi \, dx \, dt + \sum_{i=1}^\infty \int_0^T \int_D g^i v \, dx \varphi \, dW_i(t)$$

$$+ \int_0^T \int_D \vartheta u \dot{\varphi} v \, dx \, dt. \tag{7.57}$$

Hence,

$$-\text{div}\, \xi(x,t) dt = f(x,t) dt + g(x,t) dW - \vartheta \, du \quad \text{in} \quad D_T. \tag{7.58}$$

Let us identify the limit ξ.
For any $\lambda \in \mathbb{R}^n$, let w_λ be the solution of the following elliptic system:

$$-\sum_{j=1}^n \frac{\partial}{\partial y_j} \Big(\sum_{i=1}^n a_{ij}(y) \frac{\partial w_\lambda}{\partial y_i} \Big) = 0, \quad \text{on} \quad Y^*, \tag{7.59}$$

$$w_\lambda - \lambda \cdot y \quad \text{is} \quad Y - \text{periodic}, \tag{7.60}$$

$$\frac{\partial w_\lambda}{\partial \nu_{A^1}} = 0, \quad \text{on} \quad \partial S. \tag{7.61}$$

Define

$$w_\lambda^\epsilon \triangleq \epsilon (\hat{Q} w_\lambda) \Big(\frac{x}{\epsilon} \Big), \tag{7.62}$$

where $\hat{Q}$ is in Lemma 7.1. Then we have [84, p. 139],

$$w_\lambda^\epsilon \rightharpoonup \lambda \cdot x \quad \text{weakly in} \quad H^1(D), \tag{7.63}$$

$$\nabla w_\lambda^\epsilon \rightharpoonup \lambda \quad \text{weakly in} \quad \oplus_n L^2(D). \tag{7.64}$$

Set

$$\eta_{\lambda,j}(y) \triangleq \sum_{i=1}^n a_{ij}(y) \frac{\partial w_\lambda(y)}{\partial y_i}, \quad y \in Y^*,$$

and $\eta_\lambda^\epsilon(x) \triangleq (\eta_{\lambda_j}(x/\epsilon)) = a(x/\epsilon) \nabla w_\lambda^\epsilon$. Note that

$$-\text{div} \tilde{\eta}_\lambda^\epsilon = 0 \quad \text{on} \quad D. \tag{7.65}$$

Due to (7.63) and (7.64) and periodicity,

$$\tilde{\eta}_\lambda^\epsilon \rightharpoonup \frac{1}{|Y|} \int_{Y^*} \eta_\lambda \, dy \quad \text{weakly in} \quad L^2(D). \tag{7.66}$$

The above limit is still given by the classical homogenized matrix

$$\frac{1}{|Y|} \int_{Y^*} \eta_\lambda \, dy = \bar{A}^T \lambda = (\bar{a}_{ji}) \lambda$$

with

$$\bar{a}_{ij} \triangleq \frac{1}{|Y|} \int_{Y*} \sum_{k=1}^{n} a_{kj}(y) \frac{\partial w_{e_i}}{\partial y_k} dy. \tag{7.67}$$

Now, using the test function $\varphi v w_\lambda^\epsilon$ with $\varphi \in C_0^\infty(0, T)$, $v \in C_0^\infty(D)$ in (7.56) and multiplying both sides of (7.21) by $\varphi v P_\epsilon u_\epsilon$, we obtain

$$\int_0^T \int_D \tilde{\xi}^\epsilon \cdot \nabla v \varphi w_\lambda^\epsilon \, dx \, dt + \int_0^T \int_{D_\epsilon} \xi^\epsilon \cdot \nabla w_\lambda^\epsilon v \varphi \, dx \, dt$$

$$- \int_0^T \int_D \tilde{\eta}_\lambda^\epsilon \cdot \nabla v \varphi P_\epsilon u^\epsilon \, dx \, dt - \int_0^T \int_D \tilde{\eta}_\lambda^\epsilon \cdot \nabla (P_\epsilon u^\epsilon) v \varphi \, dx \, dt$$

$$= \int_0^T \int_D \tilde{f}^\epsilon \varphi v w_\lambda^\epsilon \, dx \, dt + \sum_{i=1}^\infty \int_0^T \int_D \tilde{g}_i^\epsilon v w_\lambda^\epsilon \, dx \varphi \, dW_i(t)$$

$$+ \int_0^T \int_D P_\epsilon u^\epsilon \chi_{D_\epsilon} \dot{\varphi} v w_\lambda^\epsilon \, dx \, dt.$$

Similar to the derivation of (7.24), we conclude that

$$\vartheta \, du = (\text{div}(\bar{A} \nabla u) + f) dt + g \, dW(t). \tag{7.68}$$

We also need to determine the initial value of u. Suppose

$$\tilde{u}_0^\epsilon \rightharpoonup u_0, \quad \text{weakly in } H. \tag{7.69}$$

We determine the initial value by using suitable test functions. In fact, taking $v \in C_0^\infty(D)$ and $\varphi \in C^\infty(0, T)$ with $\varphi(T) = 0$, we have

$$\int_0^T \int_D \tilde{\xi}^\epsilon \cdot \nabla v \varphi \, dx \, dt = \int_0^T \int_D \tilde{f}^\epsilon v \varphi \, dx \, dt + \sum_{i=0}^\infty \int_0^T \int_D \tilde{g}_i^\epsilon v \, dx \varphi \, dW_i(t)$$

$$- \int_0^T \int_D \tilde{u}^\epsilon v \dot{\varphi} \, dx \, dt + \int_D \tilde{u}_0^\epsilon \varphi(0) v \, dx.$$

Then, passing the limit as $\epsilon \to 0$, noticing that

$$\int_0^T \int_D \tilde{u}^\epsilon v \dot{\varphi} \, dx \, dt = \int_0^T \int_D \chi_{D_\epsilon} P_\epsilon \tilde{u}^\epsilon v \dot{\varphi} \, dx \, dt \to \int_0^T \int_D \vartheta u v \dot{\varphi} \, dx \, dt$$

$$= - \int_0^T \int_D \vartheta \dot{u} v \varphi \, dx \, dt + \int_D \vartheta u(0) \varphi(0) v \, dx$$

and using (7.58), we get the initial value of u:

$$u(0) = \frac{u_0}{\vartheta}.$$

Therefore, we have proved the convergence of u^ϵ in the sense of $L^2(\Omega_\delta)$, i.e.,

$$\lim_{\epsilon \to 0} \mathbb{E}_\delta |P_\epsilon u^\epsilon - u|^2_{L^2(0,T;H)} = 0 \tag{7.70}$$

and

$$\lim_{\epsilon \to 0} \mathbb{E}_\delta \int_0^T \int_D (\mathcal{A}_\epsilon P_\epsilon u^\epsilon - \bar{A}\nabla u)v\varphi \, dx \, dt = 0, \tag{7.71}$$

for every $v \in C_0^\infty(D)$ and $\varphi \in C^\infty(0,T)$.

We summarize the homogenization result for the stochastic system (7.36)–(7.39) in the following theorem.

Theorem 7.13 (Homogenization on finite time interval). *For every $T > 0$, assume that (7.43), (7.44), and (7.69) hold. Let u^ϵ be the solution of (7.36)–(7.39). Then the distribution $\mathcal{L}(P_\epsilon u^\epsilon)$ converges weakly to μ in the space consisting of probability measures on $L^2(0,T;H)$ as $\epsilon \downarrow 0$, with μ being the distribution of u, which is the solution of the following homogenized effective equation:*

$$\vartheta \, du = \big(\text{div}(\bar{A}\nabla u) + f\big)dt + g \, dW(t) \quad in \quad D \times (0,T), \tag{7.72}$$

$$u = 0 \quad on \quad \partial D \times (0,T), \tag{7.73}$$

$$u(0,x) = \frac{u_0}{\vartheta} \quad on \quad D, \tag{7.74}$$

where $\bar{A} \triangleq (\bar{a}_{ij})$ with $\bar{a}_{ij}$ given by (7.67).

Proof.   First, the homogenized effective equation is well-posed [94, Theorem 5.4]. Then, by the arbitrariness of δ, the Skorohod theorem, and $L^2(\Omega_\delta)$ convergence of $P_\epsilon u^\epsilon$ on $(\Omega_\delta, \mathcal{F}_\delta, \mathbb{P}_\delta)$, we obtain the result.   □

We look at an example.

Example 7.14.   Consider the following linear SPDE on the interval $D = (-L, L)$ with $Y = [-1, 1)$ and $S = (-0.5, 0.5)$,

$$du^\epsilon(x,t) = \partial_{xx}u^\epsilon(x,t)dt + dW(x,t) \quad on \quad D_\epsilon \times (0,T),$$

$$u^\epsilon = 0 \quad on \quad \partial D \times (0,T),$$

$$\frac{\partial u_\epsilon}{\partial v} = 0 \quad on \quad \partial S_\epsilon \times (0,T),$$

$$u^\epsilon(0) = u_0^\epsilon \quad on \quad D_\epsilon.$$

The corresponding cell problem is

$$w''(y) = 0, \quad on \quad Y^* = (-1, -0.5] \cup [0.5, 1),$$

$$w(y) - y \quad is \quad Y - periodic,$$

$$w'(y) = 0, \quad on \quad \partial S = \{-0.5, 0.5\}.$$

The homogenized model is

$$\tfrac{1}{2}du(x,t) = \partial_{xx}u(x,t)dt + dW(x,t), \quad on \quad (-L,L) \times (0,T),$$

$$u(x,t) = 0, \quad on \quad \{-L, L\} \times (0,T).$$

The homogenized matrix $\bar{A}$, which is now just a real number, is given by

$$\bar{A} = \frac{1}{|Y|} \int_{Y^*} w'(y) dy = 1.$$

Effectivity on the Whole Time Interval

We have obtained the homogenized effective model on any finite time interval $[0, T]$. In fact, we can consider the long time effectivity of the homogenized equation. For simplicity, we assume that f^ϵ and g^ϵ are independent of time t and

$$\sum_{i=1}^{\infty} \|\nabla_{A^\epsilon} g_i^\epsilon(x)\|_{\oplus_n H_\epsilon}^2 < C^* \tag{7.75}$$

with C^* a positive constant independent of ϵ. Then, by the assumption on A^ϵ and the properties of $\bar{a}_{ij}$, system (7.36)–(7.39) and homogenized Equations (7.72)–(7.74) have unique stationary solutions $u^{\epsilon,*}(x, t)$ and $u^*(x, t)$, respectively. The construction of a stationary solution is classical (see Problem 6.1). We denote by $\mu^{\epsilon*}$ and μ^* the distributions of $P_\epsilon u^{\epsilon*}$ and u^* in H, respectively. Then, if $\mathbb{E}\|u_0^\epsilon\|^2 < \infty$ and $\mathbb{E}\|u^0\|^2 < \infty$,

$$\left| \int_H h \, d\mu^\epsilon(t) - \int_H h \, d\mu^{\epsilon*} \right| \leq C(u_0^\epsilon) e^{-\gamma t}, \quad t > 0, \tag{7.76}$$

$$\left| \int_H h d\mu(t) - \int_H h d\mu^* \right| \leq C(u_0) e^{-\gamma t}, \quad t > 0, \tag{7.77}$$

for some constant $\gamma > 0$ and any $h : H \to \mathbb{R}^1$ with $\sup |h| \leq 1$ and $\text{Lip}(h) \leq 1$. Here $\mu^\epsilon(t) = \mathcal{L}(P_\epsilon u^\epsilon(t, u_0^\epsilon))$, $\mu(t) = \mathcal{L}(u(t, u_0/\vartheta))$, $C(u_0^\epsilon)$, and $C(u^0)$ are positive constants depending only on the initial value u_0^ϵ and u_0, respectively. The above convergence also yields that $\mu^\epsilon(t)$ and $\mu(t)$ weakly converge to $\mu^{\epsilon*}$ and μ^*, respectively, as $t \to \infty$.

Next we show that $\mu^{\epsilon*}$ weakly converges to μ^* as $\epsilon \to 0$. For this, we first derive additional *a priori* estimates, which are uniform with respect to ϵ, to ensure the tightness of the stationary distributions. We introduce some spaces. For Banach space U and $p > 1$, we define $W^{1,p}(0, T; U)$ as the space of functions $h \in L^p(0, T; U)$ such that

$$|h|_{W^{1,p}(0,T;U)}^p = |h|_{L^p(0,T;U)}^p + \left| \frac{dh}{dt} \right|_{L^p(0,T;U)}^p < \infty. \tag{7.78}$$

For any $\alpha \in (0, 1)$, define $W^{\alpha,p}(0, T; U)$ as the space of function $h \in L^p(0, T; U)$ such that

$$|h|_{W^{\alpha,p}(0,T;U)}^p = |h|_{L^p(0,T;U)}^p + \int_0^T \int_0^T \frac{|h(t) - h(s)|_U^p}{|t - s|^{1+\alpha p}} ds \, dt < \infty. \tag{7.79}$$

For $\rho \in (0, 1)$, we denote by $C^\rho(0, T; U)$ the space of functions $h : [0, T] \to \mathcal{X}$ that are Hölder continuous with exponent ρ.

For $T > 0$, denote by $\mathbf{u}_T^{\epsilon*}$ (respectively, $\mathbf{u}_T^*$) the distribution of stationary process $P_\epsilon u^{\epsilon*}(\cdot)$ (respectively, $u^*(\cdot)$) in $L^2(0, T; H^1)$. Then we have the following result.

Lemma 7.15. *For every $T > 0$, the family $\mathbf{u}_T^{\epsilon*}$ is tight in $L^2(0, T; H^{2-\iota})$ with some $\iota > 0$.*

Proof.   Since $u^{\epsilon*}$ is stationary, by (7.52) we see that

$$\mathbb{E}|u^{\epsilon*}|^2_{L^2(0,T;H^2_\epsilon)} < C_T. \tag{7.80}$$

Now represent $u^{\epsilon*}$ in the form

$$u^{\epsilon*}(t) = u^{\epsilon*}(0) + \int_0^t \mathcal{A}_\epsilon u^{\epsilon*}(s)ds + \int_0^t f^\epsilon(x)ds + \int_0^t g^\epsilon(x)dW(s).$$

Also, by the stationarity of $u^{\epsilon*}$ and (7.52), we obtain

$$\mathbb{E}\left|\int_0^t \mathcal{A}^\epsilon P_\epsilon u^{\epsilon*}(s)ds + \int_0^t \tilde{f}^\epsilon(x)ds\right|^2_{W^{1,2}(0,T;H)} \le C_T. \tag{7.81}$$

Define $M_\epsilon(s, t) \triangleq \int_s^t \tilde{g}^\epsilon(x)dW(s)$. By Lemma 3.24 and the Hölder inequality, we derive that

$$\mathbb{E}\|M_\epsilon(s, t)\|^4_{V_\epsilon} \le c\left(\int_s^t \|\nabla_{A^\epsilon} \tilde{g}^\epsilon(x)\|^2_{\mathcal{L}_2^Q} d\tau\right)^2 \le K(t-s)\int_s^t \|\nabla_{A^\epsilon} \tilde{g}^\epsilon(x)\|^4_{\mathcal{L}_2^Q} d\tau$$

$$\le KC^{*2}|t-s|^2$$

for $t \in [s, T]$, where K is a positive constant independent of ϵ, s, and t. Therefore,

$$\mathbb{E}\int_0^T \|M_\epsilon(0, t)\|^4_{V_\epsilon} dt \le C_T \tag{7.82}$$

and

$$\mathbb{E}\int_0^T \int_0^T \frac{\|M_\epsilon(0, t) - M_\epsilon(0, s)\|^4_{V_\epsilon}}{|t-s|^{1+4\alpha}}ds\,dt \le C_T. \tag{7.83}$$

Combining (7.80)–(7.83), together with the following compact embedding

$$L^2(0, T; H^2) \cap W^{1,2}(0, T; H) \subset L^2(0, T; H^{2-\iota})$$

and

$$L^2(0, T; H^2) \cap W^{\alpha,4}(0, T; H^1) \subset L^2(0, T; H^{2-\iota}),$$

we obtain the tightness of $\mathbf{u}_T^{\epsilon*}$ in $L^2(0, T; H^{2-\iota})$. This completes the proof. $\square$

The above lemma directly yields the following result:

Corollary 7.16.   *The family $\{\mu^{\epsilon*}\}$ is tight in H^1.*

By Lemma 7.15, for any fixed $T > 0$, the Skorohod embedding theorem asserts that for any sequence $\{\epsilon_n\}_n$ with $\epsilon_n \to 0$ as $n \to \infty$, there is a subsequence $\{\epsilon_{n(k)}\}_k$, a new probability space $(\overline{\Omega}, \overline{\mathcal{F}}, \overline{\mathbb{P}})$, and random variables $\overline{u}^{\epsilon_{n(k)}*} \in L^2(0, T; V_\epsilon), \overline{u}^* \in L^2(0, T; H^1)$ such that

$$\mathcal{L}(P_\epsilon \overline{u}^{\epsilon_{n(k)}*}) = \mathbf{u}_T^{\epsilon_{n(k)}*}, \quad \mathcal{L}(\overline{u}^*) = \mathbf{u}_T^*,$$

and

$$\overline{u}^{\epsilon_{n(k)}*} \to \overline{u}^*, \quad \text{in } L^2(0, T; H^1) \quad \text{as } k \to \infty.$$

Moreover, $\overline{u}^{\epsilon_{n(k)}*}$ (respectively, $\overline{u}^*$) is the unique stationary solution of Equation (7.45) (respectively, (7.72)) with W replaced by $\overline{W}_k$ (respectively, $\overline{W}$), where $\overline{W}_k$ and $\overline{W}$ are some Wiener processes defined on $(\overline{\Omega}, \overline{\mathcal{F}}, \overline{\mathbb{P}})$ with the same distribution as W. Then, by the homogenization of (7.36)–(7.39) on a finite time interval and the uniqueness of the invariant measure,

$$\mathbf{u}_T^{\epsilon*} \rightharpoonup \mathbf{u}_T^*, \quad \text{as } \epsilon \to 0$$

for any $T > 0$.

To show long time effectivity, let $u^\epsilon(t), t \geq 0$, be a weak solution of system (7.36)–(7.39), and define $u^{\epsilon,t}(\cdot) \triangleq u^\epsilon(t + \cdot)$, which is in $L^2_{\text{loc}}(\mathbb{R}_+; V_\epsilon)$ by Theorem 7.12. Then, by (7.76),

$$\mathcal{L}(P_\epsilon u^{\epsilon,t}(\cdot)) \rightharpoonup \mathcal{L}(P_\epsilon u^{\epsilon*}(\cdot)), \quad t \to \infty$$

in the space of probability measures on $L^2_{\text{loc}}(\mathbb{R}_+; H^1)$. Having the above analysis, we have the following result, which implies the long time effectivity of the homogenized effective Equation (7.72).

Theorem 7.17 (Homogenization on the whole time interval). *Assume that $f^\epsilon \in H_\epsilon$ and $g_i^\epsilon \in V_\epsilon$ are independent of time t. Further, assume that (7.75), (7.40), and (7.44) hold. Denote by $u^\epsilon(t), t \geq 0$, the solution of (7.36)–(7.39) and u^* the unique stationary solution of (7.72). Then,*

$$\lim_{\epsilon \downarrow 0} \lim_{t \to \infty} \mathcal{L}(P_\epsilon u^{\epsilon,t}(\cdot)) = \mathcal{L}(u^*(\cdot)), \tag{7.84}$$

where the limit is understood in the sense of weak convergence of Borel probability measures in $L^2_{loc}(\mathbb{R}_+; H^1)$. That is, the solution of (7.36)–(7.39) converges to the stationary solution of (7.72), in distribution, as $t \to \infty$ and $\epsilon \to 0$.

Remark 7.18. The above result implies that the macroscopic model (7.72) is an effective approximation for the microscopic model (7.36)–(7.39) on a very long time scale. In other words, if we intend to numerically simulate the long time behavior of the microscopic model, we could simulate the macroscopic model as an approximation when ϵ is sufficiently small.

Effectivity in Energy Convergence

Here we show effectivity in energy convergence for the homogenized Equation (7.72)–(7.74). Energy convergence is in the sense of mean square on a finite time interval instead of in the sense of distribution.

Let u^ϵ be a weak solution of (7.36)–(7.39), and let u be a weak solution of (7.72)–(7.74). We introduce the following energy functionals:

$$\mathcal{E}^\epsilon(u^\epsilon)(t) \triangleq \frac{1}{2}\mathbb{E}\|\tilde{u}^\epsilon\|_H^2 + \mathbb{E}\int_0^t\int_D \chi_{D_\epsilon} A^\epsilon \nabla\big(P_\epsilon u^\epsilon(x,\tau)\big)\nabla\big(P_\epsilon u^\epsilon(x,\tau)\big)dx\,d\tau \tag{7.85}$$

and

$$\mathcal{E}^0(u)(t) \triangleq \frac{1}{2}\mathbb{E}\|u\|_H^2 + \mathbb{E}\int_0^t\int_D \bar{A}\nabla u(x,\tau)\nabla u(x,\tau)dx\,d\tau. \tag{7.86}$$

By Itô's formula,

$$\mathcal{E}^\epsilon(u_\epsilon)(t) = \frac{1}{2}\mathbb{E}\|\tilde{u}_0^\epsilon\|_H^2 + \mathbb{E}\int_0^t\int_D \tilde{f}^\epsilon(x,\tau)\tilde{u}_\epsilon(x,\tau)dx\,d\tau$$
$$+\frac{1}{2}\mathbb{E}\int_0^t\|\tilde{g}^\epsilon(x,\tau)\|_{\mathcal{L}_2^Q}^2\,d\tau$$

and

$$\mathcal{E}^0(u)(t) = \frac{1}{2}\mathbb{E}\|u^0\|_H^2 + \mathbb{E}\int_0^t\int_D f(x,\tau)u(x,\tau)dx\,d\tau + \frac{1}{2}\mathbb{E}\int_0^t\|g(x,\tau)\|_{\mathcal{L}_2^Q}^2\,d\tau.$$

Then we have the following effectivity of the macroscopic model in the sense of convergence in energy.

Theorem 7.19 (Homogenization in energy convergence).   *Assume that (7.43) and (7.44) hold. If*

$$\tilde{u}_0^\epsilon \to u^0 \quad strongly\ in \quad H, \quad as \quad \epsilon \to 0,$$

then

$$\mathcal{E}^\epsilon(u^\epsilon) \to \mathcal{E}^0(u) \quad in \quad C([0,T]), \quad as \quad \epsilon \to 0.$$

Proof.   By the homogenization approach on a finite time interval, for any $\delta > 0$, $u^\epsilon \to u$ strongly in $L^2(0,T;H)$ on Ω_δ. Then, since δ is arbitrary,

$$\mathbb{E}\int_0^t\int_D \tilde{f}^\epsilon(x,\tau)\tilde{u}^\epsilon(x,\tau)dx\,d\tau \to \mathbb{E}\int_0^t\int_D f(x,\tau)u(x,\tau)dx\,d\tau$$
$$\text{for}\quad t \in [0,T].$$

Noticing that $\tilde{g}^\epsilon \rightharpoonup g$ weakly in $L^2(0,t;\mathcal{L}_2^Q)$, we have

$$\mathcal{E}^\epsilon(u^\epsilon)(t) \to \mathcal{E}^0(u)(t) \quad \text{for any} \quad t \in [0,T]. \tag{7.87}$$

We now only need to show that $\{\mathcal{E}^\epsilon(u^\epsilon)(t)\}_\epsilon$ is equi-continuous, as in the Ascoli–Arzela theorem [113, Theorem 2.4.7], then imply the result. In fact, given any $t \in [0,T]$

and $h > 0$ small enough, we have

$$|\mathcal{E}^\epsilon(u^\epsilon)(t+h) - \mathcal{E}^\epsilon(u^\epsilon)(t)|$$

$$\leq \left| \mathbb{E} \int_t^{t+h} \int_D \tilde{f}^\epsilon(x,\tau)\tilde{u}^\epsilon(x,\tau)dx\,d\tau \right| + \mathbb{E} \int_t^{t+h} \|\tilde{g}^\epsilon(x,\tau)\|^2_{\mathcal{L}_2^Q}d\tau$$

$$\leq \mathbb{E}\left\{ |\tilde{f}^\epsilon|_{L^2(0,T;H)} \int_t^{t+h} \|\tilde{u}^\epsilon(x,\tau)\|^2_H\,dx\,d\tau \right\} + \mathbb{E} \int_t^{t+h} \|\tilde{g}^\epsilon(x,\tau)\|^2_{\mathcal{L}_2^Q}\,d\tau.$$

Noticing that $\tilde{u}^\epsilon \in L^2(0,T;H)$ a.s. and (7.42), we have

$$|\mathcal{E}^\epsilon(u^\epsilon)(t+h) - \mathcal{E}^\epsilon(u^\epsilon)(t)| \to 0, \quad \text{as} \quad h \to 0,$$

uniformly on ϵ, which implies the equicontinuity of the family $\{\mathcal{E}^\epsilon(u^\epsilon)\}_\epsilon$. This completes the proof. $\qquad\square$

7.2.2 Homogenization Under Random Dynamical Boundary Conditions

In this subsection, we consider homogenization for a microscopic heterogeneous system, modeled by a linear SPDE in a medium that exhibits small-scale obstacles, with *random dynamical boundary conditions* on the boundaries of the obstacles. Such random boundary conditions arise in the modeling of, for example, the air-sea interactions on the ocean surface [254], heat transfer in a solid in contact with a fluid [210], chemical reactor theory [211], and colloid and interface chemistry [293]. One example of such microscopic systems of interest is composite materials containing microscopic heterogeneities, under the impact of random fluctuations in the domain and on the surface of the heterogeneities [175,224]. A motivation for such a model is based on the consideration that the interaction between the atoms of the different compositions in a composite material causes thermal noise when the scale of the heterogeneity scale is small. A similar consideration appears in a microscopic stochastic lattice model [3,41] for a composite material. Here the microscopic structure is perturbed by random effect, and the complicated interactions on the boundary of the holes are dynamically and randomly evolving. So, the model can be described by an SPDE on a periodically perforated domain D_ϵ with random dynamical boundary conditions on the boundary of small holes ∂S_ϵ.

One interesting point is that, for the system with random dynamical boundary conditions, the random force on the boundary of microscopic scale holes leads, in the homogenization limit, to a random force distributed all over the physical domain, even when the model equation itself contains no stochastic influence inside the domain (§6, Remark 7.26). We could also say that the impact of small-scale random dynamical boundary conditions is quantified or carried over to the homogenized model as an extra random forcing.

Partial differential equations (PDEs) with dynamical boundary conditions have been investigated recently, as in, e.g., [16,114,119,120,157,288]. Parabolic SPDEs with noise in the Neumann boundary conditions have also been studied, such as [95, Ch. 13], [96], and [227]. In [79], the authors have studied the well-posedness of SPDEs with random dynamical boundary conditions. The present authors, with collaborators, have considered dynamical issues of SPDEs with random dynamical boundary conditions [55,106,284,300,312].

For a fixed final time $T > 0$, we consider the following Itô type nonautonomous SPDE defined on the perforated domain D_ϵ in $\mathbb{R}^n$:

$$du^\epsilon(x, t) = \left[\Delta u^\epsilon(x, t) + f(x, t, u^\epsilon, \nabla u^\epsilon)\right]dt + g_1(x, t)dW_1(x, t)$$
$$\text{on} \quad D_\epsilon \times (0, T), \tag{7.88}$$

$$\epsilon^2 du^\epsilon(x, t) = \left[-\frac{\partial u^\epsilon(x, t)}{\partial \nu_\epsilon} - \epsilon bu^\epsilon(x, t)\right]dt + \epsilon g_2(x, t) dW_2(x, t)$$
$$\text{on} \quad \partial S_\epsilon \times (0, T), \tag{7.89}$$

$$u^\epsilon(x, t) = 0 \quad \text{on} \quad \partial D \times (0, T), \tag{7.90}$$

$$u^\epsilon(0, x) = u_0(x) \quad \text{on} \quad D_\epsilon, \tag{7.91}$$

where b is a real constant, $f : D \times [0, T] \times \mathbb{R}^1 \times \mathbb{R}^n \to \mathbb{R}^1$ satisfies some properties that are described later, ν_ϵ is the exterior unit normal vector on the boundary ∂S_ϵ, and $u_0 \in L^2(D)$. Moreover, $W_1(x, t)$ and $W_2(x, t)$ are mutually independent $L^2(D)$-valued Wiener processes on a complete probability space $(\Omega, \mathcal{F}, \mathbb{P})$. Denote by Q_1 and Q_2 the covariance operators of W_1 and W_2, respectively. Here we assume that $g_i(x, t) \in \mathcal{L}(L^2(D))$, $i = 1, 2$, and that there is a constant $C_T > 0$ independent of ϵ, such that

$$\|g_i(\cdot, t)\|^2_{\mathcal{L}_2^{Q_i}} \triangleq \sum_{j=1}^{\infty} \|g_i Q_i^{\frac{1}{2}} e_j\|^2_{L^2(D)} \leq C_T, \quad i = 1, 2, \quad t \in [0, T], \tag{7.92}$$

where $\{e_j\}_{j=1}^{\infty}$ are eigenvectors of operator $-\Delta$ on D with zero Dirichlet boundary conditions, which form an orthonormal basis of $L^2(D)$. Denote by $\mathcal{L}_2^{Q_i} := \mathcal{L}^2(Q_i^{1/2}H, H)$. Recall that $\mathcal{L}^2(Q_i^{1/2}H, H)$ consists of Hilbert–Schmidt operators from $Q_i^{1/2}H$ to H [94, Ch. 4.2], and the norm is defined in §3.6.

In this section, we only consider the case when f is linear. Cases for nonlinear f's are discussed in the next section.

For a fixed $T > 0$, we always denote by C_T a constant independent of ϵ and still denote by D_T the set $D \times (0, T)$.

Basic Properties of the Microscopic Model

To homogenize the stochastic system (7.88)–(7.91), we first need some estimates for the solutions u^ϵ in an appropriate space. Suppose that the term f is independent of u^ϵ and ∇u^ϵ and that $f(\cdot, \cdot) \in L^2(0, T; L^2(D))$.

Define by $H_\epsilon^1(D_\epsilon)$ the space of elements of $H^1(D_\epsilon)$ that vanish on ∂D. Denote by $H_\epsilon^{-1}(D_\epsilon)$ the dual space of $H_\epsilon^1(D_\epsilon)$ with the usual norm, and let $\gamma_\epsilon : H^1(D_\epsilon) \to L^2(\partial S_\epsilon)$ be the continuous trace operator with respect to ∂S_ϵ [289, Ch. 4.7]. Note that $H^{\frac{1}{2}}(\partial S_\epsilon) = \gamma_\epsilon(H^1(D_\epsilon))$, and $H_\epsilon^{-\frac{1}{2}}(D_\epsilon)$ is the dual space of $H_\epsilon^{\frac{1}{2}}(D_\epsilon)$.

Introduce the following function spaces:

$$X_\epsilon^1 \triangleq \left\{(u, v) \in H_\epsilon^1(D_\epsilon) \times H_\epsilon^{\frac{1}{2}}(\partial S_\epsilon) : v = \epsilon \gamma_\epsilon u\right\}$$

and

$$X_\epsilon^0 \triangleq \left\{ L^2(D_\epsilon) \times L_\epsilon^2(\partial S_\epsilon) \right\}$$

with the usual scalar products and norms. Define an operator B^ϵ on the space $H_\epsilon^1(D_\epsilon)$ by

$$B^\epsilon u \triangleq \frac{\partial u}{\partial \nu_\epsilon} + \epsilon b u, \quad u \in H_\epsilon^1(D_\epsilon). \tag{7.93}$$

Also define an operator $\mathcal{A}^\epsilon$ on $\mathcal{D}(\mathcal{A}^\epsilon) \triangleq \{(u, v) \in X_\epsilon^1 : (-\Delta u, R_\epsilon B^\epsilon u) \in X_\epsilon^0\}$, where R^ϵ is the restriction to ∂S_ϵ, as

$$\mathcal{A}^\epsilon z \triangleq \left(-\Delta u, \frac{1}{\epsilon} R^\epsilon B^\epsilon u \right)^T, \quad z = (u, v)^T \in \mathcal{D}(\mathcal{A}^\epsilon). \tag{7.94}$$

Associated with the operator $\mathcal{A}^\epsilon$, we introduce the bilinear form on X_ϵ^1

$$a^\epsilon(z, \bar{z}) \triangleq \int_{D_\epsilon} \nabla u \nabla \bar{u}\, dx + \epsilon b \int_{S_\epsilon} \gamma_\epsilon(u)\gamma_\epsilon(\bar{u})ds \tag{7.95}$$

with $z = (u, v), \bar{z} = (\bar{u}, \bar{v}) \in X_\epsilon^1$. Noticing that $\|\gamma_\epsilon(u)\|_{L^2(\partial S_\epsilon)}^2 \leq C(S_\epsilon)\|u\|_{H_\epsilon^1(D_\epsilon)}^2$, we see that there is an $M > 0$, independent of ϵ, such that

$$a^\epsilon(z, \bar{z}) \leq M \|u\|_{H_\epsilon^1(D_\epsilon)} \|\bar{u}\|_{H_\epsilon^1(D_\epsilon)}$$

and the following coercive property of a^ϵ holds:

$$a^\epsilon(z, z) \geq \bar{\alpha}\|z\|_{X_\epsilon^1}^2 - \bar{\beta}\|z\|_{X_\epsilon^0}^2, \quad z \in X_\epsilon^1 \tag{7.96}$$

for some positive constants $\bar{\alpha}$ and $\bar{\beta}$ that are also independent of ϵ. Denote by $S_\epsilon(t)$ the C_0-semigroup generated by operator $-\mathcal{A}^\epsilon$. Then the system (7.88)–(7.91) can be rewritten as the following abstract stochastic evolutionary equation:

$$dz^\epsilon(x, t) = [-\mathcal{A}^\epsilon z^\epsilon(x, t) + F^\epsilon(x, t)]dt + G^\epsilon(x, t)dW(x, t), \quad z^\epsilon(0) = z_0, \tag{7.97}$$

where

$$F^\epsilon(x, t) = (f(x, t), 0)^T,$$
$$G^\epsilon(x, t)dW(t) = (g_1(x, t)dW_1(x, t), g_2(x, t)dW_2(x, t))^T$$

and $z_0 = (u_0, v_0)^T$. In the mild sense, this becomes

$$z^\epsilon(t) = S_\epsilon(t)z_0 + \int_0^t S_\epsilon(t - s)F^\epsilon(s)ds + \int_0^t S_\epsilon(t - s)G^\epsilon(s)dW(s). \tag{7.98}$$

Moreover, the variational or weak formulation is

$$\int_0^T \int_{D_\epsilon} \dot{u}^\epsilon \varphi \, dx \, dt + \epsilon^2 \int_0^T \int_{\partial S_\epsilon} \dot{u}^\epsilon \varphi \, dx \, dt + \epsilon b \int_0^T \int_{\partial S_\epsilon} u^\epsilon \varphi \, dx \, dt$$

$$= - \int_0^T \int_{D_\epsilon} \nabla u^\epsilon \nabla \varphi \, dx \, dt + \int_0^T \int_{D_\epsilon} f \varphi \, dx \, dt + \int_0^T \int_{D_\epsilon} g_1 \varphi \dot{W}_1 \, dx \, dt$$

$$+ \epsilon \int_0^T \int_{\partial S_\epsilon} g_2 \varphi \dot{W}_2 \, dx \, dt, \tag{7.99}$$

for $\varphi(x, t) \in C_0^\infty([0, T] \times D_\epsilon)$. Here $(\dot{\cdot})$ stands for $d \cdot /dt$.

For well-posedness of system (7.97) we have the following result.

Lemma 7.20 (Global well-posedness of microscopic model). *Assume that* (7.92) *holds for $T > 0$. If $z_0 = (u_0, v_0)^T$ is a $(\mathcal{F}_0, \mathcal{B}(X_\epsilon^0))$-measurable random variable, then the system* (7.97) *has a unique mild solution $z^\epsilon \in L^2(\Omega, C(0, T; X_\epsilon^0) \cap L^2(0, T; X_\epsilon^1))$, which is also a weak solution in the following sense:*

$$\langle z^\epsilon(t), \varphi \rangle_{X_\epsilon^1} = \langle z_0, \varphi \rangle_{X_\epsilon^1} + \int_0^t \langle -\mathcal{A}^\epsilon z^\epsilon(s), \varphi \rangle_{X_\epsilon^1} \, ds$$

$$+ \int_0^t \langle F^\epsilon, \varphi \rangle_{X_\epsilon^1} \, ds + \int_0^t \langle G^\epsilon \, dW, \varphi \rangle_{X_\epsilon^1} \tag{7.100}$$

for $t \in [0, T)$ and $\varphi \in X_\epsilon^1$. Moreover, if z_0 is independent of $W(t)$ with $\mathbb{E}|z_0|^2_{X_\epsilon^0} < \infty$, then

$$\mathbb{E}\|z^\epsilon(t)\|^2_{X_\epsilon^0} + \mathbb{E}\int_0^t \|z^\epsilon(s)\|^2_{X_\epsilon^1} \, ds \leq (1 + \mathbb{E}\|z_0\|^2_{X_\epsilon^0})C_T, \quad for \quad t \in [0, T] \tag{7.101}$$

and

$$\mathbb{E}\{\sup_{t \in [0,T]} \|z^\epsilon(t)\|^2_{X_\epsilon^0}\} \leq \left(1 + \mathbb{E}\|z_0\|^2_{X_\epsilon^0} + \mathbb{E}\int_0^T \|z^\epsilon(s)\|^2_{X_\epsilon^1} \, ds\right)C_T, \tag{7.102}$$

for some constant $C_T > 0$.

Proof. By the assumption (7.92), we have

$$\|G^\epsilon(x, t)\|^2_{\mathcal{L}_2^Q} = \|g_1(x, t)\|^2_{\mathcal{L}_2^{Q_1}} + \|g_2(x, t)\|^2_{\mathcal{L}_2^{Q_2}} < \infty.$$

Then the classical result [94, Theorem 5.4] yields the local existence of z^ϵ. By applying the stochastic Fubini theorem [94, p. 109], the local mild solution is also a weak solution.

Now we derive the following *a priori* estimates, which yield the existence of a weak solution on $[0, T]$ for any $T > 0$.

Applying Itô's formula to $\|z^\epsilon\|^2_{X_\epsilon^0}$, we have

$$d\|z^\epsilon(t)\|^2_{X_\epsilon^0} + 2\langle \mathcal{A}^\epsilon z^\epsilon, z^\epsilon \rangle_{X_\epsilon^0} dt = 2\langle F^\epsilon(x, t), z^\epsilon \rangle_{X_\epsilon^0} \, dt$$

$$+ 2\langle G^\epsilon(x, t) \, dW(t), z^\epsilon \rangle_{X_\epsilon^0}$$

$$+ \|G^\epsilon(x, t)\|^2_{\mathcal{L}_2^Q} \, dt. \tag{7.103}$$

By the coercivity (7.96) of $a^\epsilon(\cdot, \cdot)$, integrating (7.103) with respect to t yields

$$\|z^\epsilon(t)\|^2_{X^0_\epsilon} + 2\bar{\alpha} \int_0^t \|z^\epsilon(s)\|^2_{X^1_\epsilon} ds$$

$$\leq \|z_0\|^2_{X^0_\epsilon} + |F^\epsilon|^2_{L^2(0,T;X^0_\epsilon)} + (2\bar{\beta}+1) \int_0^t \|z^\epsilon(s)\|^2_{X^0_\epsilon} ds$$

$$+ \int_0^t \langle G^\epsilon(s)dW(s), z^\epsilon(s)\rangle_{X^0_\epsilon} + \int_0^t \|G^\epsilon(s)\|^2_{\mathcal{L}^0_2} ds.$$

Taking the expectation of both sides of the above inequality yields

$$\mathbb{E}\|z^\epsilon(t)\|^2_{X^0_\epsilon} + 2\bar{\alpha}\mathbb{E} \int_0^t \|z^\epsilon(s)\|^2_{X^1_\epsilon} ds$$

$$\leq \mathbb{E}\|z_0\|^2_{X^0_\epsilon} + |F^\epsilon|^2_{L^2(0,T;X^0_\epsilon)} + (2\bar{\beta}+1) \int_0^t \mathbb{E}\|z^\epsilon(s)\|^2_{X^0_\epsilon} ds$$

$$+ \int_0^t \|G^\epsilon(s)\|^2_{\mathcal{L}^0_2} ds.$$

Finally, the Gronwall inequality implies the estimate (7.101). Notice that, by Lemma 3.24,

$$\mathbb{E} \sup_{t\in[0,T]} \left\| \int_0^t S_\epsilon(t-s)G^\epsilon(s,x)ds \right\|^2_{X^0_\epsilon} \leq C_T \int_0^T \|G^\epsilon(s)\|^2_{\mathcal{L}^0_2} ds.$$

Therefore, by the assumption on f and (7.98), we have the estimate (7.102). This completes the proof.   □

By the above result and the definition of z^ϵ, we have the following corollary.

Corollary 7.21. *Assume that the conditions of Lemma 7.20 hold. Then, for $t \in [0, T]$,*

$$\mathbb{E}\left[\|u^\epsilon(t)\|^2_{L^2(D_\epsilon)} + \epsilon^2 \|\gamma_\epsilon u^\epsilon(t)\|^2_{L^2(\partial S_\epsilon)} \right]$$

$$+ \int_0^t \mathbb{E}\left[\|u^\epsilon(s)\|^2_{H^1_\epsilon(D_\epsilon)} + \epsilon^2 \|\gamma_\epsilon u^\epsilon(s)\|^2_{H^{1/2}(\partial S_\epsilon)} \right] ds \leq (1 + \mathbb{E}\|z_0\|^2_{X^0_\epsilon})C_T$$

$$(7.104)$$

and

$$\mathbb{E}\left[\sup_{t\in[0,T]} \|u^\epsilon(t)\|^2_{L^2(D_\epsilon)} + \epsilon^2 \|\gamma_\epsilon u^\epsilon(t)\|^2_{L^2(\partial S_\epsilon)} \right] \leq (1 + \mathbb{E}\|z_0\|^2_{X^0_\epsilon})C_T \qquad (7.105)$$

for some constant $C_T > 0$.

We aim at deriving an effective equation in the sense of probability. A solution u^ϵ is regarded as a random variable taking values in $L^2(0, T; L^2(D_\epsilon))$. For a solution u^ϵ of (7.88)–(7.91) defined on $[0, T]$, we focus on the behavior of the distribution of u^ϵ in $L^2(0, T; L^2(D_\epsilon))$ as $\epsilon \to 0$. For this purpose, we examine the tightness of these distributions. Here a difficult issue is that the function space changes with ϵ. Instead of using the extension operator P_ϵ, we treat $\{\mathcal{L}(u^\epsilon)\}_{\epsilon>0}$ as a family of distributions in $L^2(0, T; L^2(D))$ by extending u^ϵ to the entire domain D.

In fact, we have the following tightness result.

Lemma 7.22 (Tightness of distributions).   *Assume that $z_0 = (u_0, v_0)^T$ is a $\left(\mathcal{F}_0, \mathcal{B}(X_\epsilon^0)\right)$-measurable random variable that is independent of $W(t)$ with $\mathbb{E}\|z_0\|_{X_\epsilon^0}^2 < \infty$. Then, for any $T > 0, \{\mathcal{L}(\tilde{u}^\epsilon)\}_\epsilon$, the distributions of $\{\tilde{u}^\epsilon\}_\epsilon$, are tight in $L^2(0, T; L^2(D)) \cap C(0, T; H^{-1}(D))$.*

Proof.   Denote the projection $(u, v) \to u$ by P. By the result of Corollary 7.21,

$$\mathbb{E}|u^\epsilon|_{L^2(0,T;H_\epsilon^1(D_\epsilon))}^2 \leq C_T. \tag{7.106}$$

Write $z^\epsilon(t)$ as

$$z^\epsilon(t) = z^\epsilon(0) - \int_0^t \mathcal{A}^\epsilon z^\epsilon(s)ds + \int_0^t F^\epsilon(x, s)ds + \int_0^t G^\epsilon(x, s)dW(s).$$

Then, by (7.95) and (7.100), when $(h, 0) \in X_\epsilon^1$, we have the following estimate, for some constant $C > 0$ independent of ϵ:

$$\left|\left\langle -P \int_0^t \mathcal{A}^\epsilon z^\epsilon(s)ds + P \int_0^t F^\epsilon(x, s)ds, h\right\rangle_{L^2(D_\epsilon)}\right|$$

$$\leq \left|\int_0^t a(Pz^\epsilon(s), h)ds\right| + \left|\int_0^t \langle f(x, s), h\rangle_{L^2(D_\epsilon)}ds\right|$$

$$\leq C\left(\int_0^t \|u^\epsilon(s)\|_{H_\epsilon^1(D_\epsilon)}ds + \int_0^t \|f(s)\|_{L^2(D)}ds\right)\|h\|_{H_0^1(D_\epsilon)}.$$

Thus,

$$\mathbb{E}\left| -P \int_0^t \mathcal{A}^\epsilon z^\epsilon(s)ds + P \int_0^t F^\epsilon(x, s)ds\right|_{W^{1,2}(0,T;H^{-1}(D_\epsilon))}^2 \leq C_T. \tag{7.107}$$

Let $M_\epsilon(s, t) \triangleq \int_s^t G^\epsilon(\tau, x)dW(\tau)$. By Lemma 3.24 and the Hölder inequality, we obtain

$$\mathbb{E}\|PM_\epsilon(s, t)\|_{L^2(D_\epsilon)}^4 \leq \mathbb{E}\|PM_\epsilon(s, t)\|_{L^2(D)}^4 \leq c\mathbb{E}\left(\int_s^t \|g_1(\tau)\|_{\mathcal{L}_2^{\varrho_1}}^2 d\tau\right)^2$$

$$\leq K(t - s)\int_s^t \mathbb{E}\|g_1(\tau)\|_{\mathcal{L}_2^{\varrho_1}}^4 d\tau$$

$$\leq K'(t - s)^2,$$

for $t \in [s, T]$ and for constants $K, K' > 0$ independent of ϵ, s, and t. Therefore,

$$\mathbb{E}\int_0^T \|PM_\epsilon(0, t)\|_{L^2(D_\epsilon)}^4 dt \leq C_T, \tag{7.108}$$

and for $\alpha \in (\frac{1}{4}, \frac{1}{2})$,

$$\mathbb{E} \int_0^T \int_0^T \frac{\|PM_\epsilon(0, t) - PM_\epsilon(0, s)\|_{L^2(D_\epsilon)}^4}{|t - s|^{1+4\alpha}} ds\, dt \leq C_T. \tag{7.109}$$

Combining the estimates (7.106)–(7.109) with the Chebyshev inequality [113, p. 261], it is clear that, for any $\delta > 0$, there is a bounded set

$$K_\delta \subset \mathcal{X},$$

where $\mathcal{X} \triangleq L^2(0, T; H_\epsilon^1(D)) \cap \left(W^{1,2}(0, T; H^{-1}(D)) \oplus W^{\alpha,4}(0, T; L^2(D))\right)$ (see (7.78) and (7.79) for the definition of $W^{1,2}$ and $W^{\alpha,p}$), such that

$$\mathbb{P}\{\tilde{u}^\epsilon \in K_\delta\} > 1 - \delta.$$

Moreover, by the compact embedding

$$L^2(0, T; H^1(D)) \cap W^{1,2}(0, T; H^{-1}(D)) \subset L^2(0, T; L^2(D)) \cap C(0, T; H^{-1}(D))$$

and

$$L^2(0, T; H^1(D)) \cap W^{\alpha,4}(0, T; L^2(D)) \subset L^2(0, T; L^2(D)) \cap C(0, T; H^{-1}(D)),$$

we conclude that K_δ is compact in $L^2(0, T; L^2(D)) \cap C(0, T; H^{-1}(D))$. Thus, $\{\mathcal{L}(\tilde{u}^\epsilon)\}_\epsilon$ is tight in $L^2(0, T; L^2(D)) \cap C(0, T; H^{-1}(D))$. The proof is complete. □

Remark 7.23. When $f = f(x, t, u^\epsilon)$ is nonlinear (i.e., it depends on u^ϵ) but is globally Lipschitz in u^ϵ, the results in Lemma 7.20 and Corollary 7.21 still hold [79]. Moreover, by the Lipschitz property, we have $\|f(x, t, u^\epsilon)\|_{L^2(D)} \leq C_T$. Hence, a similar analysis as in the proof of Lemma 7.22 yields the tightness of the distribution for u^ϵ in this globally Lipschitz nonlinear case. This fact is used in the beginning of §7.3 to get the homogenized effective model when the nonlinearity $f = f(x, t, u^\epsilon)$ is globally Lipschitz. Additionally, in §7.3, we derive homogenized effective models for three types of nonlinearities $f = f(x, t, u^\epsilon)$ that are *not* globally Lipschitz in u^ϵ.

Since we consider dynamical boundary conditions, the technique of transforming surface integrals into volume integrals is useful in our approach [86,290].

For $h \in H^{-1/2}(\partial S)$ and Y-periodic, define

$$\Lambda_h \triangleq \frac{1}{|Y^*|} \int_{\partial S} h(x) dx$$

and

$$\lambda_h \triangleq \frac{1}{|Y|} \langle h, 1 \rangle_{H^{-1/2}, H^{1/2}} = \vartheta \Lambda_h.$$

Thus, in particular, $\Lambda_1 = \frac{|\partial S|}{|Y^*|}$ and

$$\lambda \triangleq \lambda_1 = \frac{|\partial S|}{|Y|}, \tag{7.110}$$

where $|\cdot|$ denotes Lebesgue measure.

For $h \in L^2(\partial S)$ and Y-periodic, define $\lambda_h^\epsilon \in H^{-1}(D)$ as

$$\langle \lambda_h^\epsilon, \varphi \rangle \triangleq \epsilon \int_{\partial S_\epsilon} h\left(\frac{x}{\epsilon}\right) \varphi(x)dx, \quad \text{for} \quad \varphi \in H_0^1(D).$$

Then we have the following result about the convergence of the integral on the boundary [86, Proposition 4.1].

Lemma 7.24. *Let φ^ϵ be a sequence in $H_0^1(D)$ such that $\varphi^\epsilon \rightharpoonup \varphi$ weakly in $H_0^1(D)$ as $\epsilon \to 0$. Then,*

$$\langle \lambda_h^\epsilon, \varphi^\epsilon|_{D_\epsilon} \rangle \to \lambda_h \int_D \varphi \, dx, \quad as \ \epsilon \to 0.$$

Homogenized Macroscopic Dynamics

We apply the two-scale convergence method to derive an effective macroscopic model for the original model (7.88)–(7.91).

By the proof of Lemma 7.22, for any $\delta > 0$ there is a bounded closed set $K_\delta \subset \mathcal{X}$, which is compact in $L^2(0, T; L^2(D))$ such that

$$\mathbb{P}\{\tilde{u}^\epsilon \in K_\delta\} > 1 - \delta.$$

Then the Prohorov theorem and the Skorohod embedding theorem [94, p. 32] assure that for any sequence $\{\epsilon_j\}_j$ with $\epsilon_j \to 0$ as $j \to \infty$, there exist a subsequence $\{\epsilon_{j(k)}\}$, random variables $\{\tilde{u}^{\epsilon_{j(k)}*}\} \subset L^2(0, T; H_{\epsilon_{j(k)}})$ and $u^* \in L^2(0, T; H)$ defined on a new probability space $(\Omega^*, \mathcal{F}^*, \mathbb{P}^*)$, such that

$$\mathcal{L}(\tilde{u}^{\epsilon_{j(k)}*}) = \mathcal{L}(\tilde{u}^{\epsilon_{j(k)}})$$

and

$$\tilde{u}^{\epsilon_{j(k)}*} \to u^* \quad \text{in} \quad L^2(0, T; H) \quad \text{as} \quad k \to \infty,$$

for almost all $\omega \in \Omega^*$. Moreover, $\tilde{u}^{\epsilon_{j(k)}*}$ solves the system (7.88)–(7.91), with W replaced by Wiener process W_k^*, defined on probability space $(\Omega^*, \mathcal{F}^*, \mathbb{P}^*)$, with the same distribution as W. In the following, we determine the limiting equation (homogenized effective equation) that u^* satisfies and show that the limit equation is actually independent of ϵ. After these are all done, we see that $\mathcal{L}(\tilde{u}^\epsilon)$ weakly converges to $\mathcal{L}(u^*)$ as $\epsilon \downarrow 0$.

For u^ϵ in set K_δ, by Lemma 7.10 there exist $u(x, t) \in H_0^1(D_T)$ and $u_1(x, y, t) \in L^2(D_T; H_{per}^1(Y))$ such that

$$\tilde{u}^{\epsilon_j}(x, t) \xrightarrow{2-s} \chi(y)u(x, t)$$

and

$$\widetilde{\nabla_x u^{\epsilon_j}} \xrightarrow{2-s} \chi(y)[\nabla_x u(x, t) + \nabla_y u_1(x, y, t)].$$

Then, by Remark 7.7,

$$\tilde{u}^{\epsilon_j}(x,t) \rightharpoonup \frac{1}{|Y|} \int_Y \chi(y) u(x,t) dy = \vartheta u(x,t), \quad \text{weakly in } L^2(D_T).$$

In fact, by the compactness of K_δ, the above convergence is strong in $L^2(D_T)$. We first determine the limit equation, which is a two-scale system that u and u_1 satisfy. Then the limit equation (homogenized effective equation) that u satisfies can be easily obtained by the relation between the weak limit and the two-scale limit.

For any $\delta > 0$, we consider $\tilde{u}^\epsilon$ in the following new probability space $(\Omega_\delta, \mathcal{F}_\delta, \mathbb{P}_\delta)$ with

$$\Omega_\delta \triangleq \{\omega \in \Omega : \tilde{u}^\epsilon(\omega) \in K_\delta\},$$

$$\mathcal{F}_\delta \triangleq \{F \cap \Omega_\delta : F \in \mathcal{F}\},$$

and

$$\mathbb{P}_\delta(F) \triangleq \frac{\mathbb{P}(F \cap \Omega_\delta)}{\mathbb{P}(\Omega_\delta)}, \quad \text{for } F \in \mathcal{F}_\delta.$$

Now we restrict the system to the probability space $(\Omega_\delta, \mathcal{F}_\delta, \mathbb{P}_\delta)$.

Replace the test function φ in (7.99) by $\varphi^\epsilon(x,t) \triangleq \varphi(x,t) + \epsilon\Phi(x, x/\epsilon, t)$ with $\varphi(x,t) \in C_0^\infty(D_T)$ and $\Phi(x,y,t) \in C_0^\infty(D_T; C_{per}^\infty(Y))$. We consider the terms in (7.99) respectively.

By the choice of φ^ϵ, and noticing that $\chi_{D_\epsilon} \rightharpoonup \vartheta$, weakly* in $L^\infty(D)$ by (7.2), we have

$$\int_0^T \int_{D_\epsilon} f(x,t)\varphi^\epsilon(x,t)dx\,dt = \int_0^T \int_D \chi_{D_\epsilon} f(x,t)\varphi^\epsilon(x,t)dx\,dt$$

$$\rightarrow \vartheta \int_0^T \int_D f(x,t)\varphi(x,t)dx\,dt, \quad \epsilon \rightarrow 0. \tag{7.111}$$

By condition (7.92), we also conclude that

$$\int_0^T \int_{D_\epsilon} g_1(x,t)\varphi^\epsilon(x,t)dx\,dW_1(t) = \int_0^T \int_D \chi_{D_\epsilon} g_1(x,t)\varphi^\epsilon(x,t)dx\,dW_1(t)$$

$$\rightarrow \vartheta \int_0^T \int_D g_1(x,t)\varphi(x,t)dx\,dW_1(t), \quad \epsilon \rightarrow 0 \text{ in } L^2(\Omega). \tag{7.112}$$

Integrating by parts, noticing that $\tilde{u}^\epsilon$ converges strongly to $\vartheta u(x,t)$ in $L^2(D_T)$, and repeating the discussion (7.29) and (7.30), we obtain

$$\int_0^T \int_{D_\epsilon} \dot{u}^\epsilon(x,t)\varphi^\epsilon(x,t)dx\,dt = -\int_0^T \int_{D_\epsilon} u^\epsilon(x,t)\dot{\varphi}^\epsilon(x,t)dx\,dt$$

$$= -\int_0^T \int_D \tilde{u}^\epsilon(x,t)\dot{\varphi}(x,t)dx\,dt - \epsilon \int_0^T \int_D \tilde{u}^\epsilon(x,t)\dot{\Phi}\left(x, \frac{x}{\epsilon}, t\right)dx\,dt$$

$$\rightarrow -\int_0^T \int_D \vartheta u(x,t)\dot{\varphi}(x,t)dx\,dt = \int_0^T \int_D \vartheta\dot{u}(x,t)\varphi(x,t)dx\,dt. \tag{7.113}$$

By the choice of φ^ϵ and Lemma 7.8, we have

$$\int_0^T \int_{D_\epsilon} \nabla u^\epsilon(x,t) \cdot \nabla \varphi^\epsilon(x,t) dx \, dt$$

$$= \int_0^T \int_{D_\epsilon} \nabla u^\epsilon(x,t) \cdot \left(\nabla_x \varphi(x,t) + \nabla_y \Phi\left(x, \frac{x}{\epsilon}, t\right) \right) dx \, dt$$

$$= \int_0^T \int_D \widetilde{\nabla u^\epsilon}(x,t) \cdot \left(\nabla_x \varphi + \nabla_y \Phi\left(x, \frac{x}{\epsilon}, t\right) \right) dx \, dt$$

$$\rightarrow \frac{1}{|Y|} \int_0^T \int_D \int_Y \chi(y) [\nabla_x u(x) + \nabla_y u_1(x,y)]$$

$$\cdot [\nabla_x \varphi(x,t) + \nabla_y \Phi(x,y,t)] dy \, dx \, dt$$

$$= \frac{1}{|Y|} \int_0^T \int_D \int_{Y^*} [\nabla_x u(x) + \nabla_y u_1(x,y)]$$

$$\cdot [\nabla_x \varphi(x,t) + \nabla_y \Phi(x,y,t)] dy \, dx \, dt. \tag{7.114}$$

Now we consider the integrals on the boundary. For a fixed $T > 0$, first,

$$\epsilon^2 \int_0^T \int_{\partial S_\epsilon} \dot{u}^\epsilon(x,t) \varphi^\epsilon(x,t) dx dt$$

$$= -\epsilon^2 \int_{\partial S_\epsilon} \int_0^T u^\epsilon(x,t) \dot{\varphi}^\epsilon(x,t) dt \, dx$$

$$= -\epsilon \left\langle \lambda_1^\epsilon, \int_0^T \tilde{u}^\epsilon(x,t) \dot{\varphi}^\epsilon(x,t) dt \Big|_{D_\epsilon} \right\rangle \rightarrow 0, \quad \epsilon \rightarrow 0, \tag{7.115}$$

and second,

$$\epsilon b \int_0^T \int_{\partial S_\epsilon} u^\epsilon(x,t) \varphi^\epsilon(x,t) dx \, dt$$

$$= \left\langle \lambda_1^\epsilon, b \int_0^T \tilde{u}^\epsilon(x,t) \varphi^\epsilon(x,t) dt \Big|_{D_\epsilon} \right\rangle$$

$$\rightarrow b \vartheta \lambda \int_0^T \int_D u(x,t) \varphi(x,t) dx \, dt, \quad \epsilon \rightarrow 0. \tag{7.116}$$

Similarly, we have the limit of the stochastic integral on the boundary

$$\epsilon \int_0^T \int_{\partial S_\epsilon} g_2(x,t) \varphi^\epsilon(x,t) dx \, dW_2(t)$$

$$\rightarrow \lambda \int_0^T \int_D g_2(x,t) \varphi(x,t) dx \, dW_2(t), \quad \epsilon \rightarrow 0, \quad \text{in } L^2(\Omega). \tag{7.117}$$

Combining the above analyses of (7.111)–(7.117) and by a density argument, we have

$$
\vartheta \int_0^T \int_D \dot{u}(x,t)\varphi(x,t)dx\,dt
$$

$$
= -\frac{1}{|Y|} \int_0^T \int_D \int_{Y^*} \big[\nabla_x u(x) + \nabla_y u_1(x,y)\big]\big[\nabla_x \varphi(x,t) + \nabla_y \Phi(x,y,t)\big]dx\,dt
$$

$$
- b\vartheta\lambda \int_0^T \int_D u(x,t)\varphi(x,t)dx\,dt + \vartheta \int_0^T \int_D f(x,t)\varphi(x,t)dx\,dt
$$

$$
+ \vartheta \int_0^T \int_D g_1(x,t)\varphi(x,t)dx\,dW_1(t) + \lambda \int_0^T \int_D g_2(x,t)\varphi(x,t)dx\,dW_2(t)
$$

$$
(7.118)
$$

for any $\varphi \in H_0^1(D_T)$ and $\Phi \in L^2(D_T; H_{\mathrm{per}}^1(Y)/\mathbb{R})$. Integrating by parts, we see that (7.118) is the variational form of the following two-scale homogenized system:

$$
\vartheta\,du = \big[\mathrm{div}_x A(\nabla_x u) - b\vartheta\lambda_1 u + \vartheta f\big]dt
$$
$$
+ \vartheta g_1\,dW_1(t) + \lambda g_2\,dW_2(t), \quad \text{on} \quad D_T \tag{7.119}
$$

$$
[\nabla_x u + \nabla u_1] \cdot \nu = 0, \quad \text{on} \quad \partial Y^* - \partial Y, \tag{7.120}
$$

where ν is the unit exterior norm vector on $\partial Y^* - \partial Y$ and

$$
A(\nabla_x u) := \frac{1}{|Y|} \int_{Y^*} [\nabla_x u(x,t) + \nabla_y u_1(x,y,t)]dy, \tag{7.121}
$$

with u_1 satisfying the following integral equation on $D \times Y^*$:

$$
\int_{Y^*} [\nabla_x u + \nabla_y u_1]\nabla_y \Phi\,dy = 0, \quad u_1 \text{ is } Y-\text{periodic}, \tag{7.122}
$$

for any $\Phi \in H_0^1(D_T; H_{\mathrm{per}}^1(Y))$. Equation (7.122) has a unique solution for any fixed u, and so $A(\nabla_x u)$ is well defined. Furthermore, $A(\nabla_x u)$ satisfies

$$
\langle A(\xi_1) - A(\xi_2), \xi_1 - \xi_2 \rangle_{L^2(D)} \geq \alpha \|\xi_1 - \xi_2\|_{L^2(D)}^2 \tag{7.123}
$$

and

$$
|\langle A(\xi), \xi \rangle_{L^2(D)}| \leq \beta \|\xi\|_{L^2(D)}^2 \tag{7.124}
$$

for some $\alpha, \beta > 0$ and any $\xi, \xi_1, \xi_2 \in H_0^1(D)$. Then, by the classical theory of the SPDEs [94, Theorem 5.4], (7.119) and (7.120) is well posed.

In fact, $A(\nabla u)$, as we have discussed for system (7.31), can be transformed to the classical homogenized matrix by

$$
u_1(x,y,t) = \sum_{i=1}^n \frac{\partial u(x,t)}{\partial x_i}(w_{\mathbf{e}_i}(y) - \mathbf{e}_i \cdot y), \tag{7.125}
$$

where $\{\mathbf{e}_i\}_{i=1}^n$ is the canonical basis of $\mathbb{R}^n$ and $w_{\mathbf{e}_i}$ is the solution of the following cell problem (problem defined on the spatial elementary cell):

$$\Delta_y w_{\mathbf{e}_i}(y) = 0 \quad \text{on} \quad Y^* \tag{7.126}$$

$$w_{\mathbf{e}_i} - \mathbf{e}_i \cdot y \quad \text{is} \quad Y \text{ periodic} \tag{7.127}$$

$$\frac{\partial w_{\mathbf{e}_i}}{\partial \nu} = 0 \quad \text{on} \quad \partial S. \tag{7.128}$$

Note that, in this case, the diffusion matrix is identity.

A simple calculation yields

$$A(\nabla u) = \bar{A} \nabla u$$

where $\bar{A} = (\bar{a}_{ij})$ is the classical homogenized matrix defined as (see 7.27)

$$\bar{a}_{ij} \triangleq \frac{1}{|Y|} \int_{Y^*} \nabla_y w_{\mathbf{e}_i}(y) \cdot \nabla_y w_{\mathbf{e}_i}(y) dy. \tag{7.129}$$

The above two-scale homogenized system (7.119) is equivalent to the following homogenized system:

$$\vartheta \, du = \left[\text{div}_x \left(\bar{A} \nabla_x u \right) - b\vartheta \lambda u + \vartheta f \right] dt$$
$$+ \vartheta g_1 \, dW_1(t) + \lambda g_2 \, dW_2(t). \tag{7.130}$$

Let $U(x, t) := \vartheta u(x, t)$. Then, we have the limiting homogenized equation

$$dU = \left[\vartheta^{-1} \text{div}_x \left(\bar{A} \nabla_x U \right) - b\lambda U \right.$$
$$\left. + \vartheta f \right] dt + \vartheta g_1 \, dW_1(t) + \lambda g_2 \, dW_2(t). \tag{7.131}$$

The limiting process u^*, as we have mentioned at the beginning of this section, satisfies (7.131) with $W = (W_1, W_2)$ replaced by a Wiener process W^* with the same distribution as W. By the classical existence result [94, Theorem 5.4], the homogenized model (7.131) is well posed. We can now formulate the following homogenization result.

Theorem 7.25 (Homogenized macroscopic model). *Assume that (7.92) holds. Let u^ϵ be the solution of (7.88)–(7.91). Then, for any fixed $T > 0$, the distribution $\mathcal{L}(\tilde{u}^\epsilon)$ converges weakly to μ in $L^2(0, T; H)$ as $\epsilon \downarrow 0$, with μ being the distribution of U, which is the solution of the following homogenized effective equation:*

$$dU = \left[\vartheta^{-1} \text{div}_x \left(\bar{A} \nabla_x U \right) - b\lambda U + \vartheta f \right] dt$$
$$+ \vartheta g_1 \, dW_1(t) + \lambda g_2 \, dW_2(t), \tag{7.132}$$

$$U = 0 \quad \text{on} \quad \partial D, \quad U(0) = \frac{u_0}{\vartheta} \tag{7.133}$$

with the effective matrix $\bar{A} = (\bar{a}_{ij})$ being determined by (7.129). Moreover, $\vartheta = \frac{|Y^*|}{|Y|}$ and $\lambda = \frac{|\partial S|}{|Y|}$.

Proof. The proof is similar to that for Theorem 7.13. $\square$

Remark 7.26. It is interesting to note the following fact. In the case that $\lambda \neq 0$, that is, $n \geq 2$, even when the original microscopic model Equation (7.88) is a deterministic PDE (i.e., $g_1 = 0$), the homogenized macroscopic model (7.132) is *still* a stochastic PDE due to the impact of random dynamical interactions on the boundary of small-scale heterogeneities.

Remark 7.27. For the macroscopic system (7.132), we see that the fast-scale random fluctuations on the boundary are recognized or quantified in the homogenized equation through the extra term $\lambda g_2 \, dW_2(t)$. The effect of random boundary evolution is thus *felt* by the homogenized system on the whole domain. This is not true for the case of a one–dimensional domain, since $\lambda = 0$ in that case. Also see Example 7.28.

Example 7.28. Consider the following system on one-dimensional perforated interval D_ϵ with $D = (-L, L)$, $Y = [-1, 1)$, and $S = (-0.5, 0.5)$:

$$du^\epsilon(x, t) = \partial_{xx} u^\epsilon(x, t) dt + dW_1(x, t)$$
$$\text{on} \quad D_c \times (0, T),$$
$$\epsilon^2 \, du^\epsilon(x, t) = -\frac{\partial u^\epsilon(x, t)}{\partial \nu} dt + \epsilon \, dW_2(x, t)$$
$$\text{on} \quad \partial S_\epsilon \times (0, T),$$
$$u^\epsilon(x, t) = 0 \quad \text{on} \quad \partial D \times (0, T),$$
$$u^\epsilon(0, x) = u_0(x) \quad \text{on} \quad D_\epsilon.$$

Noticing that $\lambda = 0$ in this one-dimensional case, we have the following two-scale homogenized equation:

$$\frac{1}{2} du(x, t) = \text{div}_x A(\nabla_x u(x, t)) dt + dW_1(x, t), \quad \text{on} \quad (-L, L) \times (0, T)$$
$$u'(x, t) + \partial_y u_1(x, y, t) = 0, \quad \text{on} \quad \{-0.5, 0.5\},$$

where

$$A(\nabla_x u) \triangleq \frac{1}{|Y|} \int_{Y^*} [u'(x, t) + \partial_y u_1(x, y, t)] dy, \tag{7.134}$$

with u_1 satisfying the following equation on $D \times Y^*$:

$$\partial_{yy} u_1(x, y, t) = 0, \quad u_1 \text{ is } Y - \text{periodic}. \tag{7.135}$$

Choose $u_1(x, y, t) = \partial_x u(x, t)(w(y) - y)$ such that

$$w''(y) = 0 \quad \text{on} \quad Y^*,$$
$$w - y \quad \text{is} \quad Y \text{periodic},$$
$$w'(y) = 0 \quad \text{on} \quad \{-0.5, 0.5\}.$$

Then we have the homogenized matrix

$$A(\nabla_x u) = \bar{A} \partial_x u$$

with

$$\bar{A} = \frac{1}{|Y|} \int_{Y^*} w'(y)w'(y)dy = 1.$$

Thus, we have the same homogenized equation as that of Example 7.14:

$$\frac{1}{2}du(x,t) = \partial_{xx}u(x,t)dt + dW_1(x,t), \quad \text{on} \quad (-L, L) \times (0, T),$$

$$u(x,t) = 0 \quad \text{on} \quad \{-L, L\} \times (0, T).$$

7.3 Homogenized Macroscopic Dynamics for Stochastic Nonlinear Microscopic Systems

In this section, we derive a homogenized macroscopic model for the microscopic system (7.88)–(7.91) with random dynamical boundary conditions when f is nonlinear, either globally or locally Lipschitz.

As in Remark 7.23, if f is a globally Lipschitz nonlinear function of u^ϵ, all the estimates in Lemma 7.20 hold. In fact, note that f satisfies $f(x, t, 0) = 0$ and the Lipschitz condition, that is,

$$|f(x, t, u_1) - f(x, t, u_2)| \le L_f |u_1 - u_2|,$$

for any $t \in \mathbb{R}$, $x \in D$, and $u_1, u_2 \in \mathbb{R}$ with some positive constant L_f. Thus if $\tilde{u}^\epsilon \to \vartheta u$ strongly in $L^2(0, T; L^2(D))$, then $f(x, t, \tilde{u}^\epsilon(x, t)) \to f(x, t, u(x, t))$ strongly in $L^2(0, T; L^2(D))$, and (7.111) still holds. Hence, we obtain the following effective macroscopic system similar to (7.132) with nonlinearity $f = f(x, t, U)$:

$$dU = \left[\vartheta^{-1}\text{div}_x(\bar{A}\nabla_x U) - b\lambda U + \vartheta f(x, t, U)\right]dt \\ + \vartheta g_1\, dW_1(t) + \lambda g_2\, dW_2(t). \tag{7.136}$$

For the rest of this section, we consider three types of nonlinear systems, with f being a locally Lipschitz nonlinear function in u^ϵ. The difficulty is in passing the limit as $\epsilon \to 0$ in the nonlinear term. These three types of nonlinearity include polynomial nonlinearity, nonlinear term that is sublinear, and nonlinearity that contains a gradient term ∇u^ϵ. We look at these nonlinearities case by case and only highlight the difference with the analysis for linear systems in the last section.

Case 1: Polynomial nonlinearity

Consider f in the following form

$$f(x, t, u) = -a(x, t)|u|^p u \tag{7.137}$$

with $0 < a_0 \le a(x, t) \le a_1$ for $t \in [0, \infty)$, $x \in D$, and assume that p satisfies the following condition:

$$p \le \frac{2}{n-2}, \quad \text{if } n \ge 3; \quad p \in \mathbb{R}^1, \text{ if }; n = 2. \tag{7.138}$$

For this case, we need the following *weak convergence lemma* from [215, Lemma 1.1.3].

Weak convergence lemma: *Let Q be a bounded region in $\mathbb{R} \times \mathbb{R}^n$. For any given functions g^ϵ and g in $L^p(Q)(1 < p < \infty)$, if*

$$|g^\epsilon|_{L^p(Q)} \le C, \quad g^\epsilon \to g \text{ in } Q \text{ almost everywhere}$$

for some constant $C > 0$, then $g^\epsilon \rightharpoonup g$ weakly in $L^p(Q)$.

Noticing that $F^\epsilon(x, t, z^\epsilon) = (f(x, t, u^\epsilon), 0)$ and $(F^\epsilon(x, t, z^\epsilon), z^\epsilon)_{X^0_\epsilon} \le 0$, the results in Lemma 7.20 can be obtained by the same method. Moreover, by assumption (7.138), $|f(x, t, u^\epsilon)|_{L^2(D_T)} \le C_T$, which, by the analysis of Lemma 7.22, yields the tightness of the distribution of $\tilde{u}^\epsilon$.

Now, we pass the limit as $\epsilon \to 0$ to $f(x, t, \tilde{u}^\epsilon)$. In fact, noticing that $\tilde{u}^\epsilon$ converges strongly to ϑu in $L^2(0, T; L^2(D))$, by the above weak convergence lemma with $g^\epsilon = f(x, t, \tilde{u}^\epsilon)$ and $p = 2$, $f(x, t, \tilde{u}^\epsilon)$ converges weakly to $f(x, t, \vartheta u)$ in $L^2(D_T)$. Therefore, by a similar analysis for linear systems, we have the following result.

Theorem 7.29 (Homogenization for Nonlinear SPDEs I). *Assume that (7.92) holds. Let u^ϵ be the solution of (7.88)–(7.91) with nonlinear term f being (7.137). Then, for any fixed $T > 0$, the distribution $\mathcal{L}(\tilde{u}^\epsilon)$ converges weakly to μ in $L^2(0, T; H)$ as $\epsilon \downarrow 0$, with μ being the distribution of U, which is the solution of the following homogenized effective equation:*

$$dU = \left[\vartheta^{-1}\text{div}_x\left(\bar{A}\nabla_x U\right) - b\lambda U + \vartheta f(x, t, U)\right]dt$$
$$+ \vartheta g_1 \, dW_1(t) + \lambda g_2 \, dW_2(t), \tag{7.139}$$

$$U = 0 \quad \text{on} \quad \partial D, \quad U(0) = \frac{u_0}{\vartheta} \tag{7.140}$$

with the effective matrix $\bar{A} = (\bar{a}_{ij})$ being determined by (7.129). Moreover, $\vartheta = \frac{|Y^*|}{|Y|}$ and $\lambda = \frac{|\partial S|}{|Y|}$.

Case 2: Nonlinear term that is sublinear

Now we consider a measurable function $f : D \times [0, T] \times \mathbb{R}^1 \to \mathbb{R}^1$, which is continuous in $(x, \xi) \in D \times \mathbb{R}^1$ for almost all $t \in [0, T]$ and which satisfies

$$\left[f(x, t, \xi_1) - f(x, t, \xi_2)\right]\left[\xi_1 - \xi_2\right] \ge 0 \tag{7.141}$$

for $t \ge 0, x \in D$ and $\xi_1, \xi_2 \in \mathbb{R}^1$. Moreover, we assume that f is sublinear,

$$|f(x, t, \xi)| \le g(t)(1 + |\xi|), \quad \xi \in \mathbb{R}^1, \ t \ge 0, \tag{7.142}$$

where $g \in L^\infty_{\text{loc}}[0, \infty)$. Notice that under assumptions (7.141) and (7.142), f may not be a Lipschitz function. By assumption (7.142), we also have the tightness of the distributions of $\tilde{u}^\epsilon$ and conclude that $\chi_{D_\epsilon} f(x, t, \tilde{u}^\epsilon)$ two-scale converges to a function denoted by $f_0(x, y, t) \in L^2(D_T \times Y)$. In the following, we need to identity $f_0(x, y, t)$.

Let $\varphi \in C^\infty_0(D_T)$ and $\psi \in C^\infty_0(D_T; C^\infty_{\text{per}}(Y))$, and for $\kappa > 0$ let

$$\xi^\epsilon(x, t) := \varphi(x, t) + \kappa \psi\left(x, \frac{x}{\epsilon}, t\right). \tag{7.143}$$

Then, by assumption (7.141) again, we have

$$0 \leq \int_0^T \int_{D_\epsilon} \left[f(x,t,u^\epsilon) - f(x,t,\xi^\epsilon) \right] \left[u^\epsilon - \xi^\epsilon \right] dx \, dt$$

$$= \int_{D_T} \chi\left(\frac{x}{\epsilon}\right) \left[f(x,t,\tilde{u}^\epsilon) - f(x,t,\xi^\epsilon) \right] \left[\tilde{u}^\epsilon - \xi^\epsilon \right] dx \, dt$$

$$\stackrel{\triangle}{=} I_\epsilon = I_{1,\epsilon} - I_{2,\epsilon} - I_{3,\epsilon} + I_{4,\epsilon},$$

with

$$I_{1,\epsilon} = \int_{D_T} \chi\left(\frac{x}{\epsilon}\right) f(x,t,\tilde{u}^\epsilon) \tilde{u}^\epsilon \, dx \, dt$$

$$\stackrel{\epsilon \to 0}{\longrightarrow} \frac{1}{|Y|} \int_{D_T} \int_Y f_0(x,y,t) \vartheta u(x,t) dy \, dx \, dt, \qquad (7.144)$$

$$I_{2,\epsilon} = \int_{D_T} \chi\left(\frac{x}{\epsilon}\right) f(x,t,\tilde{u}^\epsilon) \xi^\epsilon \, dx \, dt$$

$$\stackrel{\epsilon \to 0}{\longrightarrow} \frac{1}{|Y|} \int_{D_T} \int_Y f_0(x,y,t) [\varphi(x,t) + \kappa \psi(x,y,t)] dy \, dx \, dt, \qquad (7.145)$$

$$I_{3,\epsilon} = \int_{D_T} \chi\left(\frac{x}{\epsilon}\right) f(x,t,\xi^\epsilon) \tilde{u}^\epsilon \, dx \, dt$$

$$\stackrel{\epsilon \to 0}{\longrightarrow} \frac{1}{|Y|} \int_{D_T} \int_Y \chi(y) f(x,t,\varphi(x,t) + \kappa \psi(x,y,t))$$

$$\times \vartheta u(x,t) dy \, dx \, dt, \qquad (7.146)$$

and

$$I_{4,\epsilon} = \int_{D_T} \chi\left(\frac{x}{\epsilon}\right) f(x,t,\xi^\epsilon) \xi^\epsilon \, dx \, dt$$

$$\stackrel{\epsilon \to 0}{\longrightarrow} \frac{1}{|Y|} \int_{D_T} \int_Y \chi(y) f(x,t,\varphi(x,t) + \kappa \psi(x,y,t))$$

$$\times [\varphi(x,t) + \kappa \psi(x,y,t)] dy \, dx \, dt. \qquad (7.147)$$

In (7.144)–(7.147), we have used the fact of strong two-scale convergence of $\chi\left(\frac{x}{\epsilon}\right)$ and $f(x,t,\xi^\epsilon)$ and strong convergence of u^ϵ to ϑu.

Now we have

$$\lim_{\epsilon \to 0} I_\epsilon = \int_{D_T} \int_Y \left[f_0(x,y,t) - \chi(y) f(x,t,\varphi + \lambda \psi) \right]$$

$$\times [\vartheta u(x,t) - \varphi(x,t) - \kappa \psi] dy \, dx \, dt \geq 0$$

for any $\varphi \in C_0^\infty(D_T)$ and $\psi \in C_0^\infty(D_T; C_{per}(Y))$. Letting $\varphi \to \vartheta u$, dividing the above equation by κ on both sides, and letting $\kappa \to 0$, we get

$$\int_{D_T} \int_Y \left[f_0(x, y, t) - \chi(y) f(x, t, \vartheta u) \right] \psi \, dy \, dx \, dt \leq 0$$

for any $\psi \in C_0^\infty(D_T; C_{per}(Y))$, which means

$$f_0(x, y, t) = \chi(y) f(x, t, \vartheta u).$$

Then, by a similar analysis for linear systems, we have the following homogenized model.

Theorem 7.30 (Homogenization for Nonlinear SPDEs II). *Assume that (7.92) holds. Let u^ϵ be the solution of (7.88)–(7.91) with nonlinear term f satisfying (7.141) and (7.142). Then, for any fixed $T > 0$, the distribution $\mathcal{L}(\tilde{u}^\epsilon)$ converges weakly to μ in $L^2(0, T; H)$ as $\epsilon \downarrow 0$, with μ being the distribution of U, which is the solution of the following homogenized effective equation:*

$$dU = \left[\vartheta^{-1} \mathrm{div}_x \left(\bar{A} \nabla_x U \right) - b\lambda U + \vartheta f(x, t, U) \right] dt$$
$$+ \vartheta g_1 \, dW_1(t) + \lambda g_2 \, dW_2(t), \tag{7.148}$$
$$U = 0 \quad on \quad \partial D, \quad U(0) = \frac{u_0}{\vartheta} \tag{7.149}$$

with the effective matrix $\bar{A} = (\bar{a}_{ij})$ being determined by (7.129). Moreover, $\vartheta = \frac{|Y^*|}{|Y|}$ and $\lambda = \frac{|\partial S|}{|Y|}$.

Case 3: Nonlinearity that contains a gradient term

We finally consider f in the following form containing a gradient term:

$$f(x, t, u, \nabla u) = h(x, t, u) \cdot \nabla u, \tag{7.150}$$

here $h(x, t, u) = (h_1(x, t, u), \cdots, h_n(x, t, u))$ and each $h_i : D \times [0, T] \times \mathbb{R}^1 \to \mathbb{R}^1, i = 1, \cdots, n$, is continuous with respect to u and $h(\cdot, \cdot, u(\cdot, \cdot)) \in L^2(0, T; L^2(D))$ for $u \in L^2(0, T; L^2(D))$. Moreover, we make the following two assumptions:

1. $|\langle h(x, t, u) \cdot \nabla u, v \rangle_{L^2}| \leq C_0 \|\nabla u\|_{L^2} \|v\|_{L^2}$ with some constant $C_0 > 0$.
2. $|h_i(x, t, \xi_1) - h_i(x, t, \xi_2)| \leq k|\xi_1 - \xi_2|$ for $\xi_1, \xi_2 \in \mathbb{R}^1, i = 1, \ldots, n$ and $k > 0$ being a constant.

Now we have

$$\left| \langle F^\epsilon(x, t, z^\epsilon), z^\epsilon \rangle_{X_\epsilon^0} \right| = \left| \langle h(x, t, u^\epsilon) \cdot \nabla u^\epsilon, u^\epsilon \rangle_{L^2} \right| \leq C_0 \|z^\epsilon\|_{X_\epsilon^0} \|z^\epsilon\|_{X_\epsilon^1}. \tag{7.151}$$

By applying Itô's formula to $\|z^\epsilon\|_{X_\epsilon^0}^2$, we obtain

$$d\|z^\epsilon(t)\|_{X_\epsilon^0}^2 + 2\langle \mathcal{A}^\epsilon z^\epsilon, z^\epsilon \rangle_{X_\epsilon^0} dt$$
$$= 2\langle F^\epsilon(x, t, z^\epsilon), z^\epsilon \rangle_{X_\epsilon^0} dt$$
$$+ 2\langle G^\epsilon(x, t) dW(t), z^\epsilon \rangle_{X_\epsilon^0} + \|G^\epsilon(x, t)\|_{\mathcal{L}_2^0}^2 \, dt. \tag{7.152}$$

By (7.151), coercivity (7.96) of $a^\epsilon(\cdot, \cdot)$ and the Cauchy–Schwartz inequality, integrating (7.152) with respect to t yields

$$\|z^\epsilon(t)\|^2_{X^0_\epsilon} + \bar{\alpha} \int_0^t \|z^\epsilon(s)\|^2_{X^1_\epsilon} \, ds$$

$$\leq \|z_0\|^2_{X^0_\epsilon} + (2\bar{\beta} + \Lambda_1(\bar{\alpha})) \int_0^t \|z^\epsilon(s)\|^2_{X^0_\epsilon} \, ds$$

$$+ 2 \int_0^t \langle G^\epsilon(s) dW(s), z^\epsilon(s) \rangle_{X^0_\epsilon} + \int_0^t \|G^\epsilon(s)\|^2_{\mathcal{L}^0_2} \, ds$$

where $\Lambda_1 > 0$ is a constant depending on $\bar{\alpha}$. Then, by the Gronwall lemma, we also have that (7.101) and (7.102) hold. Moreover, the fact that

$$\|h(x, t, u) \cdot \nabla u\|_{L^2} \leq C_0 \|z^\epsilon\|_{X^1_\epsilon},$$

together with the Hölder inequality, yield

$$\mathbb{E} \left| -P \int_0^t \mathcal{A}^\epsilon z^\epsilon(s) ds + P \int_0^t F^\epsilon(x, s, z^\epsilon) ds \right|^2_{W^{1,2}(0,T;H^{-1}(D_\epsilon))} \leq C_T, \quad (7.153)$$

where P is defined in Lemma 7.22. Then, by the same discussion of Lemma 7.22, we have tightness of the distributions of $\tilde{u}^\epsilon$.

Now we pass the limit as $\epsilon \to 0$ to the nonlinear term $f(x, t, u^\epsilon, \nabla u^\epsilon)$. In fact, we restrict the system to $(\Omega_\delta, \mathcal{F}_\delta, \mathbb{P}_\delta)$. By the above assumption 2 on h and the fact that $\tilde{u}^\epsilon$ strong converges to ϑu in $L^2(D_T)$, we have

$$\lim_{\epsilon \to 0} \int_{D_T} \left[h(x, t, \tilde{u}^\epsilon(x, t)) - h(x, t, \vartheta u(x, t)) \right]^2 dx \, dt = 0.$$

For any $\psi \in C_0^\infty(D_T)$,

$$\int_{D_T} h(x, t, \tilde{u}^\epsilon) \cdot \widetilde{\nabla u^\epsilon} \psi \, dx \, dt$$

$$= \int_{D_T} \left[h(x, t, \tilde{u}^\epsilon) - h(x, t, \vartheta u) \right] \cdot \widetilde{\nabla u^\epsilon} \psi \, dx \, dt$$

$$+ \int_{D_T} h(x, t, \vartheta u) \cdot \widetilde{\nabla u^\epsilon} \psi \, dx \, dt$$

$$\xrightarrow{\epsilon \to 0} \frac{1}{|Y|} \int_{D_T} \int_Y h(x, t, \vartheta u) \cdot \chi(y) [\nabla_x u + \nabla_y u_1] \psi \, dy \, dx \, dt. \quad (7.154)$$

Combining these with a similar analysis for linear systems, we have the following result.

Theorem 7.31 (Homogenization for Nonlinear SPDEs III). *Assume that (7.92) holds. Let u^ϵ be the solution of (7.88)–(7.91) with nonlinear term (7.150). Then, for any fixed $T > 0$, the distribution $\mathcal{L}(\tilde{u}^\epsilon)$ converges weakly to μ in $L^2(0, T; H)$ as $\epsilon \downarrow 0$, with*

μ being the distribution of $U = \vartheta u$ that satisfies the following homogenized effective equation:

$$dU = \left[\vartheta^{-1}\mathrm{div}_x\left(\bar{A}\nabla_x U\right) - b\lambda U + f^*(x, t, U, \nabla_x U)\right]dt$$
$$+ \vartheta \, g_1 dW_1(t) + \lambda g_2 \, dW_2(t), \tag{7.155}$$
$$U = 0 \quad \text{on} \quad \partial D, \quad U(0) = \frac{u_0}{\vartheta} \tag{7.156}$$

where the effective matrix $\bar{A} = (\bar{a}_{ij})$ is determined by (7.129) and f^* is the following spatial average:

$$f^*(x, t, U, \nabla_x U) \triangleq \frac{1}{|Y|} \int_Y h(x, t, U) \cdot \chi(y)\left[\vartheta^{-1}\nabla_x U + \nabla_y u_1\right]dy$$

with u_1 being given by (7.125) and $\chi(y)$ the indicator function of Y^*. Moreover, $\vartheta = \frac{|Y^*|}{|Y|}$ and $\lambda = \frac{|\partial S|}{|Y|}$.

Remark 7.32. The results in this chapter hold when Δ is replaced by a more general strong elliptic operator $\mathrm{div}(A^\epsilon \nabla u)$, where A^ϵ is $Y-$ periodic and satisfies the strong ellipticity condition.

7.4 Looking Forward

In this chapter, we have discussed homogenization in space in order to extract effective dynamics for SPDEs with multiple spatial scales. We have considered both linear and several classes of nonlinear SPDEs with usual or random boundary conditions.

An open problem is the homogenization of general, nonperiodic heterogeneity in space, including, for example, media with randomly distributed holes. Another interesting problem is to investigate a system with highly oscillating coefficients in both time and space with a certain mixing property. This is in fact related to both averaging (time) and homogenization (space). Furthermore, homogenization of random attractors is worthy of further research [126].

7.5 Problems

Let $D = (-10, 10)$ and D_ϵ be the periodically perforated domain with elementary cell $Y = [-1, 1)$ and hole $S = (-0.5, 0.5)$. Denote also by S_ϵ all the holes, that is, $D_\epsilon = D \setminus S_\epsilon$. Take ϵ as a small positive parameter.

7.1. Homogenization under multiplicative noise

Consider the following system of Itô stochastic partial differential equations defined on the one-dimensional perforated open interval D_ϵ:

$$du^\epsilon(x, t) = \partial_{xx}u^\epsilon(x, t)\,dt + u^\epsilon(x, t)\,dW(t) \quad \text{on} \quad D_\epsilon \times (0, T),$$
$$\frac{\partial u^\epsilon(x, t)}{\partial \nu_\epsilon} = 0 \quad \text{on} \quad \partial S_\epsilon \times (0, T),$$

$$u^\epsilon(x, t) = 0 \quad \text{on} \quad \partial D \times (0, T),$$
$$u^\epsilon(0, x) = u_0(x) \quad \text{on} \quad D_\epsilon,$$

where $\epsilon \in (0, 1)$, and $W(t)$ is a standard scalar Wiener process. Determine the homogenized equation for u^ϵ as $\epsilon \to 0$.

7.2. Homogenization under additive noise

Consider the following system of Itô stochastic partial differential equations defined on the one-dimensional perforated open interval D_ϵ:

$$du^\epsilon(x, t) = \partial_{xx} u^\epsilon(x, t)dt + a(x/\epsilon)dW(t) \quad \text{on} \quad D_\epsilon \times (0, T),$$
$$\frac{\partial u^\epsilon(x, t)}{\partial \nu_\epsilon} = 0 \quad \text{on} \quad \partial S_\epsilon \times (0, T),$$
$$u^\epsilon(x, t) = 0 \quad \text{on} \quad \partial D \times (0, T),$$
$$u^\epsilon(0, x) = u_0(x) \quad \text{on} \quad D_\epsilon,$$

where $\epsilon \in (0, 1)$, $W(t)$ is a standard scalar Wiener process, and $a : D \to \mathbb{R}$ is bounded continuous and Y-periodic. Determine the homogenized equation for u^ϵ as $\epsilon \to 0$.

7.3. Nonlinear homogenization

Consider the following system of Itô stochastic partial differential equations defined on the one-dimensional perforated open interval D_ϵ:

$$du^\epsilon(x, t) = [\partial_{xx} u^\epsilon(x, t) + u^\epsilon(x, t) - (u^\epsilon(x, t))^3]dt + a(x/\epsilon)dW(t)$$
$$\text{on} \quad D_\epsilon \times (0, T),$$
$$\frac{\partial u^\epsilon(x, t)}{\partial \nu_\epsilon} = 0 \quad \text{on} \quad \partial S_\epsilon \times (0, T),$$
$$u^\epsilon(x, t) = 0 \quad \text{on} \quad \partial D \times (0, T),$$
$$u^\epsilon(0, x) = u_0(x) \quad \text{on} \quad D_\epsilon,$$

where $\epsilon \in (0, 1)$, $W(t)$ is a standard scalar Wiener process and $a : D \to \mathbb{R}$ is bounded, continuously differentiable, and Y-periodic. Determine the homogenized equation for u^ϵ as $\epsilon \to 0$.

7.4. Homogenization under random dynamical boundary conditions

Consider a system of stochastic partial differential equations defined on the one-dimensional perforated open interval D_ϵ,

$$du^\epsilon(x, t) = \left[\partial_{xx} u^\epsilon(x, t) + u^\epsilon(x, t) \right]dt + dW_1(x, t)$$
$$\text{on} \quad D_\epsilon \times (0, T),$$
$$\epsilon^2 du^\epsilon(x, t) = \left[-\frac{\partial u^\epsilon(x, t)}{\partial \nu_\epsilon} - \epsilon u^\epsilon(x, t) \right]dt + \epsilon \, dW_2(x, t)$$
$$\text{on} \quad \partial S_\epsilon \times (0, T),$$
$$u^\epsilon(x, t) = 0 \quad \text{on} \quad \partial D \times (0, T),$$
$$u^\epsilon(0, x) = u_0(x) \quad \text{on} \quad D_\epsilon,$$

where $\epsilon \in (0, 1)$, and W_1 and W_2 are independent $L^2(D)$-valued Wiener processes with covariance operators Q_1, Q_2, respectively. Is there a homogenized system, and if so, what is it? Under what conditions?

7.5. Nonlinear homogenization under random dynamical boundary conditions

Consider a system of stochastic partial differential equations defined on the one-dimensional perforated open interval D_ϵ:

$$du^\epsilon(x, t) = \left[\partial_{xx} u^\epsilon(x, t) + u^\epsilon - (u^\epsilon)^3 \right] dt + dW_1(x, t)$$
$$\text{on} \quad D_\epsilon \times (0, T),$$
$$\epsilon^2 du^\epsilon(x, t) = \left[-\frac{\partial u^\epsilon(x, t)}{\partial \nu_\epsilon} - \epsilon u^\epsilon(x, t) \right] dt + \epsilon \, dW_2(x, t)$$
$$\text{on} \quad \partial S_\epsilon \times (0, T),$$
$$u^\epsilon(x, t) = 0 \quad \text{on} \quad \partial D \times (0, T),$$
$$u^\epsilon(0, x) = u_0(x) \quad \text{on} \quad D_\epsilon,$$

where $\epsilon \in (0, 1)$, and W_1 and W_2 are independent $L^2(D)$-valued Wiener processes with covariance operators Q_1, Q_2, respectively. Is there a homogenized system, and if so, what is it? Under what conditions?

Hints and Solutions

Problems of Chapter 2

2.1

$$u(x,t) = \sum_{n=1}^{\infty} \frac{4}{(n\pi)^3}[1 - (-1)^n]\cos n\pi t \,\sin n\pi x.$$

2.2 (a)

$$u(x,t) = \sum_{n=1}^{\infty} u_n(t) \,\sin \frac{(2n+1)\pi x}{2l}.$$

Find u_n's.

(b)

$$u(x,t) = \sum_{n=1}^{\infty} u_n(t) \,\cos \frac{(2n+1)\pi x}{2l}.$$

Find u_n's.

2.3

$$u(x,t) = \sum_{n=1}^{\infty} \frac{4}{(n\pi)^3}[2(-1)^{n+1} - 1]e^{-4n^2\pi^2 t} \,\sin n\pi x.$$

2.4 (a)

$$u(x,t) = \sum_{n=0}^{\infty} u_n(t) \,\cos n\pi x.$$

Find u_n's.

(b)

$$u(x,t) = \sum_{n=1}^{\infty} u_n(t) \,\sin \frac{(2n+1)\pi x}{2}.$$

Find u_n's.

Effective Dynamics of Stochastic Partial Differential Equations. http://dx.doi.org/10.1016/B978-0-12-800882-9.00014-7

2.5 Taking Fourier transform with respect to x on both sides, we obtain

$$\frac{d^2}{dt^2}U(k,t) = c^2(ik)^2 U, \quad U(k,0) = F(k), \quad \frac{d}{dt}U(k,0) = G(k).$$

Thus,

$$U(k,t) = b_1 e^{-ickt} + b_2 e^{ickt}$$

for two integration constants k_1, k_2, which are to be determined by initial data $F(k)$ and $G(k)$. In fact,

$$b_1 = \frac{1}{2}\left[F(k) - \frac{1}{ick}G(k)\right],$$

$$b_2 = \frac{1}{2}\left[F(k) + \frac{1}{ick}G(k)\right].$$

By taking the Fourier inverse transform of $U(k,t)$, we get the solution $u(x,t)$. This should give us the well-known d'Alembert's formula

$$u(x,t) = \frac{1}{2}[f(x+ct) + f(x-ct)] + \frac{1}{2c}\int_{x-ct}^{x+ct} g(\xi)d\xi.$$

Problems of Chapter 3

3.1 Yes. Check the definition of a scalar random variable.

3.2 Yes. Check the definition of a scalar random variable.

3.3 (i) Define open balls in this Hilbert space, with various centers and radii. For example, the unit open ball of l^2 is

$$B \triangleq \left\{x = (x_1, \ldots, x_n, \ldots) \in l^2 : \sum_{n=1}^{\infty} |x_n|^2 < 1\right\}.$$

Then, the Borel σ-field $\mathcal{B}(l^2)$ is obtained by appropriate set operations (including unions, intersections, complement) on all open balls; it is the smallest σ-field containing all open balls.

(ii) Let $x = (x_1, \ldots, x_n, \ldots) \in l^2$. Since $x = \sum_{n=1}^{\infty} x_n e_n$ and $Qe_n = \frac{1}{n^2}e_n$, $n = 1, 2, \ldots$, we see that

$$Qx = Q\left(\sum_{n=1}^{\infty} x_n e_n\right) = \sum_{n=1}^{\infty} x_n Qe_n = \sum_{n=1}^{\infty} \frac{x_n}{n^2}e_n = \left(\frac{x_1}{1^2}, \ldots, \frac{x_n}{n^2}, \ldots\right).$$

Thus,

$$\text{Tr}(Q) = \sum_{n=1}^{\infty} \langle Qe_n, e_n \rangle = \sum_{n=1}^{\infty} \frac{1}{n^2} = \frac{\pi^2}{6} < \infty.$$

Hence, Q is a trace-class operator.

(iii) Fourier series expansion for $W(t)$:

$$W(t) = \sum_{n=1}^{\infty} \frac{1}{n} W_n(t) e_n = \left(\frac{W_1(t)}{1}, \ldots, \frac{W_n(t)}{n}, \ldots \right),$$

where $W_n(t)$'s are standard scalar independent Brownian motions.
Since

$$W(t) - W(s) = \sum_{n=1}^{\infty} \frac{1}{n} W_n(t) e_n - \sum_{n=1}^{\infty} \frac{1}{n} W_n(s) e_n$$

$$= \sum_{n=1}^{\infty} \frac{1}{n} \left(W_n(t) - W_n(s) \right) e_n$$

$$= \left(\frac{W_1(t) - W_1(s)}{1}, \ldots, \frac{W_n(t) - W_n(s)}{n}, \ldots \right), \quad t > s,$$

we see that

$$\mathbb{E}(W(t) - W(s)) = \mathbb{E} \left(\sum_{n=1}^{\infty} \frac{1}{n} \left(W_n(t) - W_n(s) \right) e_n \right)$$

$$= \sum_{n=1}^{\infty} \frac{1}{n} \mathbb{E} \left(W_n(t) - W_n(s) \right) e_n = 0,$$

and

$$\mathbb{E}\|W(t) - W(s)\|^2 = \mathbb{E} \left(\sum_{n=1}^{\infty} \frac{|W_n(t) - W_n(t)|^2}{n^2} \right)$$

$$= \sum_{n=1}^{\infty} \frac{\mathbb{E} \left(|W_n(t) - W_n(s)|^2 \right)}{n^2}$$

$$= \sum_{n=1}^{\infty} \frac{t - s}{n^2} = \frac{(t - s)\pi^2}{6} = (t - s)\mathrm{Tr}(Q).$$

Hence,

$$W(t) - W(s) \sim \mathcal{N}(0, (t - s)Q).$$

(iv) Think of this Hilbert space as an infinite dimensional "Euclidean" space, $\mathbb{R}^{\infty}$, with e_i as ith axis.
Because the sample paths $\left(\frac{W_1(t)}{1}, \ldots, \frac{W_n(t)}{n} \right)$ are continuous in $\mathbb{R}^n$, we can imagine that the sample paths of $W(t) = \left(\frac{W_1(t)}{1}, \ldots, \frac{W_n(t)}{n}, \ldots \right)$ are continuous in $\mathbb{R}^{\infty}$.

3.4 (i) Define open balls in this Hilbert space, with various centers and radii. For example, the unit open ball of $L^2(0, 1)$ is

$$B \triangleq \left\{ u \in L^2(0, 1) : \int_0^1 |u(x)|^2 \, dx < 1 \right\}.$$

Then the Borel σ-field $\mathcal{B}(L^2(0, 1))$ is obtained by appropriate set operations (including unions, intersections, complement) on all open balls; it is the smallest σ-field containing all open balls.
(ii) Let $e_n(x) \triangleq \sqrt{2} \sin(n\pi x)$ and $\lambda_n = n^2\pi^2$. Thus $-\Delta e_n = \lambda_n e_n$. Since $Q = -\Delta^{-1}$,

$$Qe_n = Q\lambda_n^{-1}(-\Delta)e_n = \lambda_n^{-1}(-\Delta^{-1})(-\Delta)e_n = \lambda_n^{-1}e_n.$$

It follows that $\{\lambda_n^{-1}\}_{n=1}^\infty$ and $\{e_n\}_{n=1}^\infty$ are the eigenvalues and eigenvectors of Q and

$$\text{Tr}(Q) = \sum_{n=1}^\infty \lambda_n^{-1} = \sum_{n=1}^\infty \frac{1}{n^2\pi^2} = \frac{1}{6}.$$

(iii) Fourier series expansion for $W(x, t)$.
Since $W(x, t) \sim \mathcal{N}(0, tQ)$,

$$W(x, t) = \sum_{n=1}^\infty \sqrt{\lambda_n} \, W_n(t)e_n = \sum_{n=1}^\infty W_n(t)\frac{\sqrt{2} \, \sin(n\pi x)}{n\pi},$$

where $W_n(t)$'s are standard scalar independent Brownian motions.
3.5 Yes. Proof by definition.
3.6 Since $F(x_0) = F(x_{01}, x_{02}) = x_{01} + x_{02}^3 + \sin(x_{01})$, we obtain

$$F_x(x_0) = \left(\frac{\partial F}{\partial x_1}(x_0), \frac{\partial F}{\partial x_2}(x_0) \right) = \left(1 + \cos(x_{01}), 3x_{02}^2 \right),$$

and

$$F_{xx}(x_0) = \begin{pmatrix} \frac{\partial^2 F}{\partial x_1^2}(x_0) & \frac{\partial^2 F}{\partial x_1 \partial x_2}(x_0) \\ \frac{\partial^2 F}{\partial x_2 \partial x_1}(x_0) & \frac{\partial^2 F}{\partial x_2^2}(x_0) \end{pmatrix} = \begin{pmatrix} -\sin(x_{01}) & 0 \\ 0 & 6x_{02} \end{pmatrix},$$

where $x_0 := (x_{01}, x_{02})^T$.
If $x_0 = (1, 2)^T$, then

$$F_x(x_0) = \left(1 + \cos(x_1), 3x_2^2 \right) = (1 + \cos(1), 12),$$

and

$$F_{xx}(x_0) = \begin{pmatrix} -\sin(x_{01}) & 0 \\ 0 & 6x_{02} \end{pmatrix} = \begin{pmatrix} -\sin(1) & 0 \\ 0 & 12 \end{pmatrix}.$$

3.7 Let $G(y, z) = \frac{1}{2}(y^2 + z^2)$. Then,

$$F(u_0) = \frac{1}{2}\int_0^1 \left(u_0^2 + u_0'^2\right) dx = \int_0^1 G(u_0, u_0')dx.$$

We also define the Jacobian matrix of G

$$DG(y, z) = \left(\frac{\partial G}{\partial y}(y, z), \frac{\partial G}{\partial z}(y, z)\right) = (y, z),$$

and the Hessian matrix of G

$$D^2G(y, z) = \begin{pmatrix} \frac{\partial^2 G}{\partial y^2}(y, z) & \frac{\partial^2 G}{\partial y \partial z}(y, z) \\ \frac{\partial^2 G}{\partial z \partial y}(y, z) & \frac{\partial^2 G}{\partial z^2}(y, z) \end{pmatrix} = \begin{pmatrix} 1 & 0 \\ 0 & 1 \end{pmatrix}.$$

Then:

$$F_u(u_0)(h) = \int_0^1 DG(u_0(x), u_0'(x))v(x)dx = \int_0^1 (u_0(x)h(x)+u_0'(x)h'(x))dx,$$

where $u_0, h \in H_0^1(0, 1)$ and $v \triangleq (h, h')^T$, and

$$F_{uu}(u_0)(h, k) = \int_0^1 w(x)^T D^2G(u_0(x)),$$

$$u_0'(x))v(x)dx = \int_0^1 (h(x)k(x) + h'(x)k'(x))dx,$$

where $u_0, h, k \in H_0^1(0, 1)$ and $v \triangleq (h, h')^T$, $w \triangleq (k, k')^T$.
If $u_0(x) = \sin^2(\pi x)$, then

$$\begin{aligned} F_u(u_0)(h) &= \int_0^1 (u_0(x)h(x) + u_0'(x)h'(x))dx \\ &= \int_0^1 (\sin^2(\pi x)h(x) + 2\pi \sin(\pi x)\cos(\pi x)h'(x))dx \\ &= \int_0^1 (\sin^2(\pi x) - 2\pi^2 \cos(2\pi x))h(x)dx, \end{aligned}$$

and

$$F_{uu}(u_0)(h, k) = \int_0^1 (h(x)k(x) + h'(x)k'(x))dx.$$

3.8 Let Q be the covariance operator of the Brownian motion $W(t)$, and also let $\{q_n\}_{n=1}^\infty$ and $\{e_n\}_{n=1}^\infty$ be the eigenvalues and eigenvectors, respectively, of Q. Then:

$$W(t) = \sum_{n=1}^\infty \sqrt{q_n}\, W_n(t)e_n,$$

where $W_n(t)$'s are standard scalar independent Brownian motions. Let $X_n(t, \omega) \triangleq \langle f(t, \omega), e_n \rangle$ for each $n \in \mathbb{N}$. Since $\mathbb{E} \int_0^T X_n^2(t)dt < \infty$,

$$\mathbb{E} \int_0^T X_n(t)dW_n(t) = 0.$$

Thus,

$$\mathbb{E} \int_0^T \langle f(t, \omega), dW(t, \omega) \rangle = \mathbb{E} \int_0^T \left\langle f(t, \omega), \sum_{n=1}^{\infty} \sqrt{q_n} dW_n(t) e_n \right\rangle$$

$$= \sum_{n=1}^{\infty} \sqrt{q_n} \mathbb{E} \int_0^T \langle f(t, \omega), e_n \rangle dW_n(t)$$

$$= \sum_{n=1}^{\infty} \sqrt{q_n} \mathbb{E} \int_0^T X_n(t) dW_n(t) = 0.$$

3.9 Note that

$$\langle e_n, e_m \rangle = \begin{cases} 1, & n \neq m, \\ 0, & n = m. \end{cases}$$

Thus:

$$\mathbb{E} \langle W(t) - W(s), W(s) \rangle$$

$$= \mathbb{E} \left\langle \sum_{n=1}^{\infty} \sqrt{q_n} W_n(t) e_n - \sum_{n=1}^{\infty} \sqrt{q_n} W_n(s) e_n, \sum_{m=1}^{\infty} \sqrt{q_m} W_m(s) e_m \right\rangle$$

$$= \sum_{n=1}^{\infty} \sum_{m=1}^{\infty} \sqrt{q_n} \sqrt{q_m} \langle e_n, e_m \rangle \mathbb{E}((W_n(t) - W_n(s)) W_m(s))$$

$$= \sum_{n=1}^{\infty} q_n \mathbb{E}((W_n(t) - W_n(s)) W_n(s)) = \sum_{n=1}^{\infty} q_n \cdot 0 = 0.$$

3.10 Without loss of generality, we assume that $t \leq s$. Note that

$$\mathbb{E}(W_n(t) W_m(s)) = \begin{cases} t, & n = m, \\ 0, & n \neq m. \end{cases}$$

Hence:

$$\mathbb{E}(\langle W(t), a \rangle \langle W(s), b \rangle)$$

$$= \mathbb{E} \left(\left\langle \sum_{n=1}^{\infty} \sqrt{q_n} W_n(t) e_n, a \right\rangle \left\langle \sum_{m=1}^{\infty} \sqrt{q_m} W_m(s) e_m, b \right\rangle \right)$$

$$= \sum_{n=1}^{\infty} \sum_{m=1}^{\infty} \sqrt{q_n} \sqrt{q_m} \langle e_n, a \rangle \langle e_m, b \rangle \mathbb{E}(W_n(t) W_m(t))$$

$$= \sum_{n=1}^{\infty} q_n \langle e_n, a \rangle \langle e_n, b \rangle t = t \sum_{n=1}^{\infty} \langle e_n, a \rangle \langle q_n e_n, b \rangle$$

$$= t \sum_{n=1}^{\infty} \langle e_n, a \rangle \langle Q e_n, b \rangle = t \sum_{n=1}^{\infty} \langle Q \langle e_n, a \rangle e_n, b \rangle$$

$$= t \left\langle Q \sum_{n=1}^{\infty} \langle e_n, a \rangle e_n, b \right\rangle = t \langle Q a, b \rangle = \min(t, s) \langle Q a, b \rangle,$$

where Q is the covariance operator of $W(t)$.

3.11 (a) Let $F(u_0) \triangleq \frac{1}{2} \int_0^l u_0^2 \, dx = \frac{1}{2} \|u_0\|^2$. Then:

$$F_u(u_0)(h) = \int_0^l u_0(x) h(x) dx, \quad F_{uu}(u_0)(h, k) = \int_0^l h(x) k(x) dx$$

for $u_0, h, k \in L^2(0, l)$. We also define $b(u) \triangleq u_{xx} + u - u^3$ and $\Phi(u) \triangleq \epsilon$. Thus:

$$du = b(u) dt + \Phi(u) dW(t), \quad u(0) = u_0.$$

Since the covariance operator of $W(t)$ is Q with eigenvalues $\left\{ \frac{1}{n^2} \right\}_{n=1}^{\infty}$ and corresponding eigenvectors $\{e_n\}_{n=1}^{\infty}$ satisfying $\|e_n\|^2 = \int_0^l e_n^2 \, dx = 1$, according to Itô's formula,

$$
\begin{aligned}
d \|u\|^2 &= 2 \, dF(u) = 2 F_u(u)(\Phi(u) dW(t)) + (2 F_u(u)(b(u)) \\
&\quad + \mathrm{Tr}(F_{uu}(u)(\Phi(u) Q^{1/2})(\Phi(u) Q^{1/2})^*)) dt \\
&= 2 F_u(u)(g(u) dW(t)) + (2 F_u(u)(b(u)) \\
&\quad + \sum_{n=1}^{\infty} \frac{1}{n^2} F_{uu}(g(u) e_n, g(u) e_n)) dt \\
&= 2 \left(\int_0^l u g(u) \sum_{n=1}^{\infty} \frac{1}{n} dW_n(t) e_n \, dx \right) \\
&\quad + \left(\int_0^l 2 u b(u) dx + \sum_{n=1}^{\infty} \frac{1}{n^2} \int_0^l g(u) e_n g(u) e_n \, dx \right) dt \\
&= \sum_{n=1}^{\infty} \frac{2\epsilon}{n} \left(\int_0^l u e_n \, dx \right) dW_n(t) \\
&\quad + \left(2 \int_0^l (u u_{xx} + u^2 - u^4) dx + \sum_{n=1}^{\infty} \frac{\epsilon^2}{n^2} \int_0^l e_n e_n \, dx \right) dt \\
&= \sum_{n=1}^{\infty} \frac{2\epsilon \langle u, e_n \rangle}{n} dW_n(t) + \left(2 \int_0^l (u u_{xx} + u^2 - u^4) dx + \sum_{n=1}^{\infty} \frac{\epsilon^2}{n^2} \right) dt \\
&= \sum_{n=1}^{\infty} \frac{2\epsilon \hat{u}_n}{n} dW_n(t) + \left(2 \int_0^l (u u_{xx} + u^2 - u^4) dx + \frac{\epsilon^2 \pi^2}{6} \right) dt,
\end{aligned}
$$

where $u = \sum_{n=1}^{\infty} \hat{u}_n e_n$ for each t.

(b) According to (a),

$$\|u\|^2 = \|u_0\|^2 + \frac{\epsilon^2 \pi^2 t}{3} + 2 \int_0^t \int_0^l (uu_{xx} + u^2 - u^4) dx \, ds + \sum_{n=1}^{\infty} \frac{2\epsilon}{n} \int_0^t \hat{u}_n \, dW_n(s).$$

3.12 (a) Let $F(u_0) \triangleq \frac{1}{2} \int_0^l u_0^2 \, dx = \frac{1}{2} \|u_0\|^2$. Then:

$$F_u(u_0)(h) = \int_0^l u_0(x)h(x)dx,$$

$$F_{uu}(u_0)(h, k) = \int_0^l h(x)k(x)dx, \quad u_0, h, k \in L^2(0, l).$$

We also define $b(u) \triangleq vu_{xx} + uu_x + \sin(u)$ and $\Phi(u) \triangleq g(u)$. Thus:

$$du = b(u)dt + \Phi(u)dw(t).$$

Since the covariance operator of $w(t)$ is $Q = 1$, according to Itô's formula,

$$\begin{aligned}
d\|u\|^2 &= 2dF(u) = 2F_u(u)(\Phi(u)dw(t)) + (2F_u(u)(b(u)) \\
&\quad + \mathrm{Tr}(F_{uu}(u)(\Phi(u)Q^{1/2})(\Phi(u)Q^{1/2})^*))dt \\
&= 2F_u(u)(g(u)dw(t)) + (2F_u(u)(b(u)) + F_{uu}(g(u), g(u)))dt \\
&= \left(\int_0^l 2ug(u)dx \right) dw(t) + \left(\int_0^l 2ub(u)dx + \int_0^l g(u)g(u)dx \right) dt \\
&= \left(\int_0^l (2vuu_{xx} + 2u^2u_x + 2u \sin(u) + g(u)^2)dx \right) dt \\
&\quad + \left(\int_0^l 2ug(u)dx \right) dw(t).
\end{aligned}$$

(b) According to (a),

$$\begin{aligned}
\|u\|^2 &= \|u(0)\|^2 + \int_0^t \int_0^l (2vuu_{xx} + 2u^2u_x + 2u \sin(u) + g(u)^2)dx \, ds \\
&\quad + \int_0^t \int_0^l 2ug(u)dx \, dw(s).
\end{aligned}$$

3.13 Follow the solution for Problem 3.12.

Problems of Chapter 4

4.1 (a) Denote by A the Laplace operator with zero Dirichlet boundary conditions on $(0, l)$. Then we have a unique mild solution

$$u(t) = \int_0^t e^{A(t-s)} dW(s).$$

Next we give a more explicit representation of the mild solution. Denote by $\{\lambda_i, e_i\}_{i=1}^{\infty}$, the eigenvalues and eigenfunctions of $-A$. Because Q commutes with $-A$, we can write the Wiener process in the following series form:

$$W(x, t) = \sum_{i=1}^{\infty} \sqrt{q_i}\, w_i(t) e_i(x)$$

for some $q_i > 0$ and for $\{w_i\}$'s being mutually independent standard Brownian motions. Expand the solution $u(t, x)$ by e_i in the following form:

$$u(x, t) = \sum_{i=1}^{\infty} u_i(t) e_i(x),$$

with u_i solving

$$du_i = -\lambda_i u_i \, dt + \sqrt{q_i}\, dw_i(t), \quad u_i(0) = 0.$$

Then:

$$u_i(t) = \sqrt{q_i} \int_0^t e^{-\lambda_i(t-s)} dw_i(s),$$

and

$$\mathbb{E}\|u(t)\|^2 = \sum_{i=1}^{\infty} \mathbb{E}|u_i(t)|^2 = \sum_{i=1}^{\infty} \frac{q_i}{2\lambda_i}(1 - e^{-2\lambda_i t}) < \infty.$$

By Definition 4.8, $u(t)$ is a mild solution that is unique.
In order to show that the mild solution is also a weak solution, by Theorem 4.4 we only need to show that it is a strong solution. In fact, we just need to show that

$$\mathbb{E}\|Au(t)\|^2 \le C_t,$$

for C_t a positive continuous function of t. Indeed, this is implied by the assumption $\mathrm{Tr}(-AQ^{1/2}) < \infty$, i.e.,

$$\sum_{i=1}^{\infty} \lambda_i^2 q_i < \infty.$$

(b) By the same notation as above and expanding $u(t, x) = \sum_{i=1}^{\infty} u_i(t) e_i(x)$ and $u_0(x) = \sum_{i=1}^{\infty} u_{0,i} e_i(x)$, we have

$$du_i(t) = -\lambda_i u_i dt + u_i dw(t), \quad u_i(t_0) = u_{0,i}.$$

Then:

$$u_i(t) = \exp\left\{-\lambda_i(t - t_0) - \frac{1}{2}(t - t_0) + w(t) - w(t_0)\right\} u_{0,i}, \quad t \ge t_0.$$

A similar discussion as in (a) yields that $u(t)$ is the unique mild, weak, and strong solution.

(c) Let $v = u_t$. Expand

$$u(x,t) = \sum_{i=1}^{\infty} u_i(t)e_i(x), \quad v(x,t) = \sum_{i=1}^{\infty} v_i(t)e_i(x).$$

Then:

$$\dot{u}_i = v_i,$$
$$\dot{v}_i = -c^2 \lambda_i u_i + \sigma \dot{w}_i.$$

Now we can follow the same discussion as in part (a).

4.2 The Kolmogorov operator is

$$\mathcal{L}\varphi(u) = \frac{1}{2}\text{Tr}(((Q^{1/2})^* D^2\varphi(u))Q^{1/2})$$
$$+ \langle Au + f(u), D\varphi(u) \rangle, \quad \text{for every } \varphi \in C_b^2(H).$$

4.3 Yes, well posed.

4.4 Yes, well posed.

4.5 (a) Since $w(t)$ is a scalar Brownian motion,

$$du = (vu_{xx} + uu_x)\,dt + g(u) \circ dw(t)$$
$$= (vu_{xx} + uu_x)\,dt + \frac{1}{2}g'(u)g(u)dt + g(u)dw(t)$$
$$= \left(vu_{xx} + uu_x + \frac{1}{2}g'(u)g(u)\right)dt + g(u)dw(t)$$

(b) Since $w(t)$ is a scalar Brownian motion,

$$du = (vu_{xx} + uu_x)\,dt + g(u)dw(t)$$
$$= (vu_{xx} + uu_x)\,dt - \frac{1}{2}g'(u)g(u)dt + g(u) \circ dw(t)$$
$$= \left(vu_{xx} + uu_x - \frac{1}{2}g'(u)g(u)\right) + g(u) \circ dw(t)$$

(c) If g is a constant or does not depend on u, then $g'(u) = 0$. Therefore, the two types of SPDEs are identical.

4.6 (a) Consider the orthonormal basis of $L^2(0,l)$ formed by eigenfunctions for $A = \partial_{xx}$ under zero Neumann boundary conditions:

$$e_n(x) \triangleq \sqrt{\frac{2}{l}} \cos\left(\frac{n\pi x}{l}\right), \quad n = 0, 1, \ldots,$$

with the corresponding eigenvalues

$$\lambda_n \triangleq -\left(\frac{n\pi}{l}\right)^2, \quad n = 0, 1, \ldots$$

We define

$$u(x, t, \omega) = \sum_{n=0}^{\infty} u_n(t, \omega) e_n(x),$$

$$W(x, t) = \sum_{n=0}^{\infty} \sqrt{q_n} W_n(t) e_n(x), \quad \text{Tr}(Q) = \sum_{n=0}^{\infty} q_n.$$

Therefore,

$$\sum_{n=0}^{\infty} \dot{u}_n(t) e_n(x) = \sum_{n=0}^{\infty} \nu \lambda_n u_n(t) e_n(x) + \sum_{n=0}^{\infty} \sigma \sqrt{q_n} \dot{W}_n(t) e_n(x),$$

or, equivalently,

$$\dot{u}_n(t) = \nu \lambda_n u_n(t) + \sigma \sqrt{q_n} \dot{W}_n(t), \quad n = 0, 1, \ldots$$

We solve this stochastic ordinary differential equation to obtain

$$u_n(t) = u_n(0) \exp(\nu \lambda_n t) + \sigma \sqrt{q_n} \int_0^t \exp(\nu \lambda_n (t - s)) dW_n(s).$$

Here $u_n(0)$ is determined from the initial condition

$$u(x, 0) = u_0(x).$$

Due to

$$u_0(x) = \sum_{n=0}^{\infty} u_n(0) e_n(x),$$

we have

$$u_n(0) = \int_0^l u_0(x) e_n(x) dx.$$

So, now we have the solution u in a Fourier series: $u = \sum_{n=0}^{\infty} u_n(t, \omega) e_n(x)$.
(b) We calculate

$$\mathbb{E}(u) = \mathbb{E} \left(\sum_{n=0}^{\infty} u_n(t) e_n(x) \right)$$

$$= \sum_{n=0}^{\infty} \mathbb{E} \left(u_n(t) \right) e_n(x) = \sum_{n=0}^{\infty} u_n(0) \exp(\nu \lambda_n t) e_n(x).$$

Since $W_m(t)$ and $W_n(t)$ are independent when $m \neq n$, and also $\nu > 0$, we have

$$
\begin{aligned}
\mathrm{Var}(u) &= \mathbb{E}\,\langle u - \mathbb{E}(u),\, u - \mathbb{E}(u)\rangle \\
&= \mathbb{E}\left\langle \sum_{n=0}^{\infty} u_n(t)e_n - \sum_{n=0}^{\infty} E\left(u_n(t)\right) e_n, \right. \\
&\qquad\qquad \left. \times \sum_{m=0}^{\infty} u_m(t)e_m - \sum_{m=0}^{\infty} E\left(u_m(t)\right) e_m \right\rangle \\
&= \mathbb{E}\left(\sum_{n=0}^{\infty} (u_n(t) - \mathbb{E}(u_n(t)))^2 \right) = \sum_{n=0}^{\infty} \mathbb{E}\left(u_n(t) - \mathbb{E}(u_n(t))\right)^2 \\
&= \sum_{n=0}^{\infty} \mathrm{Var}\left(u_n(t)\right) = \sum_{n=0}^{\infty} \int_0^t \sigma^2 q_n \exp\left(2\nu\lambda_n(t-s)\right) ds \\
&= \sigma^2 q_0 t + \sum_{n=1}^{\infty} \frac{\sigma^2 q_n}{2\nu\lambda_n} \left(\exp\left(2\nu\lambda_n t\right) - 1\right).
\end{aligned}
$$

Moreover, for $0 < t < s$,

$$
\begin{aligned}
\mathrm{Cov}(u(x,t), u(x,s)) &= \mathbb{E}\,\langle u(\cdot,t) - \mathbb{E}(u(\cdot,t)),\, u(\cdot,s) - \mathbb{E}(u(\cdot,s))\rangle \\
&= \mathbb{E}\left\langle \sum_{n=0}^{\infty} u_n(t)e_n - \sum_{n=1}^{\infty} \mathbb{E}\left(u_n(t)\right) e_n, \right. \\
&\qquad\qquad \left. \times \sum_{m=0}^{\infty} u_n(s)e_m - \sum_{m=1}^{\infty} \mathbb{E}\left(u_m(s)\right) e_m \right\rangle \\
&= \mathbb{E}\left(\sum_{n=0}^{\infty} (u_n(t) - \mathbb{E}(u_n(t)))(u_n(s) - \mathbb{E}(u_n(s))) \right) \\
&= \sum_{n=0}^{\infty} \mathbb{E}\left(u_n(t) - \mathbb{E}(u_n(t))\right)\left(u_n(s) - \mathbb{E}(u_n(s))\right) \\
&= \sum_{n=0}^{\infty} \mathrm{Cov}\left(u_n(t), u_n(s)\right) \\
&= \sum_{n=0}^{\infty} \left(\mathbb{E}(u_n(t)u_n(s)) - \mathbb{E}(u_n(t))\mathbb{E}(u_n(s))\right) \\
&= \sum_{n=0}^{\infty} \mathrm{Var}\left(u_n(t)\right) \\
&= \sigma^2 q_0 t + \sum_{n=1}^{\infty} \frac{\sigma^2 q_n}{2\nu\lambda_n} \left(\exp\left(2\nu\lambda_n t\right) - 1\right).
\end{aligned}
$$

In the case of the zero Dirichlet boundary conditions

$$u(0, t) = u(l, t) = 0, \quad t \geq 0,$$

we choose the orthonormal basis of $L^2(0, l)$ to be

$$e_n(x) \triangleq \sqrt{\frac{2}{l}} \sin\left(\frac{n\pi x}{l}\right), \quad n = 1, 2, \ldots$$

Then:

$$u(x, t, \omega) = \sum_{n=1}^{\infty} u_n(t, \omega) e_n(x),$$

where $u_n(t)$ is the same as in (a) for each $n \in \mathbb{N}$.

4.7 Define $v(t) = \dot{u}(t)$. We rewrite the second-order SDE as an equivalent system of first-order SDEs:

$$\begin{cases} \dot{u}(t) = v(t), \\ \dot{v}(t) = c^2 \lambda u + \epsilon \sqrt{q} \dot{w}(t). \end{cases} \tag{7.157}$$

In the matrix form, this becomes

$$\begin{pmatrix} \dot{u}(t) \\ \dot{v}(t) \end{pmatrix} = \begin{pmatrix} 0 & 1 \\ c^2 \lambda & 0 \end{pmatrix} \begin{pmatrix} u(t) \\ v(t) \end{pmatrix} + \begin{pmatrix} 0 \\ \epsilon \sqrt{q} \dot{w}(t) \end{pmatrix}. \tag{7.158}$$

Let

$$A = \begin{pmatrix} 0 & 1 \\ c^2 \lambda & 0 \end{pmatrix}.$$

We solve (7.158) to obtain

$$\begin{pmatrix} u(t) \\ v(t) \end{pmatrix} = e^{At} \begin{pmatrix} u(0) \\ v(0) \end{pmatrix} + \int_0^t e^{A(t-s)} \begin{pmatrix} 0 \\ \epsilon \sqrt{q} \end{pmatrix} dw(s). \tag{7.159}$$

Noticing that

$$A^2 = \begin{pmatrix} 0 & 1 \\ c^2 \lambda & 0 \end{pmatrix}\begin{pmatrix} 0 & 1 \\ c^2 \lambda & 0 \end{pmatrix} = \begin{pmatrix} c^2 \lambda & 0 \\ 0 & c^2 \lambda \end{pmatrix} = c^2 \lambda I,$$

$$A^3 = c^2 \lambda A, \quad A^4 = c^4 \lambda^2 I, \quad A^5 = c^4 \lambda^2 A, \quad \ldots,$$

we have

$$
\begin{aligned}
e^{At} &= I + At + \frac{c^2 \lambda I}{2!} t^2 + \frac{c^2 \lambda A}{3!} t^3 + \frac{c^4 \lambda^2 I}{4!} t^4 + \frac{c^4 \lambda^2 A}{5!} t^5 + \cdots \\
&= \begin{pmatrix} 1 - \frac{(c\sqrt{-\lambda})^2}{2!} t^2 + \frac{(c\sqrt{-\lambda})^4}{4!} t^4 + \cdots & t + \frac{c^2 \lambda t^3}{3!} + \frac{c^4 \lambda^2 t^5}{5!} + \cdots \\ c^2 \lambda t + \frac{c^4 \lambda^2 t^3}{3!} + \frac{c^6 \lambda^3 t^5}{5!} + \cdots & 1 + \frac{c^2 \lambda t^2}{2!} + \frac{c^4 \lambda^2 t^4}{4!} + \cdots \end{pmatrix} \\
&= \begin{pmatrix} 1 - \frac{(c\sqrt{-\lambda})^2}{2!} t^2 + \frac{(c\sqrt{-\lambda})^4}{4!} t^4 + \cdots & \frac{1}{c\sqrt{-\lambda}}\left[c\sqrt{-\lambda} t - \frac{(c\sqrt{-\lambda})^3 t^3}{3!} + \frac{(c\sqrt{-\lambda})^5 t^5}{5!} + \cdots \right] \\ -c\sqrt{-\lambda}\left[c\sqrt{-\lambda} t - \frac{(c\sqrt{-\lambda})^3 t^3}{3!} + \frac{(c\sqrt{-\lambda})^5 t^5}{5!} + \cdots \right] & 1 - \frac{(c\sqrt{-\lambda})^2 t^2}{2!} + \frac{(c\sqrt{-\lambda})^4 t^4}{4!} + \cdots \end{pmatrix} \\
&= \begin{pmatrix} \cos c\sqrt{-\lambda} t & \frac{1}{c\sqrt{-\lambda}} \sin c\sqrt{-\lambda} t \\ -c\sqrt{-\lambda} \sin c\sqrt{-\lambda} t & \cos c\sqrt{-\lambda} t \end{pmatrix}.
\end{aligned}
$$

Substituting this e^{At} into (7.159), we obtain that

$$u(t) = \left[u_0 - \epsilon \frac{1}{c\sqrt{-\lambda}} \sqrt{q} \int_0^t \sin c\sqrt{-\lambda} s \, dw(s) \right] \cos c\sqrt{-\lambda} t$$
$$+ \left[\frac{1}{c\sqrt{-\lambda}} u_1 + \epsilon \frac{1}{c\sqrt{-\lambda}} \sqrt{q} \int_0^t \cos c\sqrt{-\lambda} s \, dw(s) \right] \sin c\sqrt{-\lambda} t.$$

4.8 (a) Consider the orthonormal basis of $L^2(0, l)$ formed by the eigenfunctions of $A = \partial_{xx}$

$$e_n(x) \triangleq \sqrt{\frac{2}{l}} \cos\left(\frac{n\pi x}{l}\right), \quad n = 0, 1, \ldots,$$

with corresponding eigenvalues

$$\lambda_n \triangleq -\left(\frac{n\pi}{l}\right)^2, \quad n = 0, 1, \ldots$$

We expand

$$u(x, t, \omega) = \sum_{n=0}^{\infty} u_n(t, \omega) e_n(x),$$

$$W(t) = \sum_{n=0}^{\infty} \sqrt{q_n} W_n(t) e_n(x), \quad \text{Tr}(Q) = \sum_{n=0}^{\infty} q_n.$$

Thus,

$$\sum_{n=0}^{\infty} \ddot{u}_n(t) e_n(x) = \sum_{n=0}^{\infty} c^2 \lambda_n u_n(t) e_n(x) + \sum_{n=1}^{\infty} \sigma \sqrt{q_n} \dot{W}_n(t) e_n(x),$$

and hence,

$$\ddot{u}_n(t) = c^2 \lambda_n u_n(t) + \sigma \sqrt{q_n} \dot{W}_n(t), \quad n = 0, 1, \ldots$$

This equation is rewritten as

$$\begin{cases} \dot{u}_n(t) = v_n(t), \\ \dot{v}_n(t) = c^2 \lambda_n u_n(t) + \sigma \sqrt{q_n} \dot{W}_n(t), \end{cases}$$

or

$$\begin{pmatrix} \dot{u}_n(t) \\ \dot{v}_n(t) \end{pmatrix} = A \begin{pmatrix} u_n(t) \\ v_n(t) \end{pmatrix} + h \dot{W}_n(t), \quad A = \begin{pmatrix} 0 & 1 \\ c^2 \lambda_n & 0 \end{pmatrix}, \quad h = \begin{pmatrix} 0 \\ \sigma \sqrt{q_n} \end{pmatrix}.$$

We solve this equation to obtain

$$\begin{pmatrix} u_n(t) \\ v_n(t) \end{pmatrix} = \exp(tA) \begin{pmatrix} u_n(0) \\ v_n(0) \end{pmatrix} + \int_0^t \exp((t-s)A) h \, dW_n(s).$$

Note that

$$\exp(tA) = \sum_{m=0}^{\infty} \frac{t^m}{m!} A^m$$

$$= \begin{pmatrix} \sum_{m=0}^{\infty} \frac{t^{2m}}{(2m)!} \left(c^2\lambda_n\right)^m & \sum_{m=0}^{\infty} \frac{t^{2m+1}}{(2m+1)!} \left(c^2\lambda_n\right)^m \\ * & * \end{pmatrix}$$

$$= \begin{pmatrix} \cosh\left(\frac{icn\pi}{l}t\right) & \sinh\left(\frac{icn\pi}{l}t\right) \\ * & * \end{pmatrix},$$

$$u_n(t) = u_n(0)\cosh\left(\frac{icn\pi}{l}t\right) + v_n(0)\sinh\left(\frac{icn\pi}{l}t\right)$$

$$+ \sigma\sqrt{q_n} \int_0^t \sinh\left(\frac{icn\pi}{l}(t-s)\right) dW_n(s).$$

Using the initial conditions

$$u_0(x) = \sum_{n=0}^{\infty} u_n(0)e_n(x), \quad v_0(x) = \sum_{n=0}^{\infty} v_n(0)e_n(x),$$

we obtain

$$u_n(0) = \int_0^l u_0(x)e_n(x)dx, \quad v_n(0) = \int_0^l v_0(x)e_n(x)dx.$$

So, we have the solution u.

(b)

$$\mathbb{E}(u) = \mathbb{E}\left(\sum_{n=0}^{\infty} u_n(t)e_n(x)\right) = \sum_{n=0}^{\infty} \mathbb{E}\left(u_n(t)\right) e_n(x)$$

$$= \sum_{n=0}^{\infty} \left(u_n(0)\cosh\left(\frac{icn\pi}{l}t\right) + v_n(0)\sinh\left(\frac{icn\pi}{l}t\right)\right) e_n(x).$$

Since $W_m(t)$ and $W_n(t)$ are independent when $m \neq n$,

$$\text{Var}(u) = \mathbb{E}\langle u - \mathbb{E}(u), u - \mathbb{E}(u)\rangle$$

$$= \mathbb{E}\left\langle \sum_{n=0}^{\infty} u_n(t)e_n - \sum_{n=1}^{\infty} \mathbb{E}\left(u_n(t)\right) e_n, \right.$$

$$\left. \times \sum_{m=0}^{\infty} u_m(t)e_m - \sum_{m=0}^{\infty} \mathbb{E}\left(u_m(t)\right) e_m \right\rangle$$

$$= \mathbb{E}\left(\sum_{n=0}^{\infty} (u_n(t) - \mathbb{E}(u_n(t)))^2\right) = \sum_{n=0}^{\infty} \mathbb{E}\left(u_n(t) - \mathbb{E}(u_n(t))\right)^2$$

$$= \sum_{n=0}^{\infty} \mathrm{Var}\left(u_n(t)\right) = \sum_{n=0}^{\infty} \int_0^t \sigma^2 q_n \left(\sinh\left(\frac{icn\pi}{l}(t-s)\right)\right)^2 ds$$

$$= \sum_{n=0}^{\infty} \sigma^2 q_n \int_0^t \frac{\cosh\left(\frac{i2cn\pi}{l}(t-s)\right) - 1}{2} ds$$

$$= \sum_{n=1}^{\infty} \frac{1}{2}\sigma^2 q_n \left(\frac{l}{i2cn\pi}\sinh\left(\frac{i2cn\pi}{l}t\right) - t\right).$$

For $0 < t < s$,

$$\mathrm{Cov}(u(x,t), u(x,s))$$

$$= \mathbb{E}\left\langle u(\cdot,t) - \mathbb{E}(u(\cdot,t)), u(\cdot,s) - \mathbb{E}(u(\cdot,s))\right\rangle$$

$$= \mathbb{E}\left\langle \sum_{n=0}^{\infty} u_n(t)e_n - \sum_{n=0}^{\infty} \mathbb{E}\left(u_n(t)\right)e_n, \sum_{m=0}^{\infty} u_n(s)e_m - \sum_{m=0}^{\infty} \mathbb{E}\left(u_m(s)\right)e_m\right\rangle$$

$$= \mathbb{E}\left(\sum_{n=0}^{\infty} (u_n(t) - \mathbb{E}(u_n(t)))(u_n(s) - \mathbb{E}(u_n(s)))\right)$$

$$= \sum_{n=0}^{\infty} \mathbb{E}\left(u_n(t) - \mathbb{E}(u_n(t))\right)\left(u_n(s) - \mathbb{E}(u_n(s))\right)$$

$$= \sum_{n=0}^{\infty} \mathrm{Cov}\left(u_n(t), u_n(s)\right) = \sum_{n=0}^{\infty} \left(\mathbb{E}(u_n(t)u_n(s)) - \mathbb{E}(u_n(t))\mathbb{E}(u_n(s))\right)$$

$$= \sum_{n=0}^{\infty} \mathrm{Var}\left(u_n(t)\right) = \sum_{n=1}^{\infty} \frac{1}{2}\sigma^2 q_n \left(\frac{l}{i2cn\pi}\sinh\left(\frac{i2cn\pi}{l}t\right) - t\right).$$

In the case of zero Dirichlet boundary conditions

$$u(0,t) = u(l,t) = 0, \quad t \geq 0,$$

we take an orthonormal basis of $L^2(0,l)$ to be

$$e_n(x) := \sqrt{\frac{2}{l}}\sin\left(\frac{n\pi x}{l}\right), \quad n = 1, 2, \ldots$$

Then:

$$u(x,t,\omega) = \sum_{n=1}^{\infty} u_n(t,\omega)e_n(x),$$

where $u_n(t)$ is the same as in (a) for each $n \in \mathbb{N}$.

4.9 We define

$$F(u_0) \triangleq \frac{1}{2}\|u_0\|^2 = \frac{1}{2}\int_0^l |u_0|^2\, dx, \quad u_0 \in L^2(0,l).$$

Then the first Fréchet derivative of F is

$$F_u(u_0)(h) = \int_0^l u_0 h \, dx, \quad u_0, \, h \in L^2(0, l),$$

and the second Fréchet derivative of F is

$$F_{uu}(u_0)(h, k) = \int_0^l h k \, dx, \quad u_0, \, h, \, k \in L^2(0, l).$$

(a) Let

$$b(u) \triangleq u_{xx} - \sin(u), \quad \Phi(u) := \sigma u.$$

Since $du = b(u)dt + \Phi(u)dW(t)$, where $W(t)$ is a scalar Brownian motion and $u(0, t) = u(l, t) = 0$ for $t \geq 0$, by Itô's formula

$$d\mathbb{E}(F(u))$$

$$= \frac{1}{2} d \|u\|^2 = 0 + \left(\mathbb{E}\left(\int_0^l u(u_{xx} - \sin(u)) dx \right) \right.$$

$$\left. + \frac{1}{2} \mathbb{E}\left(\int_0^l |\sigma u|^2 \, dx \right) \right) dt$$

$$= \mathbb{E}\left(\int_0^l u \, du_x \right) dt - \mathbb{E}\left(\int_0^l u \sin(u) dx \right) dt + \frac{\sigma^2}{2} \mathbb{E}\left(\int_0^l |u|^2 \, dx \right) dt$$

$$= -\mathbb{E} \|u_x\|^2 dt - \mathbb{E}\left(\int_0^l u \sin(u) dx \right) dt + \frac{\sigma^2}{2} \mathbb{E} \|u\|^2 dt.$$

Since $C(l) \|u\|^2 \leq \|u_x\|^2$, where $C(l) > 0$ for $t \geq 0$, according to Poincaré's inequality, and $u(x, 0) = u_0(x)$ for $x \in [0, l]$, we obtain

$$\mathbb{E} \|u\|^2$$

$$= \mathbb{E} \|u_0\|^2 - 2 \int_0^t \mathbb{E} \|u_x\|^2 \, ds - 2 \int_0^t \mathbb{E}\left(\int_0^l u \sin(u) dx \right) ds$$

$$+ \sigma^2 \int_0^t \mathbb{E} \|u\|^2 \, ds$$

$$\leq \mathbb{E} \|u_0\|^2 - 2C(l) \int_0^t \mathbb{E} \|u\|^2 \, ds + 2 \int_0^t \mathbb{E} \|u\| \, ds + \sigma^2 \int_0^t \mathbb{E} \|u\|^2 \, ds$$

$$\leq \mathbb{E} \|u_0\|^2 - 2C(l) \int_0^t \mathbb{E} \|u\|^2 \, ds + 2 \int_0^t \mathbb{E} \|u\|^2 \, ds + \sigma^2 \int_0^t \mathbb{E} \|u\|^2 \, ds.$$

Let

$$\alpha \triangleq \mathbb{E} \|u_0\|^2, \quad \beta(l) \triangleq 2 + \sigma^2 - 2C(l).$$

By the Gronwall inequality,

$$\mathbb{E} \|u\|^2 \leq \alpha \exp\{\beta t\}, \quad t \geq 0.$$

(b) According to (a), we see that the upper bound of the mean energy $\mathbb{E}\|u\|^2$ increases with the intensity σ of multiplicative noise. However, this does not mean that $u(t)$ increases with σ. In fact, if we apply Itô's formula to $F(u) = \log \|u\|^2$, and by the same discussion as that for the stochastic Burgers' equation with multiplicative noise in §4.7.1, we have the Lyapunov exponent estimate

$$\lambda \leq -2C(l) + 2 - 3\sigma^2, \quad \text{a.s.}$$

This shows that for larger σ the solution is almost surely decaying exponentially, although the upper bound of mean energy is increasing.

4.10 See Theorem 4.19 and Example 4.20.

4.11 See Theorem 4.19 and Example 4.20.

Problems of Chapter 5

5.1 Let $A = \partial_{xx}$ with zero Dirichlet boundary conditions and denote by $\{\lambda_i, e_i\}_{i=1}^{\infty}$ the eigenvalues and eigenfunctions for $-A$. Then we can write $\eta(x,t) = \sum_{i=1}^{\infty} \eta_i(t) e_i(x)$ and $W(x,t) = \sum_{i=1}^{\infty} w_i(t) e_i(x)$. By analyzing the equation for η_i, $i = 1, 2, \ldots$, we have a unique stationary solution $\bar{\eta}$, which is a Gaussian process with normal distribution

$$\mathcal{N}\left(-A^{-1}a(x), -\tfrac{\sigma^2}{2} A^{-1} Q\right)$$

in Hilbert space $L^2(0, l)$. Moreover,

$$\mathbb{E}(\bar{\eta}(t) \otimes \bar{\eta}(s)) = -\frac{\sigma^2}{2} e^{A|t-s|} A^{-1} Q - A^{-1}(1 - e^{A|t-s|}) a \otimes a.$$

5.2 By the discussion for random equation (5.20), we have the tightness of u^ϵ in $C(0, T; H)$ for every $T > 0$. Then, by Theorem 5.11, the averaged equation in the sense of distribution is

$$du = u_{xx} \, dt + a(x) dB(t), \quad u(0) = u_0,$$

where $B(t)$ is a standard scalar Brownian motion.

5.3 Let $A = \partial_{xx}$ with zero Dirichlet boundary conditions. The averaged equation is

$$\dot{\bar{u}} = \bar{u}_{xx} - A^{-1}a\bar{u}.$$

Let $z^\epsilon \triangleq (u^\epsilon - \bar{u})/\sqrt{\epsilon}$. Then, z is the unique limit of z^ϵ in the sense of convergence in distribution, and it solves the following SPDE:

$$dz = (z_{xx} + az)dt + \bar{u}\sqrt{-A^{-1}Q} \, d\bar{W},$$

where $\bar{W}$ is $L^2(0, l)$-valued Q-Wiener process with covariance operator $Q = Id_H$. The averaged equation together with deviation up to the order of $\mathcal{O}(\epsilon)$ is

$$d\bar{u}^\epsilon = [\bar{u}_{xx}^\epsilon - A^{-1}a\bar{u}^\epsilon]dt + \sqrt{\epsilon}\bar{u}^\epsilon\sqrt{-A^{-1}Q} \, d\bar{W}.$$

5.4 The averaged equation is the same as that in Problem 5.3. The random PDE can be seen as the random slow manifold reduced model of the slow-fast SPDE in Problem 5.3.

5.5 The first result follows directly from Theorem 5.20. For the second result, introduce $v(x, t) \triangleq \dot{u}(x, t)$. Thus,

$$du = v \, dt,$$
$$dv = \frac{1}{\epsilon}[-v + u_{xx}] \, dt + \frac{1}{\sqrt{\epsilon}} \, dW(t).$$

Using the same notation as in Problem 5.1, write $u(x, t) = \sum_{i=1}^{\infty} u_i(t)e_i(x)$ and $v(x, t) = \sum_{i=1}^{\infty} v_i(t)e_i(x)$. Then, for each i,

$$du_i = v_i \, dt,$$
$$dv_i = \frac{1}{\epsilon}[-v_i - \lambda_i u_i] \, dt + \frac{1}{\sqrt{\epsilon}} \, dw_i(t).$$

One can follow the same discussion as for the first result.

5.6

$$B(u) = 18u^4 + 54u^2 + 17/2.$$

The averaged equation together with deviation is

$$du = [-4u - u^3] \, dt + \sqrt{\epsilon} \sqrt{18u^4 + 54u^2 + 17/2} \, d\tilde{w}(t)$$

with $\tilde{w}$ being another standard Brownian motion.

Problems of Chapter 6

6.1 Denote by A the Laplace operator on $(0, l)$ with zero Dirichlet boundary conditions. Let e_i be the eigenfunction corresponding to λ_i, $i = 1, 2, \ldots$ Denote by P_N, the projection from H to span$\{e_1, e_2, \ldots, e_N\}$, and $Q_N = I - P_N$. Then denote by $W_1(t, \omega) \triangleq P_N W(t, \omega)$ and $W_2(t, \omega) \triangleq Q_N W(t, \omega)$. By the same discussion as for (6.24), the stationary solution is

$$z(\omega) = \left(\int_0^{\infty} e^{-P_N(A+\lambda)s} \, dW_1(s, \omega), \int_{-\infty}^0 e^{-Q_N(A+\lambda)s} \, dW_2(s, \omega) \right).$$

6.2 Denote by A the Laplace operator on $(0, l)$ with zero Dirichlet boundary conditions. Assume that $\lambda_{N+1} > \lambda \geq \lambda_N$ for some N. Then, by the stationary solution $z(\theta_t \omega)$ obtained in Exercise 6.1 and introducing $v(t, \omega) \triangleq u(t, \omega) - z(\theta_t \omega)$, we have

$$v_t(t) = Av(t) + f(v(t) + z(\theta_t \omega)).$$

This is a random evolutionary equation, for which the following spectrum gap condition holds:

$$L_f \left(\frac{1}{\lambda_N - \lambda - \eta} + \frac{1}{\lambda_{N+1} - \lambda + \eta} \right) < 1,$$

for some $-\lambda_{N+1} + \lambda < \eta < \lambda_N - \lambda$. Then, applying the Lyapunov–Perron method to the above random evolutionary equation, we can construct a N-dimensional random invariant manifold and reduce the original system.

6.3 Denote by A the Laplace operator on $(0, l)$ with zero Dirichlet boundary conditions. The random manifold is the graph of the mapping $h^\epsilon : L^2(0, l) \rightarrow L^2(0, l)$ with

$$h^\epsilon(u, \omega) = \frac{1}{\epsilon} \int_{-\infty}^0 e^{-As/\epsilon} g(u^{\epsilon*}(s, \omega)) ds + \eta^\epsilon(\omega),$$

where $u^{\epsilon*}(t, \omega)$ solves

$$u^{\epsilon*}(t, \omega) = e^{at} u + \int_0^t e^{a(t-s)} f(u^{\epsilon*}(s, \omega), V^{\epsilon*}(s, \omega) + \eta^\epsilon(\theta_s \omega)) ds.$$

Moreover, V^ϵ solves

$$V^{\epsilon*}(t, \omega) = \frac{1}{\epsilon} \int_\infty^t e^{-As/\epsilon} g(u^{\epsilon*}(s, \omega)) ds,$$

and η^ϵ is a stationary solution of

$$\epsilon d\eta^\epsilon = \partial_{xx} \eta^\epsilon dt + \sqrt{\epsilon} dW.$$

Up to the order of $\mathcal{O}(\epsilon)$, Lemma 6.32 we have the following approximation

$$h^\epsilon(u, \omega) = -A^{-1} g(u) + \eta^\epsilon(\omega) + \text{h.o.t.}$$

Then the random slow manifold reduced system is

$$\dot{\bar{u}}^\epsilon = a\bar{u}^\epsilon + f(\bar{u}^\epsilon, -A^{-1} g(\bar{u}^\epsilon) + \eta^\epsilon(\theta_t \omega)).$$

6.4 (a) This can be obtained by Theorem 5.34 or Theorem 5.11. In fact, write $W(t) = \sum_i \sqrt{q_i} \beta_i(t)$, where $\beta_i(t)$ are mutually independent standard scalar Brownian motions, and also expand η^ϵ as $\sum_i \eta_i^\epsilon e_i$, where $\{e_i\}$ are the eigenfunctions corresponding to the eigenvalues λ_i of ∂_{xx} on $(0, l)$ with zero Dirichlet boundary conditions. Then:

$$d\eta_i^\epsilon = -\frac{1}{\epsilon} \lambda_i \eta_i^\epsilon + \frac{1}{\sqrt{\epsilon}} \sqrt{q_i} \beta_i(t).$$

Now, denoting by $u^\epsilon(t) = \frac{1}{\sqrt{\epsilon}} \int_0^t \eta^\epsilon(s) ds$ and expanding $u^\epsilon(t) = \sum_i u_i^\epsilon(t) e_i$, we have

$$\dot{u}_i^\epsilon(t) = \frac{1}{\sqrt{\epsilon}} \eta_i^\epsilon(t), \quad u_i^\epsilon(0) = 0.$$

Theorem 5.11 thus yields the result.

(b) By the assumption on f and the above random slow manifold reduction, we have

$$\dot{\bar{u}}^\epsilon = a\bar{u}^\epsilon + \bar{u}^\epsilon - A^{-1}g(\bar{u}^\epsilon) + \eta^\epsilon(\theta_t\omega).$$

Then the result in (a) yields the following further reduced system for small ϵ:

$$\dot{\bar{u}}^\epsilon = a\bar{u}^\epsilon + \bar{u}^\epsilon - A^{-1}g(\bar{u}^\epsilon) + \sqrt{\epsilon}\dot{\bar{W}}(t),$$

for some Wiener process $\bar{W}$.

6.5 Applying random slow manifold reduction twice yields the following reduced system:

$$\partial_t\bar{u} = \gamma\bar{u} - \mathcal{P}_1\bar{u}^3 - \frac{3\sigma^2}{6}\mathcal{P}_1(\bar{u}\sin^2 2x).$$

6.6 Let $0 > \lambda_1 \geq \lambda_2 \geq \cdots$ be the eigenvalues of ∂_{xx} on $(0, l)$ with zero Dirichlet boundary conditions. For every $N > 0$, denote by P_N the linear projection to the space $\text{span}\{e_1, \ldots, e_N\}$ and $Q_N \triangleq I - P_N$. Let $u \triangleq u_N + v_N$. Then:

$$du_N = [\partial_{xx}u_N + P_N f(u_N + v_N)]\,dt + P_N dW,$$
$$dv_N = [\partial_{xx}v_N + Q_N f(u_N + v_N)]\,dt + Q_N dW.$$

We now apply Theorem 6.10 with $\alpha = \lambda_N$ and $\beta = \lambda_{N+1}$. If L_f, the Lipschitz constant of f, is small enough, then the SPDE is reduced to the following N-dimensional system:

$$d\tilde{u}_N = [\partial_{xx}\tilde{u}_N + P_N f(\tilde{u}_N + h_N(\tilde{u}_N, \theta_t\omega))]\,dt + P_N dW,$$

where $h_N(\cdot, \omega) : P_N H \to Q_N H$ is Lipschitz continuous. We can see that, compared to the Galerkin equation, there is an extra term $h_N(\tilde{u}, \omega)$ in the random invariant manifold reduced equation, which includes the influence from higher modes. Then we obtain the approximation $\tilde{u} = \tilde{u}_N + h_N(\tilde{u}_N, \omega)$, which is more accurate than the Galerkin approximation u_N^G. However, we can show that for almost all ω and u_N, $h_N(u_N, \omega) \to 0$, which implies that $\tilde{u}_N + h_N(u_N, \omega) - u_N^G \to 0$ as $N \to \infty$.

6.7 Rewrite the stochastic wave equation as

$$u_t = v, \quad u(0) = u_0,$$
$$dv = \frac{1}{\nu}\left[-v + \Delta u + f(u)\right]dt + \frac{1}{\nu}dW, \quad v(0) = u_1.$$

This can be put in the following abstract form, for $Z = (u, v)^T$,

$$dZ = [\mathcal{A}Z + F(Z)]dt + dW,$$

where

$$\mathcal{A} = \begin{pmatrix} 0 & 1 \\ \frac{1}{\nu}\Delta & \frac{-1}{\nu} \end{pmatrix}, \quad F(Z) = \begin{pmatrix} 0 \\ \frac{1}{\nu}f(u) \end{pmatrix}, \quad \mathcal{W} = \begin{pmatrix} 0 \\ \frac{1}{\nu}W \end{pmatrix}.$$

A simple calculation yields the eigenvalues of the operator $\mathcal{A}$:

$$\delta_k^{\pm} = -\frac{1}{2\nu} \pm \sqrt{\frac{1}{4\nu^2} - \frac{\lambda_k}{\nu}},$$

where $\{\lambda_k, k \in \mathbb{Z}^+\}$, with $\lambda_k > 0$, are the eigenvalues of ∂_{xx} on $(0, l)$ with zero Dirichlet boundary conditions. Therefore, if ν is small enough so that for some N

$$\nu \leq \frac{1}{4\lambda_N},$$

then the spectral gap condition holds for f with small Lipschitz constant.

Problems of Chapter 7

7.1 First show the tightness of $\{P_\epsilon u^\epsilon\}$ in $L^2(0, T; L^2(D))$. Then by the properties of stochastic integrals, we can pass the limit as $\epsilon \to 0$ in the following stochastic integral:

$$\int_0^T \int_D P_\epsilon u^\epsilon v \, dx \varphi \, dB(s) \to \int_0^T \int_D uv \, dx \varphi \, dB(s)$$

in $L^2(\Omega)$.

7.2 Having the tightness of $\{P_\epsilon u^\epsilon\}$, we then pass the limit as $\epsilon \to 0$ in the sense of $L^2(\Omega)$:

$$\int_0^T \int_D \chi_{D_\epsilon} a(x/\epsilon) v \, dx \varphi \, dB(t) \to \int_0^T \int_D \vartheta \bar{a} v \, dx \varphi \, dB(t),$$

where $\bar{a} = \frac{1}{|D|} \int_D a(y) dy$.

7.3 Similar to Exercise 7.2.

7.4 The covariance operators Q_1 and Q_2 need to satisfy certain conditions.

7.5 Examine and follow the proofs in this chapter.

Notations

$\triangleq$: is defined to be

a. s.: Almost sure or almost surely

$a \wedge b \triangleq \min\{a, b\}$

$a \vee b \triangleq \max\{a, b\}$

$a^+ \triangleq \max\{a, 0\}$

$a^- \triangleq \max\{-a, 0\}$

B_t: Brownian motion

$\mathcal{B}(\mathbb{R}^n)$: Borel σ-field of $\mathbb{R}^n$

$\mathcal{B}(S)$: Borel σ-field of space S

$\mathrm{Supp}(f) \triangleq$ Closure of$\{x \in S : f(x) \neq 0\}$: The support of function f defined on space S

$\mathrm{Cov}(X, Y)$: Covariance of X and Y

$C(\mathbb{R}^n)$: Space of continuous functions on $\mathbb{R}^n$

$C^k(\mathbb{R}^n)$: Space of continuous functions, which have up to kth order continuous derivatives, on $\mathbb{R}^n$

$C^\infty(\mathbb{R}^n)$: Space of continuous functions, which have derivatives of all orders, on $\mathbb{R}^n$

$C_c^\infty(\mathbb{R}^n)$: Space of continuous functions on $\mathbb{R}^n$, which (i) have derivatives of all orders, and (ii) have compact support

$\delta(\xi)$: Dirac delta function

δ_{mn}: Kronecker delta function

$\mathbb{E}(X)$: Expectation (or mean) of a random variable X

$F_X(x)$: Distribution function of a random variable X

$\mathcal{F}^X$ or $\sigma(X)$: σ-field generated by a random variable X; the smallest σ-field with which X is measurable

$\mathcal{F}^{X_t} = \sigma(X_s, s \in \mathbb{R})$: σ-field generated by a stochastic process X_t; the smallest σ-field with which X_t is measurable for every t

$\mathcal{F}_\infty \triangleq \sigma(\bigcup_{t \geq 0} \mathcal{F}_t)$

$\mathcal{F}_{t+} \triangleq \bigcap_{\varepsilon > 0} \mathcal{F}_{t+\varepsilon}$

$\mathcal{F}_{t-} \triangleq \sigma(\bigcup_{s < t} \mathcal{F}_s)$

$\mathcal{F}_t^X \triangleq \sigma(X_s : 0 \leq s \leq t)$: Filtration generated by a stochastic process X_t

$\mathcal{F}_t^W \triangleq \sigma(W_s : 0 \leq s \leq t)$: Filtration generated by Wiener process W_t

$\mathcal{F}_{-\infty}^t \triangleq \sigma(\cup_{s \leq t} \mathcal{F}_s^t)$: Also denoted as $\bigvee_{s \leq t} \mathcal{F}_s^t$

$\mathcal{F}_s^\infty \triangleq \sigma(\cup_{t \geq s} \mathcal{F}_s^t)$: Also denoted as $\bigvee_{t \geq s} \mathcal{F}_s^t$

$H(\xi)$: Heaviside function

$H^k(D)$: Sobolev space; see §2.5

$H_0^k(D)$: Sobolev space of functions with compact support; see §2.5

$\| \cdot \|_k$: Sobolev norm in $H^k(D)$ or $H_0^k(D)$; see §2.5

Id_H: The identity operator in the space H

Effective Dynamics of Stochastic Partial Differential Equations. http://dx.doi.org/10.1016/B978-0-12-800882-9.00015-9

lim in m.s.: Convergence in mean square, i.e., convergence in $L^2(\Omega)$

$\mathcal{L}(X)$ or P^X: Law of a random variable X; also called the probability distribution measure induced by a random variable X. See §3.1

$L^2(\mathbb{R}^n)$: Space of square-integrable functions defined on $\mathbb{R}^n$

$L^p(\mathbb{R}^n)$: Space of p-integrable functions defined on $\mathbb{R}^n$, with $p \geq 1$

$L^2(\Omega)$ or $L^2(\Omega, \mathbb{R}^n)$: Space of random variables, taking values in Euclidean space $\mathbb{R}^n$, with finite variance

$L^2(\Omega, H)$: Space of random variables, taking values in Hilbert space H, with finite variance

$L^2(\Omega, C([0, T]; H))$: See §4.1

$L^2(\Omega \times [0, T]; H)$: See §4.1

$\mathcal{L}(U, H)$: Space of bounded linear operators $A : U \to H$

$\mathcal{L}(H) \triangleq \mathcal{L}(H, H)$

$\mathcal{L}^1(U, H)$: Space of bounded linear operators $A : U \to H$ that are of trace class (i.e., $\text{Tr}(A) < \infty$)

$\mathcal{L}^1(H) \triangleq \mathcal{L}^1(H, H)$

$\mathcal{L}^2(U, H)$: Space of Hilbert–Schmidt operators $A : U \to H$

$\mathcal{L}^2(H) := \mathcal{L}^2(H, H)$

L_t: Lévy motion

L_t^α: α-stable Lévy motion

$\mathbb{N}$: Set of the natural numbers

$\mathcal{N}(\mu, \sigma^2)$: Normal (or Gaussian) distribution with mean μ and variance σ^2

$\nu(dy)$: Lévy jump measure

$\mathbb{P}$: Probability or probability measure

$\mathbb{P}(A)$ or $\mathbb{P}\{A\}$: Probability of an event A

P^X or $\mathcal{L}(X)$: Probability distribution measure (also called law) induced by a random variable X

$P(\lambda)$: Poisson distribution with parameter $\lambda > 0$

$\mathbb{R}$: Two-sided time set

$\mathbb{R}^1$: One dimensional Euclidean space; the set of real numbers

$\mathbb{R}^n$: n-dimensional Euclidean space

$\sigma(X)$ or $\mathcal{F}^X$: σ-field generated by a random variable X; the smallest σ-field with which X is measurable

$\text{Tr}(A)$: Trace of an operator or matrix A

$U(a, b)$: Uniform distribution on the interval $[a, b]$

$UC^\gamma(H, \mathbb{R})$: See §5.3

$\text{Var}(X)$: Variance of a random variable X

$\bigvee_{s \leq t} \mathcal{F}_s^t \triangleq \sigma(\cup_{s \leq t} \mathcal{F}_s^t)$: Also denoted as $\mathcal{F}_{-\infty}^t$

$\bigvee_{t \geq s} \mathcal{F}_s^t \triangleq \sigma(\cup_{t \geq s} \mathcal{F}_s^t)$: Also denoted as $\mathcal{F}_s^\infty$

$W^{k,p}(D)$: Sobolev space; see §2.5

$W_0^{k,p}(D)$: Sobolev space of functions with compact support; see §2.5

$\|\cdot\|_{k,p}$: Sobolev norm in $W^{k,p}(D)$ or $W_0^{k,p}(D)$; see §2.5

$W_t(\omega)$: Wiener process (or Brownian motion)

$w_t(\omega)$: Scalar Wiener process (or Brownian motion)

References

[1] Abraham R, Marsden JE, Ratiu T. Manifolds, tensor analysis and applications. 2nd ed. New York: Springer–Verlag; 1988.

[2] Adams RA. Sobolev spaces. New York: Academic Press; 1975.

[3] Albeverioa S, Bernabeib MS, Röckner M, Yoshida MW. Homogenization with respect to Gibbs measures for periodic drift diffusions on lattices. C R Math 2005;341(11):675–8.

[4] Allaire G. Homogenization and two-scale convergence. SIAM J Math Anal 1992;23(6):1482–518.

[5] Allaire G, Murat M, Nandakumar A. Appendix of Homogenization of the Neumann problem with nonisolated holes. Asym Anal 1993;7(2):81–95.

[6] Allen EJ. Derivation of stochastic partial differential equations. Stoc Anal Appl 2008;26:357–78.

[7] Allen EJ. Derivation of stochastic partial differential equations for size- and age-structured populations. J Biol Dynam 2009;3(1):73–86.

[8] Allouba H. A differentiation theory for Itô's calculus. Stoch Anal Appl 2006;24(2):367–80.

[9] Allouba H. Sddes limits solutions to sublinear reaction-diffusion spdes. Electron J Differ Equations 2003;111:21.

[10] Allouba H. Uniqueness in law for the Allen–Cahn spde via change of measure. C R Acad Sci Paris Ser I Math 2000;330(5):371–6.

[11] Allouba H. Different types of spdes in the eyes of Girsanov's theorem. Stoch Anal Appl 1998;16(5):787–810.

[12] Allouba H. Brownian-time, Brownian motion SIEs on $\mathbb{R}_+ \times \mathbb{R}^d$: ultra regular btrw sies limits solutions, the K-martingale approach, and fourth order spdes. Disc Cont Dyna Syst A 2013;33(2):413–63. Available from: arxiv.org/abs/0708.3419v3.

[13] Allouba H. Spdes law equivalence and the compact support property: applications to the Allen–Cahn spde. C R Acad Sci Paris Ser I Math 2000;331(3):245–50.

[14] Anantharaman A, Le Bris C. A numerical approach related to defect-type theories for some weakly random problems in homogenization. Multiscale Model Simul 2011;9:513–44.

[15] Anosov DB. Averaging in systems of ordinary differential equations with fast oscillating solutions. Izv Acad Nauk SSSR Ser Mat 1960;24:731–42. [in Russian].

[16] Antontsev SN, Kazhikhov AV, Monakhov VN. Boundary value problems in mechanics of nonhomogeneous fluids. Amsterdam, NY: North-Holland; 1990.

[17] Apostol TM. Mathematical analysis. 2nd ed. Reading, Mass: Addison–Wesley; 1974.

[18] Arnold L. Stochastic differential equations. New York: John Wiley & Sons; 1974.

[19] Arnold L. Random dynamical systems. New York: Springer–Verlag; 1998.

[20] Arnold L. Hasselmann's program revisited: the analysis of stochasticity in deterministic climate models. In: Imkeller P, von Storch J-S, editors. Stochastic climates models. Progress in probability. Basel: Birkhäuser; 2001.

[21] Arnold L, Scheutzow M. Perfect cocycles through stochastic differential equations. Probab Theor Relat Field 1995;101:65–88.

Effective Dynamics of Stochastic Partial Differential Equations. http://dx.doi.org/10.1016/B978-0-12-800882-9.00016-0

[22] Arnold VI, Kozlov VV, Neishtadt AI. Mathematical aspects of classical and celestial mechanics. Dynamical systems III. In: Encyclopaedia of mathematical sciences. 3rd ed. Berlin: Springer; 2006.

[23] Artstein Z, Linshiz J, Titi ES. Young measure approach to computing slowly advancing fast oscillations. Multiscale Model Simul 2007;6:1085–97.

[24] Ash RB. Probability and measure theory. 2nd ed. New York: Academic Press; 2000.

[25] Avellaneda M. Iterated homogenization, differential effective medium theory and applications. Commun Pure Appl Math 1987;40(5):527–54.

[26] Bal G. Convergence to homogenized or stochastic partial differential equations. Appl Math Res Express 2011;2:215–41.

[27] Bal G. Central limits and homogenization in random media. Multiscale Model Simul 2008;7(2):677–702.

[28] Bates PW, Jones CKRT. Invariant manifolds for semilinear partial differential equations. Dyn Rep 1989;2:1–38.

[29] Bates PW, Lu K, Zeng C. Existence and persistence of invariant manifolds for semiflows in Banach space. Mem Amer Math Soc 1998;645.

[30] Baxendale P, Lototsky S, editors. Stochastic differential equations: theory and applications, Interdisciplinary mathematical sciences 2, a volume in honor of Professor B. L. Rozovskii. Hackensack, NJ: World Scientific; 2007.

[31] Ben Arous G, Owhadi H. Multiscale homogenization with bounded ratios and anomalous slow diffusion. Commun Pure Appl Math 2003;56(1):80–113.

[32] Bensoussan A, Lions JL, Papanicolaou G. Asymptotic analysis for periodic structure. Amsterdam, NY: North-Holland; 1978.

[33] Bensoussan A, Temam R. Equations stochastiques du type Navier–Stokes. J Funct Anal 1973;13(2):195–222.

[34] Berger MS. Nonlinearity and functional analysis. New York: Academic Press; 1977.

[35] Berglund N, Gentz B. Noise-induced phenomena in slow-fast dynamical systems: a sample-paths approach. New York: Springer; 2005.

[36] Bessaih H. Martingale solutions for stochastic Euler equations. Stoch Anal Appl 1999;17(5):713–25.

[37] Bhattacharya R, Denker M, Goswami A. Speed of convergence to equilibrium and to normality for diffusions with multiple periodic scales. Stoch Proc Appl 1999;80(1):55–86.

[38] Billingsley P. Weak convergence of probability measures. 2nd ed. New York: John Wiley&Sons; 1999.

[39] Billingsley P. Probability and measure. 3rd ed. New York: John Wiley&Sons; 1995.

[40] Birnir B, Svanstedt N. Existence and homogenization of the Rayleigh–Benard problem. J Non Math Phys 2000;2:136–69.

[41] Blanc X, Le Bris C, Loins PL. On the energy of some microscopic stochastic lattices. Arch Rational Mech Anal 2007;184:303–39.

[42] Blömker D. Nonhomogeneous noise and Q-Wiener process on bounded domain. Stoch Anal Appl 2005;23(2):255–73.

[43] Blömker D. Amplitude equations for stochastic partial differential equations (interdisciplinary mathematical sciences). Hackensack, NJ: World Scientific; 2007.

[44] Blömker D. Amplitude equations for locally cubic nonautonomous nonlinearities. SIAM J Appl Dyn Syst 2003;2(3):464–86.

[45] Blömker D, Duan J. Predictability of the Burgers' dynamics under model uncertainty. In: Baxendale P, Lototsky S, editors. Boris Rozovsky 60th birthday volume, stochastic

differential equations: theory and applications. Hackensack, NJ: World Scientific; 2007. p. 71–90.

[46] Blömker D, Mohammed W. Amplitude equation for spdes with quadratic nonlinearities. Electron J Probab 2009;14:2527–50.

[47] Blömker D, Wang W. Qualitative properties of local random invariant manifolds for spdes with quadratic nonlinearity. J Dyn Differ Equations 2010;22:677–95.

[48] Bogolyubov NN, Mitropolskii YA. Asymptotic methods in the theory of nonlinear oscillations. New York: Gordon & Breach; 1961.

[49] Bolthausen E, Sznitman AS. Ten lectures on random media. DMV-lectures 32. Basel: Birkhäuser; 2002.

[50] Boñgolan-Walsh VP, Duan J, Fischer P, Özgökmen T, Iliescu T. Impact of boundary conditions on entrainment and transport in gravity currents. Appl Math Model 2007;31:1338–50.

[51] Boñgolan-Walsh VP, Duan J, Fischer P, Özgökmen T. Dynamics of transport under random fluxes on the boundary. Commun Nonlinear Sci Numer Simul 2008;13(8):1627–41.

[52] Boxler P. A stochastic version of center manifold theory. Probab Theor Relat Field 1989;83:509–45.

[53] Brahim-Otsmane S, Francfort GA, Murat F. Correctors for the homogenization of the wave and heat equations. J Math Pure Appl 1998;71:197–231.

[54] Briane M, Mazliak L. Homogenization of two randomly weakly connected materials. Port Math 1998;55:187–207.

[55] Brune P, Duan J, Schmalfuß B. Random dynamics of the Boussinesq system with dynamical boundary conditions. Stoch Anal Appl 2009;27:1096–116.

[56] Brzezniak Z, Li Y. Asymptotic compactness and absorbing sets for 2D stochastic Navier–Stokes equations on some unbounded domains. Trans AMS 2006;358:5587–629.

[57] Brzezniak Z, Peszat S. Hyperbolic equations with random boundary conditions. Recent development in stochastic dynamics and stochastic analysis, vol. 8. World Scientific; 2010. p. 1–21.

[58] Cabana EM. The vibrating string forced by white noise. Z Wahrscheinlichkeitstheor Verw Geb 1970;15:111–30.

[59] Caffarelli LA, Souganidis PE, Wang L. Homogenization of fully nonlinear, uniformly elliptic and parabolic partial differential equations in stationary ergodic media. Commun Pure Appl Math 2005;58:319–61.

[60] Caraballo T, Duan J, Lu K, Schmalfuß B. Invariant manifolds for random and stochastic partial differential equations. Adv Nonlinear Stud 2009;10:23–52.

[61] Caraballo T, Langa J, Robinson JC. A stochastic pitchfork bifurcation in a reaction-diffusion equation. Proc R Soc A 2001;457:2041–61.

[62] Caraballo T, Kloeden P, Schmalfuß B. Exponentially stable stationary solutions for stochastic evolution equations and their perturbation. Appl Math Optim 2004;50:183–207.

[63] Carmona RA, Rozovskii B, editors. Stochastic partial differential equations: six perspectives. Providence, RI: Amer Math Soc; 1999.

[64] Cazenave T, Haraux A. An introduction to semilinear evolution equations. Oxford: Oxford University Press; 1998.

[65] Chekroun M, Ghil M, Roux J, Varadi F. Averaging of time-periodic systems without a small parameter. Disc Cont Dynam Syst A 2006;14(4):753–82.

[66] Cerrai S. Second order pdes in finite and infinite dimension: a probabilistic approach. Lecture notes in mathematics, vol. 1762. Heidelberg: Springer; 2001.

[67] Cerrai S. Normal deviations from the averaged motion for some reaction-diffusion equations with fast oscillating perturbation. J Math Pure Appl 2009;91:614–47.

[68] Cerrai S. A Khasminskii type averaging principle for stochastic reaction-diffusion equations. Ann Appl Probab 2009;19(3):899–948.

[69] Cerrai S, Freidlin M. Averaging principle for a class of stochastic reaction-diffusion equations. Probab Theor Relat Field 2009;144:137–77.

[70] Chen G, Duan J, Zhang J. Geometric shape of invariant manifolds for a class of stochastic partial differential equations. J Math Phys 2011;52:072702.

[71] Chen X, Duan J, Scheutzow M. Evolution systems of measures for stochastic flows. Dyn Syst 2011;26:323–34.

[72] Chicone C, Latushkin Yu. Evolution semigroups in dynamicals systems and differential equations. Providence, RI: American Mathematical Society; 1999.

[73] Chechkin GA, Piatnitski AL, Shamaev AS. Homogenization: methods and applications. Providence, RI: American Mathematical Society; 2007.

[74] Chepyzhov VV, Vishik MI. Nonautonomous 2D Navier–Stokes system with a simple global attractor and some averaging problems. ESAIM: Control Optim Calc Var 2002;8:467–87.

[75] Cherkaev A, Kohn RV. Topics in the mathematical modeling of composite materials. Boston: Birkhäuser; 1997.

[76] Chow PL. Stochastic partial differential equations. New York: Chapman & Hall/CRC; 2007.

[77] Chow PL. Thermoelastic wave propagation in a random medium and some related problems. Int J Eng Sci 1973;11:953–71.

[78] Chow SN, Lin XB, Lu K. Smooth invariant foliations in infinite-dimensional spaces. J Differ Equations 1991;94(2):266–91.

[79] Chueshov I, Schmalfuß B. Parabolic stochastic partial differential equations with dynamical boundary conditions. Diff Integ Equa 2004;17:751–80.

[80] Chueshov I, Schmalfuß B. Qualitative behavior of a class of stochastic parabolic pdes with dynamical boundary conditions. Disc Cont Dyn Syst A 2007;18(2–3):315–38.

[81] Chueshov I, Schmalfuß B. Averaging of attractors and inertial manifolds for parabolic PDE with random coefficients. Adv Nonlinear Stud 2005;5:461–92.

[82] Chung Y, Titi ES. Inertial manifolds and Gevrey regularity for the Moore–Greitzer model of turbo-machine engine. J Nonlinear Sci 2003;13:1–26.

[83] Cialenco I, Lototsky SV, Pospisil J. Asymptotic properties of the maximum likelihood estimator for stochastic parabolic equations with additive fractional Brownian motion. Stoch Dynam 2009;9(2):169–85.

[84] Cioranescu D, Donato P. An introduction to homogenization. New York: Oxford University Press; 1999.

[85] Cioranescu D, Donato P. Exact internal controllability in perforated domains. J Math Pure Appl 1989;68:185–213.

[86] Cioranescu D, Donato P. Homogenization of the Stokes problem with nonhomogeneous slip boundary conditions. Math Method Appl Sci 1996;19:857–81.

[87] Cioranescu D, Donato P, Murat F, Zuazua E. Homogenization and correctors results for the wave equation in domains with small holes. Ann Scuola Norm Sup Pisa 1991;18:251–93.

[88] Coddington EA, Levinson N. Theory of ordinary differential equations. New York: McGraw Hill; 1955.

[89] Constantin P, Foias C, Nicolaenko B, Temam R. Integral manifolds and inertial manifolds for dissipative partial differential equations. New York: Springer; 1988.

[90] Cranston M, Mountford TS. Lyapunov exponent for the parabolic Anderson model in $\mathbb{R}^d$. J Funct Anal 2006;236(1):78–119.

[91] Crauel H, Debussche A, Flandoli F. Random attractors. J Dynam Differ Equations 1997;9:307–41.

[92] Crauel H, Flandoli F. Attractors for random dynamical systems. Probab Theor Relat Field 1994;100:1095–113.

[93] Dalang R, Mueller C. Some nonlinear spdes that are second order in time. Electron J Probab 2003;8(paper 1):1–21.

[94] Da Prato G, Zabczyk J. Stochastic equations in infinite dimensions. Cambridge: Cambridge University Press; 1992.

[95] Da Prato G, Zabczyk J. Ergodicity for infinite dimensional systems. Cambridge: Cambridge University Press; 1996.

[96] Da Prato G, Zabczyk J. Evolution equations with white-noise boundary conditions. Stoch Stoch Rep 1993;42:167–82.

[97] Da Prato G, Zabczyk J. Second order partial differential equations in hilbert spaces. London mathematical society lecture notes, vol. 293. Cambridge: Cambridge University Press; 2002.

[98] Diop MA, Iftimie B, Pardoux E, Piatnitski AL. Singular homogenization with stationary in time and periodic in space coefficients. J Funct Anal 2006;231:1–46.

[99] Doering CR. A stochastic partial differential equation with multiplicative noise. Phys Lett A 1987;122(3–4):133–9.

[100] Doering CR. Nonlinear parabolic stochastic differential equations with additive colored noise on $\mathbb{R}^d \times \mathbb{R}^+$: a regulated stochastic quantization. Commun Math Phys 1987;109(4):537–61.

[101] Doering CR. Microscopic spatial correlations induced by external noise in a reaction-diffusion system. Phys A 1992;188(1–3):386–403.

[102] Doering CR, Mueller C, Smereka P. Interacting particles, the stochastic Fisher–Kolmogorov–Petrovsky–Piscounov equation, and duality. Phys A 2003;325(1–2):243–59.

[103] Du A, Duan J. Invariant manifold reduction for stochastic dynamical systems. Dynam Syst Appl 2007;16:681–96.

[104] Du A, Duan J. A stochastic approach for parameterizing unresolved scales in a system with memory. J Algor Comput Tech 2009;3:393–405.

[105] Duan J. Predictability in spatially extended systems with model uncertainty I. Eng Simul 2009;31(2):17–32; Predictability in spatially extended systems with model uncertainty II. Eng Simul 2009;31(3):21–35.

[106] Duan J, Gao H, Schmalfuß B. Stochastic dynamics of a coupled atmosphere-ocean model. Stoch Dynam 2002;2:357–80.

[107] Duan J, Goldys B. Ergodicity of stochastically forced large-scale geophysical flows. Int J Math Math Sci 2001;28:313–20.

[108] Duan J, Holmes P, Titi ES. Regularity, approximation and asymptotic dynamics for a generalized Ginzburg–Landau equation. Nonlinearity 1993;6:915–33.

[109] Duan J, Lu K, Schmalfuß B. Invariant manifolds for stochastic partial differential equations. Ann Probab 2003;31:2109–35.

[110] Duan J, Lu K, Schmalfuß B. Smooth stable and unstable manifolds for stochastic evolutionary equations. J Dynam Differ Equations 2004;16:949–72.

[111] Duan J, Nadiga B. Stochastic parameterization of large eddy simulation of geophysical flows. Proc Am Math Soc 2007;135:1187–96.

[112] Duan J, Millet A. Large deviations for the Boussinesq equations under random influences. Stoch Proc Appl 2009;119:2052–81.

[113] Dudley RM. Real analysis and probability. Cambridge: Cambridge University Press; 2002.

[114] Duncan TE, Maslowski B, Pasik-Duncan B. Ergodic boundary/point control of stochastic semilinear systems. SIAM J Control Optim 1998;36:1020–47.

[115] Durrett R. Stochastic calculus: a practical introduction. Boston: CRC Press; 1996.

[116] Weinan E. Principles of multiscale modeling. Cambridge: Cambridge University Press; 2011.

[117] Weinan E, Mattingly JC, Sinai Ya. Gibbsian dynamics and ergodicity for the stochastically forced Navier–Stokes equation. Commun Math Phys 2001;224(1):83–106.

[118] Weinan E, Li X, Vanden-Eijnden E. Some recent progress in multiscale modeling. In: Multiscale modeling and simulation. Lecture notes in computer science and engineering, vol. 39. Berlin: Springer; 2004. p. 3–21.

[119] Escher J. Quasilinear parabolic systems with dynamical boundary. Commun Part Differ Equations 1993;18:1309–64.

[120] Escher J. On the qualitative behavior of some semilinear parabolic problem. Diff Integ Equa 1995;8(2):247–67.

[121] Evans LC. Partial differential equations. Providence, RI: American Mathematical Society; 1998.

[122] Fannjiang A, Papanicolaou G. Convection enhanced diffusion for random flows. J Stat Phys 1997;88:1033–76.

[123] Faris WG, Jona-Lasinio G. Large fluctuations for a nonlinear heat equation. J Phys A: Math Gen 1982;15:3025–55.

[124] Fenichel N. Persistence and smoothness of invariant manifolds for flows. Ind University Math J 1971;21:193–225.

[125] Feng J, Kurtz T. Large deviation for stochastic processes. Providence, RI: American Mathematical Society; 2006.

[126] Fiedler B, Vishik MI. Quantitative homogenization of global attractors for reaction-diffusion systems with rapidly oscillating terms. Asym Anal 2003;34:159–85.

[127] Flandoli F. Regularity theory and stochastic flows for parabolic Spdes. USA: Gordon and Breach; 1995.

[128] Flandoli F, Lisei H. Stationary conjugation of flows for parabolic spdes with multiplicative noise and some applications. Stoch Anal Appl 2004;22(6):1385–420.

[129] Flandoli F, Schaumloffel KU. Stochastic parabolic equations in bounded domains: random evolution operator and Lyapunov exponents. Stoch Stoch Rep 1990;29:461–85.

[130] Flandoli F, Schaumloffel KU. A multiplicative ergodic theorem with applications to a first order stochastic hyperbolic equation in a bounded domain. Stoch Stoch Rep 1991;34:241–55.

[131] Flandoli F.. Stochastic flow and Lyapunov exponents for abstract stochastic pdes of parabolic type. Lecture notes in mathematics, vol. 2. Springer; 1991. p. 196–205.

[132] Flandoli F, Maslowski B. Ergodicity of the 2D Navier–Stokes equation under random perturbations. Commun Math Phys 1995;171:119–41.

[133] Fouque J-P, Garnier J, Papanicolaou G, Solna K. Wave propagation and time reversal in randomly layered media. New York: Springer; 2007.

[134] Freidlin MI, Wentzell AD. Random perturbations of dynamical systems. 2nd ed. Springer-Verlag; 1998.

[135] Freidlin MI, Wentzell AD. Reaction-diffusion equations with randomly perturbed boundary conditions. Ann Probab 1992;20:963–86.

[136] Freidlin MI. Random perturbations of reaction-diffusion equations: the quasi-deterministic approximation. Trans Amer Math Soc 1988;305:665–97.

[137] Freidlin MI. Markov processes and differential equations: asymptotic problems. Basel: Birkhäuser; 1996.

[138] Fu H, Duan J. An averaging principle for two time-scale stochastic partial differential equations. Stoch Dynam 2011;11:353–67.

[139] Fu H, Cao D, Duan J. A sufficient condition for nonexplosion for a class of stochastic partial differential equations. Interdisciplinary Math Sci 2010;8:131–42.

[140] Fu H, Liu X, Duan J. Slow manifolds for multi-time-scale stochastic evolutionary systems. Commun Math Sci 2013;11(1):141–62.

[141] Fusco N, Moscariello G. On the homogenization of quasilinear divergence structure operators. Ann Math Pure Appl 1987;164(4):1–13.

[142] Garcia-Ojalvo J, Sancho JM. Noise in spatially extended systems. Springer–Verlag; 1999.

[143] Givon D, Kupferman R, Stuart A. Extracting macroscopic dynamics: model problems and algorithms. Nonlinearity 2004;17:R55–R127.

[144] Gilbarg D, Trudinger NS. Elliptic partial differential equations of second order. 2nd ed. New York: Springer; 1983.

[145] Goldys B, Maslowski B. Lower estimates of transition densities and bounds on exponential ergodicity for stochastic pdes. Ann Probab 2006;34(4):1451–96.

[146] Guckenheimer J, Holmes P. Nonlinear oscillations, dynamical systems and bifurcations of vector fields. New York: Springer–Verlag; 1983.

[147] Haberman R. Applied partial differential equations. 4th ed. Prentice Hall; 2003.

[148] Hale JK, Verduyn Lunel SM. Averaging in infinite dimensions. J Int Equa Appl 1990;2:463–94.

[149] Hale JK. Ordinary differential equations. New York: Wiley; 1969.

[150] Hanggi P, Marchesoni F, editors. Stochastic systems: from randomness to complexityPhys A 2003;325(1–2):1–296. [special issue]

[151] Hadamard J. Sur l'iteration et les solutions asymptotiques des equations differentielles. Bull Soc Math France 1901;29:224–8.

[152] Hairer M. An introduction to stochastic Pdes. Unpublished lecture notes. 2009. Available from: arXiv:0907.4178v1.

[153] Hairer M, Mattingly JC. Ergodicity of the 2D Navier–Stokes equations with degenerate stochastic forcing. Ann Math 2006;164(3):993–1032.

[154] Hairer M, Pardoux E. Homogenization of periodic linear degenerate pdes. J Funct Anal 2008;255:2462–87.

[155] Hasselmann K. Stochastic climate models: part I. Theory. Tellus 1976;28:473–85.

[156] Henry D. Geometric theory of semilinear parabolic equations. Berlin: Springer–Verlag; 1981.

[157] Hintermann T. Evolution equations with dynamic boundary conditions. Proc Roy Soc Edinburgh Sect A 1989;113:43–60.

[158] Hochberg D, Zorzano M-P. Reaction-noise induced homochirality. Chem Phys Lett 2006;431:185–9.

[159] Holden H, Oksendal B, Uboe J, Zhang T. Stochastic partial differential equations: a modeling, white noise approach. New York: Springer; 1996.

[160] Horsthemke W, Lefever R. Noise-induced transitions. Berlin: Springer–Verlag; 1984.

[161] Hou TY, Yang DP, Wang K. Homogenization of incompressible Euler equations. J Comp Math 2004;22(2):220–9.

[162] Huang Z, Yan J. Introduction to infinite dimensional stochastic analysis. Beijing/New York: Science Press/Kluwer Academic Publishers; 1997.

[163] Huebner M, Rozovskii BL. On asymptotic properties of maximum likelihood estimators for parabolic stochastic pdes. Prob Theor Relat Field 1995;103(2):143–63.

[164] Huisinga W, Schutte C, Stuart AM. Extracting macroscopic stochastic dynamics: model problems. Commun Pure Appl Math 2003;56:234–69.

[165] Hunter JK, Nachtergaele B. Applied analysis. Hackensack, NJ: World Scientific; 2001.

[166] Ibragimov IA, Hasminskii RZ. Statistical estimation–asymptotic theory. New York: Springer–Verlag; 1981.

[167] Ibragimov IA, Hasminskii RZ. Estimation problems for coefficients of stochastic partial differential equations. Theory Probab Appl Part I 1999;43:370–87; Part II 2000;44: 469–94; Part III 2001;45:210–32.

[168] Ichihara N. Homogenization problem for stochastic partial differential equations of Zaikai type. Stoch Stoch Rep 2004;76:243–66.

[169] Iftimie B, Pardoux E, Piatnitski AL. Homogenization of a singular random one-dimensional pde. AIHP–Probab Stat 2008;44:519–43.

[170] Ikeda N, Watanabe S. Stochastic differential equations and diffusion processes. New York: North-Holland; 1981.

[171] Imkeller P, Lederer C. On the cohomology of flows of stochastic and random differential equations. Probab Theor Relat Field 2001;120(2):209–35.

[172] Imkeller P, Monahan A, editors. Stochastic climate dynamicsStoch Dynam 2002;2(3). [special issue]

[173] Jacod J, Shiryaev AN. Limit theorems for stochastic processes. New York: Springer; 1987.

[174] Jentzen A, Kloeden PE. Taylor approximations for stochastic partial differential equations. Providence, RI: SIAM; 2011.

[175] Jikov VV, Kozlov SM, Oleinik OA. Homogenization of differential operators and integral functionals. Berlin: Springer–Verlag; 1994.

[176] Jost J. Partial differential equations. 2nd ed. New York: Springer; 2007.

[177] Just W, Kantz H, Rodenbeck C, Helm M. Stochastic modelling: replacing fast degrees of freedom by noise. J Phys A: Math Gen 2001;34:3199–213.

[178] Kabanov Y, Pergamenshchikov S. Two-scale stochastic systems: asymptotic analysis and control. New York: Springer; 2003.

[179] Kallenberg O. Foundations of modern probability. 2nd ed. Applied Probability Trust; 2002.

[180] Karatzas I, Shreve SE. Brownian motion and stochastic calculus. 2nd ed. New York: Springer; 1991.

[181] Khasminskii RZ. On the principle of averaging the Itô's stochastic differential equations (Russian). Kibernetika 1968;4:260–79. [in Russian].

[182] Kelley A. The stable, center-stable, center, center-unstable, unstable manifolds. J Differ Equations 1967;3:546–70.

[183] Kesten H, Papanicolaou GC. A limit theorem for turbulent diffusion. Commun Math Phys 1979;65:79–128.

[184] Khoshnevisan D, Rassoul-Agha F, editors. A minicourse on stochastic partial differential equations. Lecture notes in mathematics, vol. 1962. Berlin–Heidelberg: Springer–Verlag; 2009.

[185] Khoshnevisan D. A primer on stochastic partial differential equations. In: Khoshnevisan D, Rassoul-Agha F, editors. A Minicourse on Stochastic Partial Differential Equations. Lecture notes in mathematics, vol. 1962. Berlin–Heidelberg: Springer–Verlag; 2009. p. 1–38.

[186] Kifer Y. Diffusion approximation for slow motion in fully coupled averaging. Probab Theor Relat Field 2004;129:157–81.

[187] Kifer Y. Some recent advance in averaging. In: Brin M, Hasselblatt B, Pesin Ya, editors. Modern dynamical systems and applications. Cambridge: Cambridge University Press; 2004.

[188] Kifer Y. L^2 diffusion approximation for slow motion in averaging. Stoch Dynam 2003;3:213–46.

[189] Kifer Y. Averaging in difference equations driven by dynamical systems. Astérisque 2003;287:103–23.

[190] Klebaner FC. Introduction to stochastic calculus with applications. 2nd ed. London: Imperial College Press; 2005.

[191] Kleptsyna ML, Piatnitski AL. Homogenization of a random non-stationary convection-diffusion problem. Russian Math Surveys 2002;57:729–51.

[192] Kline M. Mathematical thought from ancient to modern times 1–4. Oxford: Oxford University Press; 1990.

[193] Kloeden PE, Platen E. Numerical solution of stochastic differential equations. Springer–Verlag; 1992. Second corrected printing, 1995.

[194] Koksch N, Siegmund S. Cone invariance and squeezing properties for inertial manifolds of nonautonomous evolution equations. Banach Center Publ 2003;60:27–48.

[195] Kotelenez P. Stochastic ordinary and stochastic partial differential equations: transition from microscopic to macroscopic equations. New York: Springer; 2008.

[196] Kotelenez P, Kurtz TG. Macroscopic limits for stochastic partial differential equations of McKean–Vlasov type. Probab Theor Relat Field 2010;146:189–222.

[197] Kreiss HO. Problems with different time scales. In: Iserles A, editor. Acta numer. Cambridge: Cambridge University Press; 1922.

[198] Kreyszig E. Introductory functional analysis with applications. New York: John Wiley & Sons; 1989.

[199] Krishnan V. Nonlinear filtering and smoothing: an introduction to martingales, stochastic integrals and estimation. New York: John Wiley & Sons; 1984.

[200] Krylov NM, Bogoliubov NN. Introduction to nonlinear mechanics. Princeton: Princeton University Press; 1947.

[201] Krylov NV. An analytic approach to stochastic partial differential equations. In: Stochastic partial differential equations: six perspectives. Mathematical surveys and monographs, vol. 64. American Mathematical Society; 1999. p. 185–242.

[202] Krylov NV, Rozovskii BL. Stochastic evolution equations. J Sov Math 1981;16:1233–77.

[203] Kukavica I, Ziane M. Uniform gradient bounds for the primitive equations of the ocean. Diff Integral Equa 2008;21:837–49.

[204] Kuksin S, Shirikyan A. Ergodicity for the randomly forced 2D Navier–Stokes equations. Math Phys Anal Geom 2001;4(2):147–95.

[205] Kuksin SB, Piatnitski AL. Khasminskii–Whitham averaging for randomly perturbed KdV equation. J Math Pure Appl 2008;89:400–28.

[206] Kunita H. Stochastic flows and stochastic differential equations. Cambridge: Cambridge University Press; 1990.

[207] Kurtz T, Xiong J. A stochastic evolution equation arising from the fluctuation of a class of interacting particle systems. Commun Math Sci 2004;2:325–58.

[208] Kushner HJ, Huang H. Limits for parabolic partial differential equations with wide band stochastic coefficients and an application to filtering theory. Stochastics 1985;14(2):115–48.

[209] Lam R, Vlachos DG, Katsoulakis MA. Homogenization of mesoscopic theories: effective properties of model membranes. Amer Inst Chem Eng J 2002;48:1083–92.

[210] Langer RE. A problem in diffusion or in the flow of heat for a solid in contact with a fluid. Tohoku Math J 1932;35:260–75.

[211] Lapidus L, Amundson N, editors. Chemical reactor theory. Prentice–Hall; 1977.

[212] Levermore CD, Pomraning GC, Sanzo DL, Wong J. Linear transport theory in a random medium. J Math Phys 1986;27(10):2526–36.

[213] Lian Z, Lu K. Lyapunov exponents and invariant manifolds for random dynamical systems in a banach space. Mem AMS 2010;206(967).

[214] Liggett TM. Interacting particle systems. New York: Springer; 1985.

[215] Lions JL. Quelques mé thodes de ré solution des problèmes non liné aires. Paris: Dunod; 1969.

[216] Lions PL, Masmoudi N. Homogenization of the Euler system in a 2D porous medium. J Math Pure Appl 2005;84:1–20.

[217] Lochak P, Meunier C. Mutiphase averaging for classical systems: with applications to adiabatic theorems. New York: Springer–Verlag; 1988.

[218] Lototsky SV, Rozovskii BL. A unified approach to stochastic evolution equations using the skorokhod integral. Theory Probab Appl 2010;54:189–202.

[219] Lv Y, Roberts AJ. Averaging approximation to singular perturbed nonlinear stochastic wave equations. J Math Phys 2012;53:062702.

[220] Lv Y, Wang W. Limit dynamics of stochastic wave equations. J Differ Equations 2008;244(1):1–23.

[221] Majda AJ, Timofeyev I, Vanden Eijnden E. A mathematical framework for stochastic climate models. Commun Pure Appl Math 2001;54:891–974.

[222] MacKay RS. Slow manifolds. In: Dauxois T, Litvak-Hinenzon A, MacKay RS, Spanoudaki A, editors. Energy localisation and transfer. Hackensack, NJ: World Scientific; 2004. p. 149–92.

[223] Mao X, Markus L. Wave equation with stochastic boundary values. J Math Anal Appl 1993;177:315–41.

[224] Marchenko VA, Khruslov EYa. Homogenization of partial differential equations. Boston: Birkhäuser; 2006.

[225] Marcus R. Stochastic diffusion on an unbounded domain. Pacific J Math 1979;84(4):143–53.

[226] Mariano AJ, Chin TM, Özgökmen TM. Stochastic boundary conditions for coastal flow modeling. Geophys Res Lett 2003;30(9):1457–60.

[227] Maslowski B. Stability of semilinear equations with boundary and pointwise noise. Ann Scuola Norm Sup Pisa Cl Sci 1995;22:55–93.

[228] Maso GD, Modica L. Nonlinear stochastic homogenization and ergodic theory. J Rei Ang Math B 1986;368:27–42.

[229] Mazenko GF. Nonequilibrium statistical mechanics. Wiley-VCH; 2006.

[230] Mikelić A, Paloi L. Homogenization of the inviscid incompressible fluid flow through a 2D porous medium. Proc Amer Math Soc 1999;127:2019–28.

[231] McOwen RC. Partial differential equations: methods and applications. NJ: Pearson Education; 2003.

[232] Metivier M. Stochastic partial differential equations in infinite dimensional spaces. Pisa: Quaderni Scuola Normale Superiore; 1988.

[233] Mielke A, Timofte A. Two-scale homogenization for evolutionary variational inequalities via the energetic formulation. SIAM J Math Anal 2007;39(2):642–68.

[234] Mielke A. Locally invariant manifolds for quasilinear parabolic equations. Rocky Mountain J Math 1991;21:707–14.

[235] Mohammed S-EA, Zhang T, Zhao H. The stable manifold theorem for semilinear stochastic evolution equations and stochastic partial differential equations. Mem AMS 2008;196(917):1–105.

[236] Moss F, McClintock PVE, editors. Noise in nonlinear dynamical systems Volume 1: theory of continuous Fokker–Planck systems. Cambridge: Cambridge University Press; 2007; Volume 2: theory of noise induced processes in special applications. Cambridge: Cambridge University Press; 2009; Volume 3: experiments and simulations. Cambridge: Cambridge University Press; 2009.

[237] Mueller C. Some tools and results for parabolic stochastic partial differential equations. In: Khoshnevisan D, Rassoul-Agha F, editors. A minicourse on stochastic partial differential equations. Lecture notes in mathematics, vol. 1962. Berlin–Heidelberg: Springer–Verlag; 2009. p. 111–44.

[238] Mueller C, Tribe R. A phase transition for a stochastic dpe related to the contact process. Probab Theor Relat Field 1994;100(2):131–56.

[239] Myint-U T, Debnath L. Linear partial differential equations for scientists and engineers. 4th ed. Boston: Birkhäuser; 2007.

[240] Najafi M, Sarhangi GR, Wang H. Stabilizability of coupled wave equations in parallel under various boundary conditions. IEEE Trans Auto Cont 1997;42(9):1308–12.

[241] Nandakumaran AK, Rajesh M. Homogenization of a parabolic equation in a perforated domain with Neumann boundary condition. Proc Indian Acad Sci (Math Sci) 2002;112:195–207.

[242] Nandakumaran AK, Rajesh M. Homogenization of a parabolic equation in a perforated domain with Dirichlet boundary condition. Proc Indian Acad Sci (Math Sci) 2002;112:425–39.

[243] Nguetseng G. A general convergence result for a functional related to the theory of homogenization. SIAM J Math Anal 1989;20:608–23.

[244] Øksendal B. Stochastic differential equations. 6th ed. New York: Springer–Verlag; 2003.

[245] Olivieri E, Vares ME. Large deviation and metastability. Cambridge: Cambridge University Press; 2004.

[246] Papanicolaou GC. Introduction to the asymptotic analysis of stochastic equations. In: Modern modeling of continuum phenomena. Lectures in applied mathematics, vol. 16. Providence, RI: American Mathematical Society; 1977. p. 109–47.

[247] Papoulis A. Probability, random variables, and stochastic processes. 2nd ed. McGraw-Hill; 1984.

[248] Pardoux E. Stochastic partial differential equations and filtering of diffusion processes. Stochastics 1979;3:127–67.

[249] Pardoux E. Homogenization of linear and semilinear second order parabolic PDEs with periodic coefficients: a probabilistic approach. J Funct Anal 1999;167(2):498–520.

[250] Pardoux E, Piatnitski AL. Homogenization of a nonlinear random parabolic partial differential equation. Stoch Proc Appl 2003;104:1–27.

[251] Pardoux E, Piatnitski AL. Homogenization of a singular random one dimensional pde with time varying coefficients. Ann Probab 2012;40(3):893–1376.

[252] Pavliotis GA, Stuart AM. Multiscale methods: averaging and homogenization. Texts in applied mathematics, vol. 53. New York: Springer; 2009.

[253] Pazy A. Semigroups of linear operators and applications to partial differential equations. Berlin: Springer; 1985.

[254] Peixoto JP, Oort AH. Physics of climate. New York: Springer; 1992.

[255] Perko L. Differential equations and dynamical systems. New York: Springer–Verlag; 2000.

[256] Perron O. Die stabilitatsfrage bei differentialgleichungssysteme. Math Zeit 1930;32:703–28.

[257] Peszat S, Zabczyk J. Stochastic partial differential equations with Levy noise. Cambridge: Cambridge University Press; 2007.

[258] Pinsky MA. Partial differential equations and boundary value problems with applications. Waveland Press Inc.; 2003.

[259] Polacik P. Parabolic equations: asymptotic behavior and dynamics on invariant manifolds. Handbook of dynamical systems, vol. 2. Elsevier; 2002. [chapter 16].

[260] Prevot C, Rockner M. A concise course on stochastic partial differential equations. Lecture notes in mathematics, vol. 1905. New York: Springer; 2007.

[261] Razafimandimby PA, Sango M, Woukeng JL. Homogenization of nonlinear stochastic partial differential equations in a general ergodic environment. Stoch Anal Appl 2013;31(5):755–84. [preprinted, 2011]. Available from: arXiv:1109.1977.

[262] Reed M, Simon B. Methods of modern mathematical physics: I. Functional analysis. San Diego: Academic Press; 1980.

[263] Ren J, Fu H, Cao D, Duan J. Effective dynamics of a coupled microscopic-macroscopic stochastic system. Acta Math Sci 2010;30:2064–76.

[264] Renardy M, Rogers R. Introduction to partial differential equations. New York: Springer–Verlag; 1993.

[265] Röckner M, Sobol Z. Kolmogorov equations in infinite dimensions: well-posedness and regularity of solutions, with applications to stochastic generalized Burgers equations. Ann Probab 2006;34(2):663–727.

[266] Roberts AJ. Resolving the multitude of microscale interactions accurately models stochastic partial differential equations. LMS J Comp Math 2006;9:193–221.

[267] Roberts AJ. Computer algebra compares the stochastic superslow manifold of an averaged spde with that of the original slow-fast spdes. Technical report. University of Adelaide; 2010 [unpublished].

[268] Robinson JC. The asymptotic completeness of inertial manifolds. Nonlinearity 1996;9:1325–40.

[269] Robinson JC. Infinite-dimensional dynamical systems. Cambridge: Cambridge University Press; 2001.

[270] Rogers LCG, Williams D. Diffusions, Markov processes and martingales 1: foundations. 2nd ed. Cambridge: Cambridge University Press; 2000.

[271] Rozovskii BL. Stochastic evolution equations. Boston: Kluwer Academic Publishers; 1990.

[272] Sanchez-Palencia E. Non homogeneous media and vibration theory. Lecture notes in physics, vol. 127. Berlin: Springer–Verlag; 1980.

[273] Sauders JA, Verhulst F, Murdock J. Averaging methods to nonlinear dynamical systems. 2nd ed. Springer; 2007.

[274] Schmalfuß B. A random fixed point theorem and the random graph transformation. J Math Anal Appl 1998;225(1):91–113.

[275] Schmalfuß B, Schneider KR. Invariant manifold for random dynamical systems with slow and fast variables. J Dyn Diff Equa 2008;20(1):133–64.

[276] Sell GR, You Y. Dynamics of evolutionary equations. New York: Springer; 2002.

[277] Shiga T. Two contrasting properties of solutions for one-dimension stochastic partial differential equations. Canadian J Math 1994;46(2):415–37.

[278] Skorokhod AV. Asymptotic methods in the theory of stochastic differential equations. Providence, RI: American Mathematical Society; 1989.

[279] Souganidis PE. Stochastic homogenization of Hamilton–Jacobi equations and some applications. Asym Anal 1999;20(1):1–11.

[280] Souza J, Kist A. Homogenization and correctors results for a nonlinear reaction-diffusion equation in domains with small holes. In: the seventh workshop on partial differential equations II. Contemporary Mathematic, vol. 23. American Mathematical Society; 2002. p. 161–83.

[281] Sowers RB. Large deviation for a reaction-diffusion equation with non-Gaussian perturbations. Ann Probab 1992;20:504–37.

[282] Stroock DW. An introduction to the theory of large deviations. New York: Springer–Verlag; 1984.

[283] Stroock DW. Probability theory: an analytic view. Revised ed. Cambridge: Cambridge University Press; 1993.

[284] Sun C, Gao H, Duan J, Schmalfuß B. Rare events in the Boussinesq system with fluctuating dynamical boundary conditions. J Differ Equations 2010;248:1269–96.

[285] Sznitman A-S. Topics in random walks in random environment. In: School and conference on probability theory. ICTP lecture notes series, vol. 17. Trieste: ICTP; 2004. p. 203–66.

[286] Taghite MB, Taous K, Maurice G. Heat equations in a perforated composite plate: influence of a coating. Int J Eng Sci 2002;40:1611–45.

[287] Temam R. Infinite–dimensional dynamical systems in mechanics and physics. 2nd ed. New York: Springer–Verlag; 1997.

[288] Timofte C. Homogenization results for parabolic problems with dynamical boundary conditions. Romanian Rep Phys 2004;56:131–40.

[289] Triebel H. Interpolation theory, function spaces, differential operators. Amsterdam: North-Holland; 1978.

[290] Vanninathan M. Homogenization of eigenvalues problems in perforated domains. Proc Indian Acad Sci 1981;90:239–71.

[291] Varadhan SRS. Large deviations and applications. Philadelphia: SIAM; 1984.

[292] Veretennikov AYu. On large deviations in the averaging principle for SDEs with a "full dependence". Ann Probab 1999;27(1):284–96.

[293] Vold R, Vold M. Colloid and interface chemistry. Reading, MA: Addison–Wesley; 1983.

[294] Volosov VM. Averaging in systems of ordinary differential equations. Russ Math Surv 1962;17:1–126.

[295] Walsh JB. An introduction to stochastic partial differential equations. Lecture notes in mathematics, vol. 1180. Springer; 1986.

[296] Wang B, Barcilon A, Fang Z. Stochastic dynamics of El Nino–Southern oscillation. J Atmos Sci 1999;56(1):5–23.

[297] Wang W, Duan J. Homogenized dynamics of stochastic partial differential equations with dynamical boundary conditions. Commun Math Phys 2007;275:163–86.

[298] Wang W, Cao D, Duan J. Effective macroscopic dynamics of stochastic partial differential equations in perforated domains. SIAM J Math Anal 2007;38:1508–27.

[299] Wang W, Duan J. A dynamical approximation for stochastic partial differential equations. J Math Phys 2007;48(10):102701.

[300] Wang W, Duan J. Reductions and deviations for stochastic partial differential equations under fast dynamical boundary conditions. Stoch Anal Appl 2008;27:431–59.

[301] Wang W, Roberts AJ. Average and deviation for slow-fast stochastic partial differential equations. J Differ Equations 2012;253(5):1265–86.

[302] Wang W, Roberts AJ. Slow manifold and averaging for slow-fast stochastic system. J Math Anal Appl 2013;398(2):822–39.

[303] Wang W, Roberts AJ. Macroscopic reduction for stochastic reaction-diffusion equations. IMA J Appl Math 2013;78(6):1237–64.

[304] Wang W, Roberts AJ, Duan J. Large deviations and approximations for slow–fast stochastic reaction-diffusion equations. J Differ Equations 2012;253(12):3501–22.

[305] Watanabe H. Averaging and fluctuations for parabolic equations with rapidly oscillating random coefficients. Probab Theor Relat Field 1988;77:359–78.

[306] Waymire E, Duan J, editors. Probability and partial differential equations in modern applied mathematics. Springer–Verlag; 2005.

[307] Wayne CE. Invariant manifolds for parabolic partial differential equations on unbounded domains. Arch Rational Mech Anal 1997;138:279–306.

[308] Weits E. Stationary freeway traffic flow modelled by a linear stochastic partial differential equation. Trans Res B: Meth 1992;26(2):115–26.

[309] Wright S. Time-dependent Stokes flow through a randomly perforated porous medium. Asym Anal 2000;23(3–4):257–72.

[310] Xu C, Roberts AJ. On the low-dimensional modelling of Stratonovich stochastic differential equations. Phys A 1996;225:62–80.

[311] Xu XF, Graham-Brady L. Computational stochastic homogenization of random media elliptic problems using Fourier Galerkin method. Finite Elem Anal Des 2006;42(7):613–22.

[312] Yang D, Duan J. An impact of stochastic dynamic boundary conditions on the evolution of the Cahn–Hilliard system. Stoch Anal Appl 2007;25:613–39.

[313] Yosida K. Functional analysis. 6th ed. Berlin: Springer–Verlag; 1980.

[314] Zabczyk J. A mini course on stochastic partial differential equations, Progress in probability, vol. 49. Springer; 2001. p. 257–84.

[315] Zambotti L. An analytic approach to existence and uniqueness for martingale problems infinite dimensions. Probab Theor Relat Field 2002;118:147–68.

[316] Zeidler E. Applied functional analysis: applications to mathematical physics. New York: Springer; 1995.

[317] Zeidler E. Applied functional analysis: main principles and their applications. New York: Springer; 1995.

[318] Zhang D. Stochastic methods for flow in porous media: coping with uncertainties. New York: Academic Press; 2001.

[319] Zhikov VV. On homogenization in random perforated domains of general type. Matem Zametki 1993;53:41–58.

[320] Zhikov VV. On homogenization of nonlinear variational problems in perforated domains. Russian J Math Phys 1994;2:393–408.

[321] Zhu W. Nonlinear stochastic dynamics and control (Chinese). Beijing: Science Press; 2003.

[322] Zwanzig R. Nonequilibrium statistical mechanics. Oxford: Oxford University Press; 2001.

Printed and bound by CPI Group (UK) Ltd, Croydon, CR0 4YY

03/10/2024

01040422-0006